TB. Lloc

Unhastening Science

STUDIES IN SOCIAL AND POLITICAL THOUGHT 7

STUDIES IN SOCIAL AND POLITICAL THOUGHT
Editor: Gerard Delanty, *University of Liverpool*

This series publishes peer-reviewed scholarly books on all aspects of social and political thought. It will be of interest to scholars and advanced students working in the areas of social theory and sociology, the history of ideas, philosophy, political and legal theory, anthropological and cultural theory. Works of individual scholarship will have preference for inclusion in the series, but appropriate co- or multi-authored works and edited volumes of outstanding quality or exceptional merit will also be included. The series will also consider English translations of major works in other languages.

Challenging and intellectually innovative books are particularly welcome on the history of social and political theory; modernity and the social and human sciences; major historical or contemporary thinkers; the philosophy of the social sciences; theoretical issues on the transformation of contemporary society; social change and European societies.

It is not series policy to publish textbooks, research reports, empirical case studies, conference proceedings or books of an essayist or polemical nature.

Discourse and Knowledge: The Making of Enlightenment Sociology
Piet Strydom

Social Theory after the Holocaust
edited by Robert Fine and Charles Turner

The Moment: Time and Rupture in Modern Thought
edited by Heidrun Friese

Essaying Montaigne
John O'Neill

The Protestant Ethic Debate: Max Weber's Replies to his Critics, 1907–1910
edited by David Chalcraft and Austin Harrington

Breeding Superman: Nietzsche, Race and Eugenics in Edwardian and Interwar Britain
Dan Stone

Unhastening Science

Autonomy and Reflexivity in the Social Theory of Knowledge

DICK PELS

LIVERPOOL UNIVERSITY PRESS

First published 2003 by
Liverpool University Press
4 Cambridge Street
Liverpool
L69 7ZU

Copyright © Dick Pels 2003

The right of Dick Pels to be identified as the author
of this work has been asserted by him in accordance
with the Copyright, Design and Patents Act, 1988

All rights reserved. No part of this book may be reproduced,
stored in a retrieval system, or transmitted, in any form or
by any means, electronic, mechanical, photocopying, recording
or otherwise without the prior written permission of the publishers.

British Library Cataloguing-in-Publication Data
A British Library CIP Record is available

ISBN 0–85323–638–0 hardback
      0–85323–598-8 paperback

Typeset in Plantin by Koinonia, Bury
Printed and bound in Great Britain by MPG Books Ltd, Bodmin

In memory of Pierre Bourdieu, *maître à penser*
(1930–2002)

Contents

	Acknowledgments	ix
1	The Timescape of Science	1
2	What (Again) is So Special about Science?	25
3	Two Traditions in the Social Theory of Knowledge	51
4	The Natural Proximity of Facts and Values	74
5	Knowledge Politics and Anti-Politics: Bourdieu on Science and Intellectuals	108
6	The Politics of Symmetry	130
7	Reflexivity: One Step Up	157
8	Intellectual Autonomy and the Politics of Slow Motion	179
	Epilogue: Weak Social Theory	217
	Notes	222
	Bibliography	246
	Index	269

Acknowledgments

Portions of this book have been previously published, in slightly different form, as follows. Chapter 3 is a revised version of 'Karl Mannheim and the Sociology of Scientific Knowledge: Toward a New Agenda', *Sociological Theory* 14(1) (1996): 30–48 and 'Indifference or Critical Difference? Reply to Bogen', *Sociological Theory* 14 (2) (1996): 195–98. Chapter 4 is a reworked and expanded version of a Dutch article, 'De natuurlijke saamhorigheid van feiten en waarden', *Kennis & Methode* 14(1) (1990): 14–43, which also incorporates a section from 'Values, Facts and the Social Theory of Knowledge', *Kennis & Methode* 15(3) (1991): 274–84. Chapters 5 and 6 are substantially revised and expanded versions of 'Knowledge Politics and Anti-Politics: Toward a Critical Appraisal of Bourdieu's Concept of Intellectual Autonomy', *Theory and Society* 24 (1995): 79–104, and 'The Politics of Symmetry', *Social Studies of Science* 26(2) (May 1996): 277–304. A shorter version of Chapter 7 has appeared in *Theory, Culture and Society* 17(3) (2000): 1–25. I am grateful to the editors and publishers of these journals (Blackwell, Boom, Elsevier and Sage) for their permission to draw on this material.

I would also like to record my gratitude to Malcolm Ashmore, David Bogen, Trudy Dehue, Nil Disco, René Gabriëls, Emilie Gomart, Hans Harbers, Paul ten Have, Kevin Hetherington, Martin Jay, Sjaak Koenis, Mike Lynch, Frédéric Vandenberghe, Hans Radder, and Steven Yearley for various critical comments and suggestions about the papers that have supplied the raw materials for this book.

I wish to dedicate this book to the memory of Pierre Bourdieu, who died just after the manuscript was completed. He may have been too relentless and intense about the calling of social science, and would not have appreciated my predilection for a weaker social theory. Taking over in a sense from Alvin

Acknowledgments

Gouldner, another imperious sociological mentor, he has nevertheless inspired much of my writing over the past decade. I salute and cherish his unfaltering commitment to autonomy and reflexivity in social science. To simultaneously agree and disagree with him about both issues, as I venture to do in the following pages, is the finest compliment I can pay to someone who always invited others to think with and against the great social scientists of the past, into whose ranks he has now been admitted with a flourish. I like to think that, in that celebrated company, he will now finally be able to slow down (*ralentir*) a little.

Dick Pels, Amsterdam, February 2002

CHAPTER 1

The Timescape of Science

Time travels in divers paces with divers persons. I'll tell you who Time ambles withal, who Time trots withal, who Time gallops withal and who he stands still withal.

William Shakespeare, *As You Like It*

Time does not pass. Times are what are at stake between forces.

Bruno Latour

Il faut écrire, il faut écrire et de temps en temps il faut parler.

Pierre Bourdieu

Temporal Differences

This book offers a new account of what makes science special in the company of other human pursuits.[1] It engages critically with a variety of current approaches, especially constructivist and relativist studies of science and technology, which tend to level science down to an ordinary enterprise that is just as idiosyncratic, localized, interest-ridden, and 'political' as other forms of social endeavour. This view is now widely shared by a large tribe of those whom I frivolously call the 'nothing-specialists'. Taking up the challenge of their 'knowledge-political' demystification of science, I shall mount a defence of scientific autonomy, albeit in a reflexive and self-critical way, without reverting to traditional foundational principles such as transcendental truth, universalist method, disinterestedness, or value-neutrality. In this sense, this book reclaims the venerable pursuit of intellectual autonomy and academic freedom in a way that transcends conventional essentialist criteria of demarcation between science and society.

Unhastening Science

This specificity of science is primarily located in its studied 'lack of haste', its relative stress-freeness, or its socially sanctioned withdrawal from the swift pace of everyday life and alternative professional cultures. This 'unhastened' quality defines science's peculiar 'delaying tactics', which systematically slow down and objectify ordinary conversations and contests (e.g. by means of reading and writing technologies, publishing conventions, peer review procedures, flexible time economies), and which tend to attract 'slow' personalities who read more (books rather than reports or memos) and talk less (e.g. in board meetings, on mobile phones, in video conferences) than the 'fast cats' who are attracted to and recruited by more decisionist, stress-driven and hasty cultures. It is possible to articulate a detailed phenomenology of pragmatically rather than philosophically defined practices and work-styles (observation, fieldwork, experimentation, reading and writing, over against public speaking, participation in board and committee meetings, soundbiting and spinning in the media), which are intrinsically tied to differential workpaces (slow vs. fast), and differential workspaces (the glare of publicity vs. the stillness of private study). Compared with other professional pursuits such as journalism, politics or business management, science indulges in a typical delay and deferral of decisions about what the world is like, how to describe and explain it, and what to do about it.

Such a description, rather than claiming normative innocence, immediately implies the ambition to strengthen what it characterizes: that is, to liberate science from the stress and haste that are increasingly imposed upon it by 'external' pressures and incentives such as user relevance, managerial efficiency, cost effectiveness, employability, benchmarking, league-tabling, and audit accountability, or, more generally, by the mounting impact of consumerism, media publicity, the celebrity system, political correctness and the culture of enterprise. A more intense political prioritization of research has accompanied a new agenda of global competitiveness, introducing forms of 'academic capitalism' which critics interpret as having precipitated a shrinking of time horizons and a multiplication of deadlines, resulting in a marked loss of autonomy on the part of the institutions of higher learning (Readings 1996; Slaughter and Rhoades 1996; Slaughter and Leslie 1997; Currie 1998; Robins and Webster 1999: 192ff.; Monbiot 2000: 281ff.).[2] Far from presenting a nostalgic case for academic freedom that harks back to transcendental principles, however, the present account does not aim to rebuild the ivory tower, nor does it wish to barricade the legitimate intrusion of economic and political metaphors, work-styles, and criteria of excellence into the academy and the cultural field more generally. It does however argue for the preservation of a place of quiet, stillness, and unhastened reflection, which must be incessantly negotiated for and demarcated against the speedy cultures of the 'outside' world. In this sense, the book's main title simultaneously encapsulates a new factual description and a new normative/political project to enhance the autonomy of science.

Traditionally, the argument for intellectual autonomy and academic freedom has tended to rely on spatial and territorial metaphors that emphasized distance, decentredness, and marginality, thriving on the topological imagery of the republic of letters, the ivory tower, or the ivy-covered walls and hallowed groves of academe. From Socrates to Descartes and Nietzsche, philosophical and scientific doubt have been seen as requiring a spiritual/corporeal *askèsis* and a 'renunciation of the world', which invariably took the form of a spatial withdawal from the hustle and bustle of the *agora*. Reflective stillness typically breathed from the 'dreamy spires of academia', ethereal niches that were set aside from the hectic pursuit of power, profit, and fame that swept modern cosmopolitan life. More generally, a long-standing topographical imagination in social theory has articulated the social universe in terms of fields, spheres, domains, regions, systems, levels, centres, and peripheries, which are marked off by boundaries, limits, or thresholds, or are instead interconnected by networks, circuits, interfaces, and transit zones. In this manner, social ordering and sorting has conventionally privileged a discourse of cognitive mapping and spatial zoning, in which ontological issues of similarity and difference were preferably dealt with in topological terms.

This cartographic and spatial impulse has persisted even when social ontology, in its Foucaldian, Deleuzian, constructivist or postmodernist incarnations, has grown critical of rigid Euclidean divisions and Cartesian grids, discovering the constructed contingency of boundary-work (Gieryn 1983; 1990; 1999), shifting attention from homogeneous to heterogeneous spaces or 'heterotopias' (e.g. Shields 1991; Thrift 1996; Hetherington 1997), favouring a new metaphorics of folding, fluidity, and flow (Mol and Law 1994; Law 1999; 2002; Urry 2000: 36ff.), celebrating the hybrid 'in-betweenness' or 'liminality' of third spaces and third ways (e.g. Bhabha 1994; Soja 1996), or recognizing the 'blank' figure as representing 'the presence of absence' in social (dis)orders (Hetherington and Lee 2000). These approaches to social complexity still privilege a territorial vocabulary and a politics of location in which even the much-admired mobility of 'de-territorialized', 'migratory', or 'nomadic' subjects (see Pels 2000: 176ff. for a critical discussion) is considered in terms of covering distances across finite but permeable cultural spaces.

In this book, I experiment with a more pointedly temporal perspective on social differentiation and autonomy. This temporal ontology is of course immediately continuous with and constitutively interwoven with the topological one, as expressed, for example, by the increasingly converging vocabulary of 'fluid spaces' and 'time flows' and, more crucially, by the idea of the inextricable co-production of envelopes of time/space (Giddens 1984; Massey 1994: 2–4; 260–64; Urry 1995: 4, 18ff.; 2000: 103ff.; Castells 1996: 407ff.; Thrift 1996: 285–86; May and Thrift 2001).[3] These approaches suggest that our perspective on social differentiation is considerably enriched by introducing a time-geography or political chronology of social relations and of the temporal edges and boundaries that separate them. This angle enables an alternative ordering of social activity in terms of a multiplicity of temporal codes and settings that define different social time patterns or institutional speeds, and hence favours a retranslation of the seemingly topological variables of distanciation, estrangement and marginality as *temporal* experiences and categories. The liberal view of social differentiation, which partitioned the domains of the state from those of religion and culture, polity from economy, and private from public spheres, traditionally found a territorial expression in, for instance, Weber's 'value spheres', Parsons' 'pattern variables', Luhmann's 'systems', or Bourdieu's 'fields'. But it could simultaneously transcribe a diversity of *time cultures* and advocate a *temporal* pluralism according to which every social activity operates its own velocity or follows its own distinctive rhythm (Gurvitch 1964: 20–25; Giddens 1987: 148; Zerubavel 1981; Young 1988; Adam 1990; Bessin and Gasparini 2000; May and Thrift 2001).

In Gurvitch's pioneering approach, the individual macrosocial units and their distinctive social times are classifiable by their differential rhythms, enabling distinctions to be drawn between groupings of slow, medium and rapid cadence. His extensive but somewhat speculative typology of social temporalities separates, for example, the 'enduring time' of slowed-down long duration that characterizes kinship and rural locality groupings from the 'surprise time' of social organizations that alternate between slow and agitated velocities (cf. Saint-Simon's idea of an alternation between normal and crisis times, or Durkheim's notion of an oscillation between moments of collective effervescence and stretches of routine behaviour). The latter is in turn distinguished from the 'erratic time' of accentuated contingency and discontinuity that is deemed characteristic of technological advances, non-political publics, and 'global societies in transition', and from the 'time in advance of itself' that belongs to moments of collective effervescence, active mass behaviour, and communities in revolt (Gurvitch 1964: 30–33, 66ff.).[4] Schools and universities are thrown in with factories, unions, small cities, professions and guilds as hovering intermediately between precipitation and deceleration, and are doubly contrasted with the precipitated rhythm of economic strata and large cities, and with the slow rhythm of isolated villages and 'permanent groupings' such

The Timescape of Science

as state and church (which unexpectedly also include colleges and university *faculties*). In an equally curious discussion of the different time-scales of social classes, Gurvitch also sketches a broad historical acceleration that develops from slow-motion 'peasant time' via the more rapid 'bourgeois time' towards the 'explosive' proletarian time-scale, which is typically 'time in advance of itself', because it projects the future into the present (1964: 86ff., cf. 103ff.).

The Devil of Speed

It is this broad notion about the 'acceleration of history' that provides the sociological and chronopolitical backdrop to my present reflections about 'unhastening science'.[5] It stands in particular tension with the idea of temporal differentiation and multiple cultures of time, because the universal synchronization and time compression enforced by the ever-increasing velocity of the 'fast systems' tendentially erases or at least threatens a peaceful coexistence (or peaceful competition) between different social rhythms. In close parallel to the 'territorial' dilemma of social ontology (which will be extensively reviewed throughout this book), which must reconcile a modernist-liberal insistence on differentiation, boundary maintenance and autonomy with a postmodernist and post-liberal emphasis on de-differentiation, boundary confusion and hybridity, the *temporal* version of this dilemma addresses the difficulty of balancing a liberal notion of time differentiation (e.g. the varying tempi of life in private and public spheres) against a postmodernist one that, in spite of its sustained talk of difference, tends to project a fusion of time horizons in a homogenized culture of accelerated individualization, globalization, technologization, and mediatization. As in the spatial or topological version, 'temporal postmodernism', at least in this aspect, verges remarkably and disconcertingly close to the totalitarian utopia of a fusion of times in a uniform symphony of total mobilization (cf. Jauréquiberry 2000: 258–59). The theoretical challenge here is to develop a vision that neither essentializes the temporal ontologies of different social fields, nor is swept along in a universal mix of total mobilization and synchronization. How are we to reconcile the liberal defence of temporal difference with a post-liberal acknowledgment that (time) boundaries are permeable, that temporal intrusions and confusions are to some extent legitimate, and that all social activities are moving along a faster track than ever before?

Let me venture a thumbnail sketch of this epochal acceleration of history, with the purpose of detailing a politics of time that may begin to resist the 'speed fetishism' of modern ultrafast societies (Adam 1995: 100), and to delineate intellectual and scientific autonomy in terms of temporal as well as

Unhastening Science

spatial withdrawal and estrangement. It is fair to introduce Adam Smith's pin-maker as iconically personifying the dramatic increase of speed that was accomplished through the perfected division of labour and the mechanical inventions that launched the Industrial Revolution. In Smith's calculation, the rationalized division of labour and the technological innovations deriving from it vastly increased the productivity (i.e. the speed) of work, enabling a single pin-maker to jump from one pin per day in 'non-industrial' conditions to 4800 pins per day in 'industrial' ones. Characteristically, for a utility-minded political economist, Smith did not exempt the trades of philosophy and science from his analysis, arguing that both the institution of philosophy as a separate branch of activity and the internal division of labour between philosophers themselves had considerably speeded up the pace of productivity in science (Smith 1976 [1776]: 15–19, 21–22).[6] It was the technology that Smith described as the 'fire-engine' (an early version of the steam engine) that revolutionized speed by substituting machine time (especially rail travel time) for the 'natural' times of walking, riding a horse, or being propelled by wind and water. Time management and technologically engineered (i.e. steam-powered) mobility emerged as critical success factors in the developing system of intercontinental economic competition. Time became money, and laziness, delay, and time wasted were increasingly considered sources of evil. The French *Encyclopédie*, as its subtitle 'A Rational Dictionary of Arts and Crafts' promised, followed Adam Smith in systematically linking occupational rationalization to the effort to gain time (Gurvitch 1964: 131). Weber's account of Puritan asceticism similarly identified time thrift as an essential element of the developing spirit of capitalism. In order to cut time losses, social time was increasingly quantified to uniform standards of industrial clock time. Accelerated communications and transportation mobilities, afforded by the telegraph, the telephone, the railways, general electrification, the fuel-powered car and finally the aeroplane went on to define the essential experience of modernity. In the early twentieth century, Frederick Taylor (nicknamed 'Speedy Taylor') summarized his time-and-motion studies of manual labour routines into a system of scientific management which managed to drastically reduce labour times and boost productivity. By 1926, Ford's assembly-line method had cut the complete production cycle for a Model T, from the digging of the ore to the car driving away from the factory, to 81 hours.

In Sloterdijk's critical interpretation, even Marx's paradoxical hymn of capitalist acceleration ('all that is solid melts into air') revealed him as a thinker of mobilization-in-action (1998: 51); and his view of social revolution clearly marked him as an enthusiast of what Gurvitch has called 'explosive proletarian time'. In this respect, Marx was not all that different from a Futurist and future

Fascist such as Marinetti, who exalted the beauty of speed and the dynamism of machine time, hoping to replace Christianity with the new secular religion and morality of 'divine' *velocità*.[7] Speed protected humanity from the decay of slowness, memory, analysis, repose, and habit; slowness was 'naturally *unclean*' whereas speed, having as its essence 'the intuitive synthesis of every force in movement' was naturally 'pure'. The Futurists therefore had to 'destroy the museums, libraries, academies of every kind', which represented nostalgic repositories of 'slow ideas' and 'dead books' (Marinetti 1972: 41–42, 94–96, 158). Marinetti prefigured the Fascist and National Socialist aesthetics of speed, their cult of fast and intuitive decision in a 'life of danger', and their drive for the *totale Mobilmachung* of society – a dynamic that was clearly indebted to the left-totalitarian planning experiment in post-Tsarist Russia, which was led by ardent followers of both Marx and Taylor. Both the left- and right-wing totalitarian techno-bureaucracies accelerated their modes of production and geared their politics into overdrive in order to manage a warlike total mobilization of all national resources, both human and material.

But the post-totalitarian world economy with its lightning global communications, warp-speed money flows, space-annihilating mobilities and zero-time production processes has likewise, and justifiably, been described in terms of general mobilization; its economic life has been characterized as 'a continuation of war by the means of civil society' (de Gaudemar 1979: 18; cf. Virilio 1986; Sloterdijk 1998; Robins and Webster 1999). Globalization, especially that of economic and technological processes, is essentially a process of speeding up, of the displacement of slower cultures by 'an accelerated global monoculture' (Millar and Schwarz [eds] 1998: 16; Eriksen 2001). The times of international financial markets, of the information superhighway, of the globalized media industry, and of 'cities that never sleep', are times of the highest velocity. The tempo of mobility is also turned up by the inexorable advance of individualization (De Tocqueville already noted that individualistic America ran on impatience and impetuousness), which induces forms of rationalization of production and consumption that focus upon time-saving technologies. The significance of Ritzer's thesis about the McDonaldization of society is precisely to offer the fast-food restaurant as the contemporary paradigm of the rationalization process, emphasizing the *accelerations* induced by classical components of rationalization such as greater efficiency, calculability, predictability and control – the latter as primarily accomplished through the substitution of non-human for human technologies (Ritzer 1996; 1998).

A 'dromocratic' society in which speed is dominant, and the dominant are those who have easiest access to the highest speeds, unleashes a torrential motion that unsettles the uneven time-scales of different institutions, and enforces a social synchronization that tends to suck up the slower cultures into the faster ones. The 'restless society' and its ever-shrinking time horizons produces a general time famine (Young 1988: 218) in which every second is made to count.[8] Working becomes multi-tasking, watching becomes zapping, eating turns into 'grazing', petting into 'speed-dating', and people run fast even in their leisure time. This adulation of speed and stigmatization of slowness results in a general colonization of time that is homologous with and supportive of Habermas's view about the colonization of the lifeworld as a world of (*slower*) communication and interaction by the (*faster*) systems of the state and the market.

A critical phenomenology of differential time-frames, such as I want to develop here, will accordingly attempt to heighten the 'speed awareness' of a society in the grip of acceleration, and needs to elaborate a politics of time (a political 'chrono-topology' or a political kinetics) that thematizes the close co-production of the mechanics of speed and the machinery of power, which is able to resist synchronization under the sign of the highest velocity, and which explores the opportunities for reconciling high speeds with pockets of slowness in the various registers of social life (Fabian 1983; Virilio 1986; Adam 1990; Nowotny 1994; Bessin and Gasparini 2000; Eriksen 2001). Like the spatial version of the post-liberal argument that I shall develop later, this model of time differentiation emphasizes the need to preserve a plurality of temporal cultures and a variety of 'speed lanes'. These are separated by time boundaries that are admittedly weak, but still act as filtering devices and institutional hurdles that serve to dampen or brake the high velocities of the hasty systems. In this perspective, cultural speed (the speed of innovation) is not so much generated by hooking up to a uniform and universal acceleration, but by selectively preserving a variety of time gaps, breaches, intervals and niches that create open timescapes for indulging in delay and slow motion.

In the history of rationalization and modernity, science and technology have of course most often been identified as vehicles of acceleration. For Max Weber and many other social analysts, science was the chief engine of the drive for rational efficiency and means-ends calculability. In this book, I shall adopt an alternative view of scientific practice (and of scientific rationality) as embodying slowness and as inhabiting a timescape of unhastening or deceleration. In a chronotopic version of Turner's view of science as a liminoid space (1982: 33), I shall argue that science is one essential site for installing time differentials and spatio-temporal buffers that enable a high-tech and high-info society to profit from the technologies of slowness and 'indecision' that are essential for maintaining its ever-accelerating speed of innovation. Gurvitch's view about the 'intermediate' time-scale of schools and universities, which alternate between slow and fast velocities, and about the slower pace of colleges and faculties, which apparently operate within the protective ring of an externally secured academic autonomy, begins to delineate an institutional niche in which time follows a special curvature, enabling scientists to adopt forms of time management and temporal strategies that favour a culture of 'waiting' and 'stalling', and even of 'wasting time'. Good science requires a spatio-temporal disconnection that renders it legitimately and productively 'out of sync' with the more frenzied cultures. It occupies a special chronotope, a place of shelter and enclosure that favours particular forms of reflective 'unwinding' or 'chilling out', and thus provides the essential timescape for the cultivation of critical capacities. As both Plato and Nietzsche well knew, 'untimely' thoughts take time to develop, and hence can only mature if the thinker refuses to be synchronized into the fast pace of the utilitarian worlds of power, money, and worldly fame.

Scientific Patience and Scientific Truth

If truth is the daughter of time, then time must be the father of truth – a thought that can well be radicalized beyond more conventional epistemologies of truth-saying and scientific certainty. Science, indeed, has been historically and traditionally identified as a special time- and space-*transcending* enterprise in such a way as to reify its pragmatics of slow motion into the special privileges and powers of universal truth. An opening clue for discussing this critical problem can be found in Durkheim, who may incidentally be identified as the true father of the idea of 'social time' (although not so much of the acceleration of history but of the cyclical rhythms that structure recurrent social ritual). However, I shall focus not upon Durkheim's familiar 'socialization' of the Kantian transcendental categories of time and space in his late sociology of religion, but upon his early methodological views about the autonomy of the social object and the demarcation of sociology from rival sciences of the social. Here, the special temporal experience and epistemic obligation of science is touched upon almost surreptitiously and in passing, even though the idea is obviously central to Durkheim's more general justification of scientific (sociological) authority. A first remark occurs within the broad framework of his programmatic demarcation between science and 'art' (taken in the extended French sense to include practice, morality, ethics, and politics), and is deeply embedded in a traditional phalanx of essentialist epistemic purifications and polarizations, which oppose rational knowledge to everyday opinion, objective things to subjective ideas and desires, factual explanation to evaluation, and contemplation to practical decision.

Unhastening Science

A discipline such as sociology, Durkheim insists, does not deserve to be called a science and cannot secure disciplinary autonomy if it does not have a specific object to explore. If sociology wishes to become an autonomous science, it needs to focus upon social facts (which should be treated as things) rather than upon preconceived ideas, prejudices, and vital needs, thereby distancing itself from 'art', which is not oriented to what social phenomena actually *are* but to what they *should be*. Art, to put it differently, is too *hasty* in its judgments, whereas true science is intrinsically *patient*, and does not precipitate itself towards its conclusions:

> Art is action; it is impelled by urgency, and whatever science it may contain is swept along in its headlong rush. True science does not admit of such haste. The fact is that whenever we have to decide what to do – and such decisions are the concern of art – we cannot temporize for too long; we must make up our minds as quickly as possible because life goes on ... Science is so different from art that it can be true to its own nature only by asserting complete independence, that is, by applying itself, in utter disregard of utility, to a definite object with a view to knowing it. Far from public or private debate, free from any vital necessity, a scientist must pursue his endeavours in the peace and quiet of his study, where nothing impels him to press his conclusions beyond what is justified by his evidence. (Durkheim 1965 [1892]: 5–6, 7)

Evidently, in this account, the is/ought dichotomy is made to coincide with the distinction between science and art, and epistemologically 'freezes' a more pragmatic and relative contrast between two different working speeds or temporal strategies.[9] The same distinction and substitution are echoed in Durkheim's later lectures on socialism, which is contrasted with sociology along similar lines (cf. Pels 2000: 60–61).[10] An *esprit calme* and a sense of moderation are considered scientific virtues par excellence, which sharply contrast with the categorical and totalist temper that pervades socialist ideology. 'Fervour' is the main cause of inspiration of its over-hasty doctrines. Indeed, in order to know 'what the family, property, political, moral, juridical, and economic organization of the European peoples can and ought to be', it is indispensable first to have studied them in some considerable depth:

> Since each of these problems is a world in itself, the solution cannot be found in an instant, merely because the need is felt ... the bases for a rigorous prediction about the future are not established. Socialism has not taken the time; perhaps one could even say, it did not have the time. That is why, to speak precisely, there cannot be a scientific socialism. (Durkheim 1958: 40–41)

The Timescape of Science

Science needs time and cannot be improvised. The only attitude it permits is reservation and circumspection, 'and socialism can hardly maintain this without lying to itself ... Those aware of what social science must be, of the slow pace of its processes, of the laborious investigations it implies to resolve even the narrowest questions, cannot be fond of these premature solutions, these vast systems so summarily sketched out.' Socialism must therefore itself be studied as a social fact; it is not a *product* but an *object* of science. We need to distance ourselves from it, take it more slowly, approach it as a strange, unexplored phenomenon, 'as we did suicide, the family, crime, punishment, responsibility, and religion'. Rather than taking socialism at face value and adopting its propositions ready-made, we must explain its deeper causes, understand what it is 'really' like (Durkheim 1958: 41–42, 55).

It is worth noting that the fact/value dichotomy once again functions to seal off the demarcation between science and ideology. In addition, 'factuality' or 'thinghood' act as epistemological place-holders for the methodological prescription of estrangement and 'patience', in such a way as to open an immediate rift between sociological rationality and 'ordinary' consciousness. To treat social facts as things is not to place them in some category of reality, Durkheim explains, but to observe towards them a certain perplexed attitude of mind. It is to treat them as *strange* objects, which we are ignorant about, and hence need to look at more detachedly (1982 [1895]: 36–38). On this level, 'things' are simply and tautologically that which (Durkheimian) sociology is not prepared to take at face value and finds worthy of more laborious and patient investigation. The first rule of sociological method ('treat social facts as things') is then simply coincident with the Cartesian exhortation that 'one must systematically discard all preconceptions' (Durkheim 1982 [1895]: 72):

> As the sociologist penetrates into the social world he should be conscious that he is penetrating into the unknown ... In the present state of the discipline, we do not really know the nature of the principal social institutions, such as the state or the family, property rights or contract, punishment and responsibility. We are virtually ignorant of the causes upon which they depend, the functions they fulfil, and their laws of evolution. It is as if, on certain points, we are only just beginning to perceive a few glimmers of light. (Durkheim 1982 [1895]: 37–38).

This apparently modest profession of ignorance in the face of the most significant social facts conveniently channels the more grandiose claim that the sociologist is able to grasp their 'real nature' beyond the prejudices of common sense, political ideologies, and rival human sciences. In Durkheim's work, there are always two conceptions of factuality at play that operate in continual tension, defining what are effectively the beginning and end points of scientific inquiry in the positivistic mode: facts are things one is initially uncertain and ignorant about, but they are also whatever finally emerges from sociological analysis as objectively true and certain.[11]

According to the weaker epistemology that I will unfold in this book, however, Durkheim's basic postulate about scientific patience appears in principle detachable from this positivistic methodological powerhouse, which claims access to objective causes, functions and laws, separates facts from ideas and ideals, and pits the rigorous objectivity of scientific truth against the hasty prejudgments of alternative forms of knowledge. The autonomy of science is not necessarily dependent upon a Cartesian epistemology of disinterested and universal doubt (factuality type 1) which promises the certainty of universal and necessary demonstration (factuality type 2). It simply introduces the special perspective (and prejudice) of a slower pace and of doubting some things and not others, without requiring sharp demarcations between 'art' and 'science' or values and facts, and without hoping for the closure of a conventional truth machine.

I want to compare this classical positivist demarcation and defence of sociological autonomy with two recent theorizations which support sharply contrasting suggestions about the nature of scientific rationality and scientific work. Differing from Durkheim, who in his critique of pragmatism anxiously records that it 'places reason on the same plane as sensibility, truth on the same plane as sensations and instincts' and more generally 'reduces everything to the same level' (1972: 252–53), both approaches are 'pragmatist' in levelling down and dissolving most of Durkheim's hierarchical dichotomies, depriving his science of the transcendental integrity that unarguably ordained its privileges of autonomy. Both also mobilize a discourse of disenchantment that highlights basic similarities between forms of scientific rationality and the allegedly impure rationalities of other social pursuits, undercutting the strong demarcation between science and society that was still traced by Durkheim. Pierre Bourdieu grounds his thinking on the idea that science 'is a social field like any other, with its distribution of power and its monopolies, its struggles and strategies, interests and profits', where the specific issue at stake is the monopoly of scientific authority, which he defines inseparably as a technical capacity and as a social power (1981 [1975]: 257–58). Bruno Latour more radically professes that the conventional charge of 'confusing' power and reason, might and right, politics and science, rests upon a stubborn misunderstanding about the notion of force:

> 'Force' looks very different when it is considered *in opposition* to reason, and when it is seen as what designates the complete *gradient* of resistance

where reality is tested… 'Pure force' is an expression that takes meaning only because it is opposed to 'argumentation', 'rationality', 'objectivity', 'rational discussion' and so on. It is devoid of any meaning once this contrast is removed and when a gradient of forces is allowed to settle. When such is the case, arguments also have some force; logical connections are not without strength either; legal barriers exert some pressure as well; taboos seem to have quite a lot of clout also. When the two extremes – pure might and pure right – are forgotten, all the relations of force may start to unfold. Thus, in order to study the way reality is built through trials of resistance, it is necessary *not to make any* a priori *distinction between might and right.* (Latour 1989: 112)

Introducing these two modern thinkers in this manner is also to introduce the two 'inimical friends' or 'best enemies' who figure as my main intellectual sparring partners in the present argument. Theirs are the most suggestive and exciting current reconceptualizations of the practices and institutions of science, which I intend to think with and against, learn from and debate, on occasion in a rather critical fashion. Both Bourdieu's and Latour's writings centrally inform the idea of 'knowledge politics' as it is unfolded in this book. At the end of the day, however, Bourdieu's field theory still appears caught in a residual opposition between truth and power which I think we can do without. Latour abolishes this opposition with more panache, but in doing so also unfortunately appears to lose interest in the issue of demarcation. In this regard, I find Bourdieu's continued loyalty to the idea of scientific autonomy preferable to Latour's contrasting view that scientific laboratories precisely function to *destabilize* the very distinction between the micro-level of science and the macro-level of society. In a nutshell, therefore, my purpose in this book is to reinvent the Bourdieuian idea of intellectual and scientific autonomy from the vantage point of a radical Latourian (but of course also Foucaldian) epistemology of the mixture of knowledge and power. I have already suggested that introducing a critical 'time factor' and the idea of 'unhastening science' provide the first building blocks of such a theorization. Hence I am also interested in how both thinkers consider the peculiar temporality of science (which remains somewhat undertheorized in both, although less so in Bourdieu than in Latour) and how they reconcile the demand of scientific patience with the contrary 'mobilizing' suggestion that is often implied by the idea of a *politics* or *economics* of knowledge.

At the core, therefore, my argument is about the crucial connection between doubt and time, and about the critique of rationalist Cartesian principles of doubt which erase its perspectival, socially distributed and spatio-temporally fragmented nature. Following my earlier recalcitrant reading of Durkheim, I intend to show how a number of interesting current epistemologies are transposable into the more mundane and pragmatic vocabulary of timing and unhastening. In Chapter 4, for example, I shall discuss the temporal dimension of Elias's processual notions of involvement and detachment, and shall argue that Habermas's proposals about an 'ideal speech situation' are ideal precisely because they enable intersubjective communication to *slow down*, without requiring the residual idealism of transcendentalist discursive presuppositions. In Chapter 5, I shall more broadly thematize Bourdieu's views about science and the role of intellectuals, in order to lend greater pragmatic depth to his notions about scientific autonomy and the role of time. His suggestive ideas about intellectual 'anti-politics' can be fruitfully translated into the more promising *temporal* interest of intellectuals to reduce the speed of representation in fast-running political machines. In Chapter 6, I shall critically discuss the quasi-Cartesian principles of symmetry and agnosticism, which have become central methodological benchmarks in modern science and technology studies, in order similarly to recast them in more context-sensitive temporal terms. Here I shall also pay more extensive attention to Latour's political or military metaphorization of science as inducing the risk of 'following the actors' *too fast*, of frantically keeping up with their speed, and hence of tendentially losing the capacity to follow them *critically*. In Chapter 7, I shall introduce a version of the principle of reflexivity which, by acknowledging the constitutive circularity of all accounts, further radicalizes the habit of methodical doubt, and as such offers an effective way of slowing down science and of weakening our conceptions of what the world is like.

From Partisan Science to Knowledge Politics

I have already suggested that the description of science as inhabiting an envelope of slow time immediately includes a normative capacity and extends into a chronopolitical project. Due to this immediate entanglement of fact and value, the description of the spatio-temporal 'nature' of science does not reflect a natural category, but performatively acts upon (the times of the) world. It 'naturally' extends into a critical engagement to realize what it describes, i.e. to protect scientific practices from the universal acceleration imposed by current gearshifts in faster techno-practices such as politics, business, and the media. Informed by the principle of the 'natural proximity of facts and values', which will be more fully explicated in Chapter 4, this particular take on the performativity of science emphasizes the normative complexity and complicity of all empirical descriptions and classifications. It is set in deliberate contrast with more neutralist versions of the performativity and politics of knowledge which have been widely adopted in the vocabularies of the linguistic, constructivist and ethnographic turns in science studies and social theory. These have bred forms of relativism, agnosticism, and a-criticism which resonate with an intellectual mood that has been aptly described as the current 'fascination for a-morality' (Neckel and Wolf 1994).

This polemical context requires some historical framing. Retracing the origins of the constructivist turn is useful for ascertaining not merely its overall direction but also what it has actively 'turned off'. What exactly was involved in the paradigm shift from the old configuration, in which a liberal principle of value-freedom faced strong demands for political partisanship and for taking sides in social conflicts (cf. Hammersley 2000), towards the new situation in which a principle of 'knowledge politics' has apparently superseded both? What was retained and what was discarded when the old 1970s stand-off between the Weberian demand of scholarly detachment and the Marxian one of a politicized 'unity of theory and practice' yielded before the new Foucaldian problematic of the intrinsic coalition of truth and power? My sense is that it produced an intriguing paradox, which will extensively occupy us in the pages below, but which can be briefly pointed up as follows. On the one hand, the knowledge-political shift obviously promoted a more visceral politicization of science, since the political was no longer viewed as belonging to the extra-scientific realm (to be kept at bay or to be embraced by taking sides in broader social struggles) but was permitted to invade the core business of the construction of scientific statements itself. This micropolitical shift effectively undercut the grand certainties of science and emancipation and the debunking style of ideology critique that they legitimized. Simultaneously, however, it elicited a contrary depoliticizing effect, by mounting a full-scale retreat from normative concerns towards an agnostic descriptivism which applied a strict principle of symmetry to the issue of taking sides in both scientific and social struggles. While professing indifference to the traditional terms of the fact/value issue, the new epistemological matrix of knowledge politics was therefore simultaneously caught out by a residual version of it which effectively cut short its critical potential.

In order to trace this paradoxical displacement and implosion of the political in more precise terms, it is worthwhile to linger briefly with the substance of the old set-piece debate between 'Marx' and 'Weber'. Indeed, the knowledge-political principle of analysis is immediately helpful in registering deep contradictions in both the Marxian principle of partisan inquiry and the Weberian one of value-neutrality – both of which contain elements of the other, as one would readily expect of constitutive opposites in a dialectical double bind. For its part, Marx's famous Eleventh Thesis on Feuerbach ('the philosophers have so far only interpreted the world; the point is to change it') fatally misrecognizes the internal politics and performative force of scientific proclamations, which themselves already change the world by interpreting it in a particular way – a transformative force of which Marxism itself has provided a uniquely dramatic example.[12] The Nietzschean idea of the 'violence' that is naturally committed by all forms of interpretation thematizes a proximity and simultaneity between knowledge and politics and between facts and values that is epistemologically far more radical than a scientific politics that seeks to derive the hitting power of the latter from the universal certainty of the former. Thesis Eleven indeed operates on the tacit premise that the interpretation of the world is in broad outline *finished* and *available*,[13] and that it can now be historically realized by empowering itself through the external mass of the revolutionary proletariat (and, equally tacitly, through the leadership by the 'internal mass' of revolutionary intellectuals who claim exclusive ownership of this interpretation).

In this regard, Thesis Eleven displays the same vicious circularity that is the besetting sin of all standpoint epistemologies and theories about situated knowledge, including their current neo-Marxist, feminist and postcolonial updates: while practical positions are taken to situate and adjudicate theoretical positionings, it is (radical) theory that simultaneously pre-structures and dictates these practices, begging the reflexive question of how to define the situation for these situated knowledges (cf. extensively Pels 2000). The proletariat functions as a 'philosophical double' and a false shadow of the radical intellectual, who both conceals and empowers himself by identifying with the class for which he acts as an exclusive spokesperson. Who the sides to be taken (the classes) are is a matter of performative class-ification, in which values cannot be independently derived from pre-given natural facts but are so deeply entangled with the latter as to constitute the political tension and tonality of the description itself. Political partisanship is scientifically certified by making reference to an objective fissure or immanent contradiction in reality which the critic must 'enter' and whose laws of development he or she merely needs to 'display'. Reality contradicts and judges itself, an ontological fact that the scientific observer 'merely' needs to acknowledge and articulate in order to turn political. There is no need to criticize the world if, as the young Marx expresses it, social truth can be developed out of 'the conflict of the political state with itself'. If our critical task is merely to show the world what it is 'really' fighting for, and explain to it the true meaning of its own actions, nothing prevents us from tying our criticism to a definite partisanship in politics, or identifying our criticism with real struggles (Marx 1972 [1844]: 9–10). Marx's partisanship hence affords an intriguing pose of dispassionate 'cool' on the part of the scientific observer, who merely needs to 'follow the proletariat' and the necessary unfolding of the objective contradictions of capitalism in order to know whose side he must be on.

A similarly vicious circularity disturbs Weber's famous principle of value-freedom, which Dahrendorf and many others have reverently cited as 'the beginning of the scientific century of sociology', and which has indeed come to dominate the liberal conception of the relationship between science and ethics and of the place of science in society throughout the twentieth century (Dahrendorf 1968; Proctor 1991: 154). Despite its remarkable historical success, however, Weber's postulate about the 'absolute heterogeneity' of the realms of *Sein* and *Sollen* appears both self-contradictory and self-defeating, insofar as it is predicated upon an (onto)logical judgment which itself inevitably performs an immediate *mixture* of description and evaluation. In this respect, the Weberian principle of separation between facts and values offers a highly paradoxical (and hence singularly eloquent) example of its very opposite: that of their natural proximity.[14] If Weber describes the 'unbridgeable logical gap' between factual and normative truth as a simple fact (*Sachverhalt*) which is given 'in our historical situation' as a product of the rational disenchantment of the world, i.e. as part of the fate (*Schicksal*) of an epoch that has 'eaten from the tree of knowledge', he evidently presupposes the epistemological and ontological validity of his own modernist theory of rationalization and its constitutive dichotomies between values and facts, ends and means, and reason and force. If he cites the preservation of the value/fact distinction as a 'very trivial demand' and an 'elementary duty of scientific self-control', adding in a high moralistic tone that it is 'a weakness of character when we do not dare face up to this *Schicksal der Zeit*' (Weber 1949: 1–2, 11–12, 57–58, 98; 1970: 143, 149), we are attending neither an empirical generalization nor a normative decision but an inextricable *mixture* of both which performatively counteracts the stated obviousness of their formal separation as a simple demand of scientific logic and intellectual integrity.[15]

Any strict demarcation between words that act as ploughshares, to 'loosen the soil of contemplative thought', and words that function as weapons, or as 'swords against the enemies' (Weber 1970: 145), is eroded as soon as it is acknowledged that 'words are also deeds' (Wittgenstein 1998: 140) and derive a good portion of their meaning from their polemical context or adversarial energy (Schmitt 1996: 30–32; Fabian 1983: 152–53; Pels 1998: 230–31). Conservative-revolutionary intellectuals such as Schmitt, Freyer and Steding anticipated this micro-political critique by pointing out that the Weberian demarcation between science and politics was ordained not only from the vantage point of science, but more significantly, from a particular neo-Kantian conception of it that was deeply embroiled in a liberal politics of separation and neutralization (Schmitt 1993; Freyer 1932; 1964; Steding 1932).[16] Patrolling the dividing line between rationality and irrationality, and deciding between 'good' and 'bad' faith in terms of the responsible weighting of means against ends and facts against values, liberal science preferred to conceive of these divisions as etched into the nature of things (and hence available to rational scientific interpretation) rather than as products of knowledge-political decisions.[17] But the demand of neutrality could never be a purely (methodo)logical one but was necessarily rooted in ethical arguments and in strategic, political considerations (cf. Scott 1997: 45–46). As both Bourdieu and Latour would agree, genuine theory of science therefore needs to be *political theory*: 'The relationship of science and value is a political problem. That is, any particular conception of their relations involves value judgments and cannot be neutrally determined. It is not less "political" to claim these are separate than it is to claim they are indissociably joined; the relation of science and politics is something we invent, more than it is something we discover' (Proctor 1991: 152, 231; cf. Root 1993: 38, 42).[18]

In this book, I shall likewise depart from the view that the ideal of value-neutrality, instead of reflecting an ontological gulf between facticity and normative validity, must be seen as a set of changing, context-bound strategies of redrawing and stabilizing the boundary between science and society. Far from presenting a timeless or self-evident principle, Proctor has also argued, the idea of value-freedom displays a chequered history and a complex political geography, having alternately functioned as 'myth, mask, shield, and sword' in a variety of historical contexts. It has regularly succeeded in masking the broader political ideals of its advocates, who alternatively mobilized it against intrusive political movements of the left or the right (such as socialism, fascism, feminism, or racism), to adjust to shifting balances between science, the state, and the market, or to account for intrascientific developments such as discipline formation and professionalization. In this political capacity (and its active denial of it) the principle of neutrality fits into a wider liberal vision of the relationship between science and society which, in its fully developed form, opposes the value-freedom and objectivity of science to the subjectivity of moral and political values as complementary principles which together define the fundamental political ideology of modern science (Proctor 1991: 6, 65, 70, 266–69; cf. Root 1993: 10ff.).

Within this comparative framework, *both* the Marxian and the Weberian positions therefore reveal a proximity between facts and values that is far more 'natural' and visceral than is accounted for in their official views of the relationship between science and politics. Another way of putting this is that

The Timescape of Science

both positions presume a *distanciation* between the realms of the factual and the normative (which is less principled and formal in the case of Marx than in that of Weber) that breaks down upon closer analysis of their knowledge-political 'deep structure'. Marx's derivation of political certainty from scientific objectivity results not so much from a symmetrical *confusion* of facts and values as from a disciplined *separation* between them which surreptitiously asserts the directive *priority* of the former over the latter and does not question the ability of science to ascertain universal truth. While Marx might therefore be more Weberian than is traditionally assumed, Weber is more Marxian because the divorce between science and politics is executed by a discursive politics which itself extends towards a broader liberal political agenda. Although Weber repeatedly disavows any objective validation of normative and political goals, his liberal 'politics of neutrality' hence offers the same heady brew of strong passion and strong detachment which also characterizes the Marxian principle of 'cool' partisanship. Despite their stated differences, both would affirm a strong devotion to the pursuit of disinterested causes and impersonal values as the moral foundation of both science and politics as professions.

Against this backdrop, I can now more fully spell out the critical deficit of the constructivist and ethnographic turns, which have sought to supersede the old stand-off between Marxian partisanship and Weberian neutrality. The purpose of my analysis has been to suggest that both 'Marx' and 'Weber' miss the deeper performativity of their own close entanglement of factual explanation and normative evaluation. The basic quandary of constructivist ethnographies of science could then be that they have retained too much Marxian 'cool' (and too much Weberian neutrality) by insisting that there is no point in criticizing the world in an arbitrary and subjective fashion, if one can empirically 'display' its inner mechanisms and 'follow' the critical capacities of the actors themselves. In this respect, there is an intriguing resemblance between Marx's injunction to 'follow the proletariat' and the Latourian methodology of 'following the actors'.[19] While promoting a much-needed disenchantment of scientific rationality, this move towards a descriptivist ethic of research has simultaneously arrested all normative and critical appetites within the confines of what might be called a 'value-free relativism' (see further Chapter 3). Instead of dissolving the problem of value-freedom and bridging the gap between values and facts, the new analytic of power/knowledge has retained the dualism and reasserted a new version of the neutrality ideal.

Part of my argument in this book is to suggest a more radical principle of the 'natural proximity' of facts and values and a conception of performativity which derives its force precisely from their intrinsic mixture. This also implies that I no longer consider it appropriate to invoke such residual versions of the neutrality ideal to shore up the autonomy of scientific work. I have noted elsewhere that Foucault's genealogies retain a normative complexity that is tendentially lost in alternative (de)constructivist methodologies, and that his principle of power/knowledge might fruitfully be interpreted in a minimal-normative way (Pels 1995b). More broadly, it would appear that the shift towards methodologies that privilege the agnostic description of performative *effects* leaves the fact/value issue hanging, precisely because they fail to account reflexively for the normative and performative purpose of their own descriptions.[20] The task is then to reintegrate the idea of a natural proximity of facts and values with the Foucaldian idea of a natural proximity of knowledge and power, and to duplicate this integration on the level of analytic observation itself. It is to make workable a notion of performativity that is able to supersede the two dualisms in a single gesture. The point of this reflexive exercise is to salvage a workable notion of *critique*, which builds upon the knowledge-political turn but twists it back in order to recover its normative capacity.

The Structure of this Book

Having thus introduced the guiding themes of my study, I shall now provide an overview of the individual chapters that follow. Chapter 2 takes up the fundamental notion of knowledge politics in order to prepare an alternative 'normative description' of scientific autonomy. The first step in the argument introduces a 'knowledge-political continuum', which replaces the heavily guarded, impassable boundary policed by essentialist criteria of truth, logic, and method by a gradient of lesser distinctions, lower thresholds, and weaker boundaries, which range all the way from the micropolitics of science to the macropolitics of the state. The main point here is that the autonomy of science is socially performed and pragmatically accomplished by continuous investments in these smaller boundaries, which incrementally separate it from the similarly complex and graded reality of politics. The second step in the argument, already alluded to above, introduces a differential *time* dimension in addition to the traditional dimension of space, suggesting that scientific autonomy can be reinvented within this new framework of graded distinctions and permeable boundaries by attending to the specific effects of *deceleration* or *unhastening*. This temporal gradient can be filled out by means of a critical phenomenology which points up a set of pragmatic differentials between slower 'writing cultures' and faster 'talking cultures', highlighting significant variations in public exposure, issue selectivity, and relative audience size. The question about the temporal specificity of science is then recaptured on the broader canvas of what I call the 'social triangle'. This spatio-temporal model revises the classical triadic ontology of culture, polity, and economy in order to re-balance the liberal problem of the separation of social spheres with the post-liberal one of de-differentiation and boundary erosion. This new constellation defines a weaker autonomy for culture and science that no longer relies upon the core liberal postulates of neutrality and disinterestedness, and focuses the specific temporal profile of an unhastened practice which needs to be protected against the demands of urgency, immediacy, and publicity that invade it from the 'fast tracks' in the social triangle.

Chapter 3 shifts attention to some radical strands in social epistemology that have recently advanced new agnostic and relativistic ways of analysing scientific rationality. Over the past decades, a regrettable schism has developed between two currents of theorizing and research, which I shall abbreviate as the 'Mannheimian' and 'Wittgensteinian' traditions. The radical impulse of the new social studies of science in the early 1970s was initiated not by followers of Mannheim, the founder of the reflexive sociology of knowledge, but by Wittgensteinians such as Kuhn, Bloor, Barnes, and Collins, who articulated a programme of 'value-free relativism' that favoured an agnosticist and constructivist inquiry into the inner politics of scientific discourses and practices. This chapter argues that this Wittgensteinian programme has been encountering difficulties that might to some extent be settled by reverting to a more normatively sensitive research agenda. A social theory of knowledge (or social epistemology) along neo-Mannheimian lines could begin to refocus the normative complexity of sociologically situating knowledge and science and revive the vexed issues of ethical judgment and ideology critique. In this way, it would help to reconnect the social analysis of science to a broader macrosocial account of the 'knowledge society' and its democratic dilemmas.

Chapter 4 deepens my inquiry into the issue of neutrality and critical theory by focusing more substantively upon the fact/value dichotomy itself. This dichotomy, it is claimed, freezes and eternalizes a distinction that shows great historical and contextual variation, being constitutively tied to various phases and intensities of the institutional separation between science and society. The 'knowledge-political' performance of what is traditionally assumed to be a timeless and self-evident principle generates an inner tension that is close to constituting a reflexive paradox (Weber, I have already suggested, dramatically breaks his own methodological ground rule in its very moment of enunciation). This paradox might explain why the neutrality ideal stands helplessly before the multiple fact/value interactions that are increasingly evident in present-day philosophy and science, and fails to account for various forms of 'contamination' that blur the operational logics of culture, politics, and the economy. While the intellectual history of the dualism witnesses a century-long process of conceptual fissure, culminating in asymmetrical epistemologies such as those defended by Weber, Popper or Kolakowski, this presumed philosophical end point is currently surpassed by numerous epistemologies that variously mingle and (con)fuse values and facts. The resulting diamond-shaped historical matrix will be detailed and clarified by means of brief critical excursions into the epistemological worlds of Elias, Habermas, and Latour.

Chapter 5 broadens the inquiry towards the role of intellectuals and the contentious issue of professional vs. lay expertise. Being recently deprived of its high stature and its transcendental assurances, the intellectual profession nevertheless retains a particularity that warrants a more pragmatic and knowledge-political defence of its autonomy. A weaker defence of intellectual competence and professional expertise sensitizes us to their intrinsic 'duality', according to which enabling and disabling dimensions or functions and dysfunctions exist in close symbiotic collusion. Such a dual conception of intellectual autonomy is most interestingly approximated in Bourdieu's field theory of science and his defence of the 'anti-political' corporatism of the intellectuals. In critical debate with Bourdieu's rigorous 'field scientism', this chapter elaborates a normatively enriched conception of autonomy-as-duality that aims to overcome his residual distanciation between truth and power and between science and ethics. Bourdieu's views about the differential tempi of science, the mass media, and the capitalist economy need to be further articulated to this effect, effectively radicalizing his own proposals about a *Realpolitik* of reason in the direction of a more pragmatic politics of time. The notion of intellectual 'anti-politics' itself attracts an interesting chronopolitical interpretation which singles out intellectuals as go-betweens between slower (scientific) and faster (political) timescapes. Anti-politics is a politics of deceleration, of a systematic halting or stalling of decisions that *refuses* to think under the dictate of the 'extreme case'; it precisely attempts to depressurize the political cooker, to incapacitate swift action, to weaken hard facts and propositions, and thus to increase the amount of insecurity and doubt.

Chapter 6 continues my critical inquiry into the relationship between autonomy and neutrality by focusing upon a few core notions in the recent tradition of science and technology studies as already introduced in Chapter 3. While symmetry and impartiality have become ruling methodical principles throughout science and technology studies (STS), defining its core ideal of value-free relativism, their philosophical anchorage has attracted less discussion than the issue of how far their jurisdiction can be *extended* or *generalized*. This chapter argues that these principles unwarrantedly present as generalizable procedure what are in fact contingent knowledge-political attempts to reposition various fields of controversy. An inspection of the recent 'Epistemological Chicken' and 'Capturing' debates, and of some symmetrical 'translations' of the historical dispute between Hobbes and Boyle, reveals some of the epistemological and political hazards that afflict STS's convulsive forward push of the symmetry frontier. The chapter further attempts a time-sensitive translation of the principle of symmetry as a tactic of deferral or delay in the face of a particular set of modernist dichotomies. It concludes by re-examining some tensions and contradictions in Latour's political epistemology, which are focused on the question of the relative pace of practices of fact-making. One point I wish to make is that one should be careful not to identify the pragmatic materialization that is endemic to writing and coding techniques with stronger epistemological notions of reification. The style or modality in which facts are written up (as natural entities, or as reflexive constructions) makes a critical difference to our stories about the world which it is important to retain and normatively spell out.

Reflexivity, or the systematic attempt to include the spokesperson in accounts of the social world, is indeed a magnetic signature and inherent riddle of all modern thinking about knowledge and science. Turning the narrative back upon the narrator may sharpen our critical wits about the 'inescapable perspectivity' (Mannheim) of human knowledge; but self-referential accounts may also trigger endless loops of meta-theorizing and lose track of the object itself. Negotiating the twin pitfalls of spiralling meta-reflexivity and flat naturalistic accounts, Chapter 7 argues for a reflexive 'one step up', which adds one storey to the story (no more, no less). It explores, through critical discussions of the work of Latour, Harding and Bourdieu, the ways in which reflexivity may promiscuously combine with conflicting objectivist and constructivist epistemologies, and how reflexive accounts invariably appear to run in a circle. This performative circularity, I argue, does not invalidate the reflexive effort but defines its major strength (-in-weakness), gracing both world-accounts and self-accounts with a radical incompleteness and uncertainty. As such, it also highlights the *temporal* dimension of reflexivity as incarnating a more radical version of the old imperative of scientific doubt. The reflexive acknowledgment of circular reasoning extends and aggravates this doubt, and hence further diminishes the swift pace of acting and decision-making that drives more frenzied practices both within and outside science.

The concluding chapter first of all sketches a broader historical background to the pragmatic identification of science that has been previously laid out. It briefly chronicles the chronopolitical history of the ideas of scientific autonomy and academic freedom, beginning with the 'stammering' of the Socratic philosopher and the silent prayer of the medieval ascetic, continuing via Descartes' monastic quest for 'peaceful solitude', towards the 'Baconian' compromise, which was ready to accept political censorship in exchange for a limited autonomy of science. Kant's account of the 'contest of the faculties' and Humboldt's plea for academic 'solitude and freedom' restructured this compromise and laid the groundwork for the influential neo-Kantian defences of value-neutrality that subsequently dominated much of twentieth-century social and political thought. Following my critical angle, this narrative also further disentangles the idea of methodical doubt from the neutrality ideal in order to reposition it within the more contextualized and political modality of *timing*. I then recall the main parameters of the 'social triangle' as they were outlined in Chapter 2, explaining how this matrix is able to chart multi-directional processes of infiltration and boundary erosion, while simultaneously protecting the relative autonomies of the three spheres of culture, polity, and economy by means of various gradients of resistance.

My final argument is that we precisely need this post-liberal balance between differentiation and de-differentiation, or between autonomy and fluidity, in order to articulate what makes science special among other human pursuits. In other words, we need not fear the legitimate intrusion of political and economic metaphors, performance indicators, and operational styles in the inner fabric of science, if we simultaneously acknowledge that science is not simply a form of politics or economics 'continued by other means', but that it is precisely continued by the *slower* means of observation, experiment, comparison, inscription, unhastened conversation, protracted conflict, and low-gear competition. Unhastening science without the 'scourge' of internal competition, or without the 'political' jostling for recognition and a distinctive name, tendentially degenerates towards idle privilege. On the other side of the balance, however, the pace of circulation in the scientific republic (or the scientific marketplace) must be shifted down in order to resist the 'time famine' produced by the faster machinery of politics, business, sports, and the media.[21] In a 'nanoculture' that glorifies the speedway and the 'fast-forward mode', the university may still provide a resort of stillness and slow motion where 'untimely' thoughts may be cultivated in relative liberty and leisure. A foreshortening of the temporal horizon of the collective imagination as a result of universal acceleration critically undercuts this culture of waiting, and hence paradoxically risks slowing down the rate of social creativity and innovation over the longer historical term.

CHAPTER 2

What (Again) is So Special about Science?

Well, we've got time, haven't we, Socrates?

Plato, *Theaetetus*

Self-Interested Science

Twenty-five years of irreverent thinking and thick empirical description have done much to dislodge the long-standing philosophical conviction that science has a special, singularly compelling, and context-spanning rationality that legitimately dominates and adjudicates ordinary, local forms of reasoning (what used to be called 'common sense'). It is no longer seen as the supreme legislator of all human knowledge, setting standards of truth and logic that automatically bridge disparate social and historical experiences, and defining universal principles of right reasoning and rules of proper method that explain its unique capacity to produce a truthful picture of the world. Increasingly, also, 'science' in the singular has come to be seen as bad shorthand for a vast plurality of practices, which are fragmented across many disciplines, niches, paradigms, and approaches. More dramatically, science has come to be viewed as just one culture of rationality among others, 'just another story', one among a plurality of perspectives, information bases, and interpretive communities, none of which can lay claim to a totalizing, overarching, or foundational status. The shockwaves that were generated by Feyerabend's rhetorical question 'What is so special about science?' (1978: 73) have gradually subsided, because the efforts of an entire generation of sceptical students of scientific and technological success have meanwhile been invested in arguing the implied answer: 'Nothing really'.

This new emphasis upon the ordinariness of a core element of intellectual culture is not an isolated phenomenon but partakes of broader cultural mutations and social realignments, which have tended to desacralize high culture and brought it closer to the world of everyday meaning. A tradition has been building up from Benjamin's (1973 [1939]) ground-breaking proposition that the cultural realm was shedding its 'auratic' character and could no longer be separated on principle from the broader social realm. Indeed, some twenty years before Feyerabend mooted his iconoclastic question, Raymond Williams had already established a firm baseline for British cultural studies by emphasizing that 'culture is ordinary'. In Williams' estimate, culture was not confined to the arts and high learning, to the dazzling peaks of discovery and creativity, but included the 'most ordinary common meanings' next to the 'finest individual meanings'. One needed to avoid the customary polarization between 'self-gracious sophistication' and 'doped mass', and emphasize 'not the ladder but the common highway', since both elite and popular culture belonged to a 'good common culture' (Williams 1989 [1958]: 4, 15, 17, 34–35; cf. Willis 1978). In close parallel, the new sceptical approaches to science maintained that science had lost its transcendental aura, that it was social through and through (as messy, interest-driven, power-ridden, and opportunistic as more mundane social practices), and that conventional hierarchies opposing the high rationality of philosophy and the low rationality of lay and non-scientific thought were progressively crumbling away. This reconfiguration shifted the direction of research from philosophical issues of demarcation, boundary maintenance and division of labour towards a more sociological attention to issues of continuity, symmetry, boundary *work* (Gieryn 1983; 1995) and the 'labour of division' (Hetherington and Munro [eds] 1997).

If cultural studies exchanged the idea of the autonomy of bourgeois culture for a converging culturalization of society and societalization of culture, incipient science studies similarly emphasized the mundanity and social embeddedness of its subject matter by letting in politics and economics through the front door. The unexceptional, down-to-earth character of scientific rationality, like that of other practices and products of high culture, was preferably affirmed by highlighting the salience of everyday *interests* that derived from the mundane competition for power, money, and prestige. These interests imparted to scientific practices a quasi-political or quasi-economic character that was previously considered foreign to and pollutant of the very essence of science, since they imported the supposed perspectival blinkers of social location (standpoints in time and place) into the heart of a practice that was considered to be free-floating and universal. If cognitive and social elements in science, knowledge and power (or knowledge and 'capital') were to be reconceived as proximate, intimately interlaced and reciprocally determinant – as high art and low commerce appeared intriguingly to mesh for students of the new consumer

What (Again) is So Special about Science?

culture – venerable philosophical dichotomies separating truth from interest, morality from politics, or values from facts would lose whatever force they still retained in policing the Great Wall between science and society.

In this fashion, the levelling discourses that revalorized the everyday (of lay rationality, practical morality, commercial and pop art) while 'devaluing' the culturally unique (high science, high ethics, classical art) effectively co-opted and combined two venerable vocabularies of disenchantment: the Marxian one of economics and capital and the Nietzschean one of power and politics. Registering something analogous to the search for profit or the will to power as intrinsic features of the professional quest for knowledge, culture and technology, these metaphoric repertoires expressed the inseparable duality of cognitive and strategic interests and the resulting agonistic structure of scientific endeavour, revealing the intense competition, the scramble for reputation, and the daily commerce of turning soft claims into hard facts that characterized science as a this-worldly practice. In this fashion, the two metaphors appeared equally useful in hastening the desecration of what Nietzsche dubbed the 'ascetic' ideal of philosophical truth, and in disburdening science of its conventional epistemological privileges.

This generalization of economic categories to the analysis of cultural, intellectual, artistic and political practices and institutions has since become a familiar gesture in a variety of approaches, which include the 'economic theory of democracy' such as elaborated by Downs or Barry; rational choice theory as practised by Becker, Olsen, Coleman, or Elster; neo-Marxist cultural studies in the line of Williams, Jameson, Hall and Harvey; or the general 'economy of practices' which has been advanced by Pierre Bourdieu. The alternative expansion of political metaphorics has become a manifest feature of feminist studies (e.g. Harding's and Haraway's articulations of standpoint theory), ecological criticism and its call for a 'politics of nature' (e.g. Bookchin's eco-anarchism), Elias's processual sociology, Foucault's microphysics of power and discipline, the sociology of the 'risk society' and reflexive modernization as elaborated by Beck, Giddens, and Lash; and the recent enthusiasm in constructivist studies of science and technology about a 'politics of things'. Even though economic and political metaphors are increasingly 'synonymized' in recent analyses of culture and science (cf. extensively Pels 1998), there is still a lingering tendency to prefer either the Marxian idiom of property and accumulation, capital and market, credit and investment, or the Nietzschean idiom of power and negotiation, strategy and conflict, negotiation and network-building.[1]

As Rouse has indicated (1987: 12ff.), two types of encounter are usually thought to occur between knowledge and power (a parallel argument applies to the interaction between knowledge and capital). In the received or 'extrinsic' view, knowledge and power are essentially separate and antagonistic: power impedes or distorts the acquisition of knowledge, imposes false beliefs and suppresses the truth. True knowledge uncovers the distortions and unmasks the disguises, and is therefore capable of liberating us from power's repressive effects. Knowledge can also be applied to achieve power: knowing how things are or how they function creates opportunities for manipulation and control. The alternative 'intrinsic' view, which is clearly inspired by Foucault's slogan *pouvoir/savoir*, radically reappraises this relation, in describing power itself as the mark of knowledge. Dismissing the understanding of power as primarily repressive or censorious, this view suggests that power relations permeate the most ordinary activities in scientific research, and that scientific knowledge arises out of such relations rather than in opposition to them. Modern scientific practices are political in ways that centrally define their epistemic success (Latour and Woolgar 1986 [1979]; Latour 1987; 1988a).

In a slightly different phrasing, this distinction turns on two radically opposite interpretations of the Baconian dictum that 'knowledge is power'. Its 'linear' Enlightenment articulation, which has informed classical utopian thought, French positivism (cf. Comte's slogan *savoir pour prévoir pour pouvoir*), classical liberalism, and Marxian political economy, typically specifies scientific knowledge as a body of true, well-founded, and universally valid propositions, which ideally mirror (social) nature in undistorted fashion. The opposite, 'dual' interpretation, which characterizes more recent pragmatist, constructivist and performativist approaches, simultaneously reverses the arrow of influence. The idea of a 'knowledge politics' or a 'politics of truth' evokes the double notion that knowledge is never 'outside' power and that power itself is always discursively and symbolically constituted. Knowledge does not copy reality but actively intervenes in it, construes it, performs it, even if (especially if) it claims to do nothing but transparent copying. Power, politics, and interests do not so much blind the knower and pollute the clear source of truth; they are also 'productive', both in highlighting sections of reality and opening up perspectives, and in actively co-producing the realities they describe or explain.

Knowledge Politics and Time Economy

In the following, I want to take this fundamental idea of the mutual determination of power and knowledge (or the intrinsic entanglement of reason and force) and differentiate it by spreading it across a 'knowledge-political continuum'.[2] If power is indeed invested into the heart of (scientific) knowledge, and science simultaneously enters into the thick of the political, traditional liberal demarcations between the two realms are destined to become blurred.

What (Again) is So Special about Science?

The heavy, impassable boundary which is marked by essentialist criteria of truth, logic, and method, and which is shored up by Popperian and Mertonian values such as neutrality, disinterestedness, community and universalism, is replaced by a gradient of lesser distinctions, weaker boundaries, and softer demarcations, which range all the way from the micropolitics of knowledge to the macropolitics of government, passing through all sorts of practices and institutions which mix and mediate them, and which define different forms of coalition and articulation.

In this fashion, the topology of a knowledge-political continuum excludes two less attractive epistemological options, dutifully offering itself as a supervening third position. One of these focuses the classical quest for truth 'for its own sake', which requires science to jealously guard the autonomy and elevated neutrality of its special location (the ivory tower). Contrary to this view, I shall (rather abruptly and performatively) maintain that there 'can be no such thing' as a search for knowledge that is purely interest-free, curiosity-driven, or value-neutral, or a form of dialogue or discussion that can (or, for that matter, should) be liberated from power, interested negotiation, or strategic calculation. But I simultaneously want to distinguish this view from 'identitarian' theories that simply posit the coincidence of knowledge and power and tendentially collapse the realms of science and politics into one another (an erasure that can originate from either the scientific or the political pole, issuing in forms of scientific imperialism or political totalitarianism respectively). These approaches advocate the notion of a 'seamless web' of relations where differentiations are no longer legible, where mobility and complexity are rampant, where all distinctions between inside and outside become fluid and evaporate, and where the ideas of autonomy and bounded practice lose much of their meaning.[3]

In this book, I propose to recapture the idea of scientific autonomy within this framework of graded distinctions and permeable boundaries, placing it at an equal remove from the notion of disinterested, value-free science and that of 'other-interested', externally driven, and 'politically correct' science. If autonomy can no longer be sustained in absolute terms, on the basis of the unique epistemological status of science, it can be more realistically described as an interactive and variable process: as the ever-precarious outcome of negotiations about flexible and shifting boundaries between science and society (Cozzens 1990; Gieryn 1995). It is a distributed and dynamic state that is as much a product of relatedness as of delineation (Fox Keller 1985: 98), demanding incessant definitional performance in order to produce a stability that can only be contingent and provisional. A non-neutral, knowledge-political defence of scientific autonomy might therefore begin from the more promising notion of

Unhastening Science

self-interested science, which finds itself incessantly negotiating the weak boundaries that separate the various institutional forms and locations of knowledge politics and 'big' politics. Self-interested science may serve as an appropriately demythified reformulation of the idea of the pursuit of knowledge 'for its own sake'; highlighting both the need for its professional independence and the risks of corporatist insulation, monopolistic control of resources, and collective arrogance that are coincident with it. If science is (to be) 'nothing special', one might still defend its autonomy in the same sceptical and ambivalent mood in which one would support the relative independence of other ordinary occupations and communities of skill, including the professional conduct of politics.

How can we rethink the relative autonomy of science (and that of the political), starting from the continuity and mediation rather than the principled opposition between truth and power? I have already begun to draft a continuum of weak demarcations (and weak autonomies) that exchanges the normative problem of how to specify universal and invariant criteria distinguishing science from non-science for a more descriptive problem of how to follow a variety of distinctions across a range of institutions which are not separated by a Great Divide but are more equally balanced across many small ones. Focusing upon the mixtures rather than the oppositions, this continuum follows the entire span from the extreme of 'blue sky' academic research to that of 'big' professional politics and state administration, detailing a complex range of mixed institutional logics which – travelling from the left-hand to the right-hand pole – may include university-based policy-oriented research, academic administration, professional auditing boards, independent consultancy, science journalism, 'movement intellectuals', semi-statal funding agencies, research units attached to political parties, and research departments in government bureaucracies. Figure 2.1 illustrates a simplified British version of this positional continuum, which simultaneously articulates a gradient of occupational attitudes, psychological predilections and career-bound beliefs.

Representations such as these, which would need much finer elaboration, begin to articulate a critical phenomenology of science–politics relations, the purpose of which is to offer an ethnographically enriched picture of what scientists and politicians (in all their different role mixtures and combinations)

Figure 2.1: The Knowledge-Political Continuum

'actually do'.[4] That is, if we are also agreed that we can no longer count upon a strict philosophical separation between facts and values, and acknowledge that all such descriptions are performative and hence carry an inherent normative intent. We are looking at a continuum of interested positions (in the strong dual sense in which cognitive and social interests are deemed inseparable), where each link in the chain between 'self-interested' science and 'big' professional politics displays a different mixture of interests. Between the extremes are located various thresholds, transit zones and liminal spaces through which individuals may pass in order to shift positions, either temporarily or more permanently (e.g. academic intellectuals 'defecting' to posts in administration). Institutions may renegotiate their identities and displace their boundaries in competition with others (a dramatic example is offered by politicization drives in totalitarian regimes). Moreover, an incessant struggle (both discursive and physical) rages about the precise location of the boundaries and the height of the thresholds (e.g. scandals involving conflicts of interest, corruption and sleaze, political correctness, or abuse of power). As one travels from one pole in the direction of the other, jumping one or several institutional fences, atmospheric changes occur that build up incremental differences in the institutional logic or ethos of different occupational fields; pushing individual players to conform to the norms, rules and habits of adjacent but different social games. The combined effect of these displacements is to proliferate weak autonomies across a multiplicity of intermediate occupational zones which act as buffers (but also as trading zones and transit areas), and which generate stronger forms of autonomy as institutional distances become more stretched. Instead of being transcendentally guaranteed by principles of right reason, universal logic, and proper method, the autonomy of science is socially performed and pragmatically accomplished by continuous investments in the many lesser boundaries that incrementally separate it from the similarly complex and graded reality of politics.

It is time to fill out this phenomenology of science–politics relations, starting 'from the ground up', by way of entering a minimal and 'naïve' description of practical differences that is no longer weighed down by such lofty principles of differentiation. Travelling from the intellectual to the political pole, one may for example witness a gradual decrease in the crucial activities of *reading* and *writing* (and hence of the incidence of silence and quietness), and a concomitant increase in *talking*, disputing and negotiating (and hence of the amount of noise, anxiety and haste). It is a crucially defining characteristic of any politician that she needs to talk a lot, and that her investment in reading and writing typically remains confined to newspapers, reports, summaries of reports, notes, briefings, and even briefer briefings. If scientists talk, they typically conduct the far slower conversations that turn upon a careful perusal of arguments proposed by partners who may be far removed in time (e.g. quoting long-dead founding fathers) and/or space (debating with disciplinary colleagues on the other side of the globe), and that follow the leisurely rhythm of written commentary and critique: the quiet exchange and unhastened turn-taking of articles and counter-articles, books and counter-books. Science is typically of the 'long breath', depending on long-term cycles of investment in human and material resources, whereas politics expects quicker returns within a much shorter time-span. Even in terms of the act of reading itself, the sheer *pace* of the exercise typically differs across the range of professional activities. Politicians (including academics in a more public or 'political' role) tend to speed through administrative memos and reports, which supposedly feed into other memos, oral summaries, or discursively presented decisions. In the intellectual mode, on the other hand, articles and books, particularly if they must fertilize the composition of other articles and books, often require a sustained effort of re-reading, rethinking and sense-making – a technique of deceleration that is equally characteristic of the act of writing itself, which requires a succession of versions and revisions before the text is permitted to see the light of day. There is a specific delay in these objectifications (words 'staring back at you') and an endless deferral of the moment of decision, which is entirely out of place in the daily conduct of more speedy enterprises such as business or politics.[5]

Textual Tactics of Delay

As Goody and others have shown, there is an intimate linkage between the technology of writing (and especially the materiality of the text as objectified speech), the effect of unhastening or delay, and the ability to scrutinize discourse in a different, potentially more abstract, generalized and critical way. Writing, by giving oral communication a semi-permanent form, enables one to stand back from and quietly study a static and rigid 'thing' rather than being swept along by the hectic immediacy and fleeting dynamics of face-to-face interaction. Reading and writing eliminate the typical redundancy of oral thought and speech; they force the mind into a slowed-down pattern that enables it to continually reorganize itself and add precision, and in this sense affords more reflective and analytic habits of thinking and communication. Written speech is no longer tied to an occasion or a speaker; it becomes 'timeless', abstract, depersonalized, distanced from lived experience. Texts can be inspected in much greater detail, in their parts as well as wholes, backwards as well as forwards, out of context as well as in their setting. Evidently, their materiality and 'objectivity' also favour a more permanent storage of information; opening up a wider range of thought for the reading public and increasing the potentiality for cumulative knowledge and collective memorization. In this respect, there exists an originary and constitutive link between the technology of writing (as a technology of reflective delay) and the emergence of scientific rationality in the modern sense of this term (Goody 1977: 11–12, 37, 44–45; 1987: 300; Ong 1982: 39–41, 81–83; Goody and Watt 1968; McLuhan 2001 [1964]). Without wishing to subscribe to Goody's more sanguine conclusions about an intrinsic connection between literacy and rationality in its high-minded Enlightenment form (cf. Street 1984: 44ff.; Finnegan 1988), we may still secure the notion of the 'delay of writing' as a minimal criterion of distinction for a science that no longer offers itself as a compulsory standard of excellence for all other traditions of thought.[6]

Despite the recent avalanche of social interpretations, practising science (including *social* science) is still much more a matter of non-verbal, solitary (if not solipsistic) interaction with nonhuman objects (such as books, articles, protocols, instruments, machines, pen and paper, keyboard and screen) in the comparative stillness of one's study or laboratory than a matter of talking to and negotiating with other human subjects – which is the daily fare of politics and other 'verbomotor lifestyles' (cf. Ong 1982: 68).[7] Incessant talk and high interaction imply that the politician is expected to be far more gregrarious and personable than the average academic, which also defines a range of differences in presentation, tone, bearing, attitude, and dress code. While the scientist typically talks a lot to herself, the politician routinely (and loudly) talks to many others (face to face, on the phone, or confronting large radio and television audiences), spending long days in committee meetings and parliamentary sessions where the art of writing (brief notes and comments) is basically subservient to the production of more effective and authoritative public speech. While politicians speak as representatives of particular interest groups, constituencies, parties, and movements (or in their more ambitious moments, of the Nation or People at large), scientists primarily speak for themselves, for their research groups or departments, and for their discoveries (which they can similarly inflate to represent Nature or Reality). Even within the academic field, more 'political' or 'entrepreneurial' functionaries such as heads of department and chairs of disciplinary and administrative committees talk (or phone) a lot more, write a lot less, and regularly mingle with many more people than their colleagues who engage in sustained research and reporting of research. Conversely, 'intellectuals' and 'professors' in the political arena (a rapidly declining species) are often mistrusted or ridiculed because they are thought incapable of making quick decisions, unnecessarily philosophize about principles, and take time out to read or even to write books, which confers upon them the 'unnatural' distinction of working long hours in sustained privacy and tranquillity. Even within the fast political field, the pace of life quickens during election years and other crisis times, when politicians become even more short-termist, preferring to work towards the next poll rather than concerning themselves with long-term issues and the ultimate 'verdict of history' about their doings and failings.[8]

Another relative difference that further delineates the time economies that gradually segregate academics from politicians concerns the selectivity and the level of attention to issues. While scientists are expected to concentrate on an few isolated topics for a long, sometimes extremely long, period of time, politicians must be ready to switch among topics and issues very rapidly. Their professional situation typically favours a broad but necessarily superficial sweep of knowledge about a plethora of subjects, while scientists reverse this logic in favouring a deep acquaintance with a highly selective and narrow set of discipline-driven questions (e.g. the PhD student's solitary confinement in a detailed research topic for a three- or four-year period). In between these extreme positions, researchers employed in commercial consultancy, policy think tanks or scientific advisory panels of political parties (such as, in the British context, Demos or the Institute for Public Policy Research), while being allowed more time to conduct research on individual issues (a few weeks, a few months), are nevertheless expected to report to an agenda that is not self-selected but is normally dictated by their commercial or political commissioners. The weekly 'updating' television chat with the prime minister, which is a customary feature in several European democracies, typically involves a swift-paced ticking off of a whole cascade of current issues. Journalists who interview politicians on a regular basis in news and current affairs programmes tend to act as legitimate interpellators and 'partners in expertise' on all these different political topics; and the interpellated politicians are required to make up their minds very quickly, in conditions of uncertainty and stressful visibility, producing credible soundbites on any number of relevant issues that currently figure in the news.

By contrast, at the opposite end of this decisional continuum, academic researchers are extremely *slow* decision-makers, who endlessly ponder and prevaricate over which words to use in which particular context, preferring to keep silent rather than saying things they are not completely confident about.[9] Occupational accidents in science, such as the Cold Fusion episode or spoof discoveries of AIDS cures, are virtually always traceable to journalistic or managerial pressures to decide the issue and go public before the time is ripe ('discovery through press conference') (Haslam and Bryman [eds] 1994; Bucchi 1998; Gieryn 1999: 183ff.).[10] Academics who closely identify with public or political causes, or who more generally wish to cut a public figure, and prefer to write for newspapers or appear on radio and TV talk shows rather than publishing academic articles in journals 'that nobody ever reads', often incur the censure and displeasure of colleagues who accuse them of sacrificing their intellect to the seductions of publicity and the demands of the large audience on the other side of the screen. The novel habit of celebrity intellectuals of airing their ideas in interviews is often interpreted (and dismissed) by their less famous (and more envious) colleagues as a genuflection to the 'quick fix', signifying an unfortunate intrusion of the habits of journalism and the logic of politics which denature the scientific field. The true scientist, they are told in Socratic fashion, has no ambition to persuade audiences as large as the politician's electorate, and is satisfied to converse with smaller circles of students and fellow professionals in a more quiet retreat. She or he loves the semi-private spaces of the academy and is not overly attracted by the glare of the spotlights and the nervous pace of public life. She or he fails to be seduced by the fame or notoriety that is the inevitable dowry of the public personality who continually operates in the vicinity of cameras and microphones.

I would like to illustrate and further focus this incipient phenomenology of differences by briefly noting some typical patterns of interaction that occur at scientific meetings and conferences. Although embedded in the normal rhythm of academic and professional life, such events nevertheless introduce a distinct spatio-temporality that effectively speeds up and 'publicizes' (and in this minimal sense: politicizes) the routines of everyday academic teaching and research. Conferences produce a Durkheimian collective effervescence by assembling a crowd of talking bodies in a liminal space (away from the home turf and ordinary discipline, which usually invites forms of festivity, tourism, and other escapades and transgressions), in some ways similar to that which politicians experience as part of their everyday professional routine. A conference is a talking shop (a parliament), where real-time conversations are struck up against the background of the slow-paced ones conducted during the 'normal' duration of research and writing. They are about meeting people, establishing and renewing contacts, negotiating and striking deals, building up networks, and hence of affirming, maintaining, attacking, or losing scientific reputations. They create situations of heightened fervour and tension, suffused with the anxiety of having to perform to large(r) audiences that more or less immediately talk back. The 'natural' slowness of academic production also reasserts itself in the convention of participants' reading without interruption (for as long as 45 minutes) from a written-out paper on which they have bestowed weeks or perhaps months of thoughtful preparation in the stillness of their study. In this sense, reading a scientific paper is very much a rehearsed performance and a 'coming out' experience rather than an improvised talk, while the ensuing conversation does not follow the noisy and haphazard pattern of public political debate, but requires the audience to hold its fire until official question time. Question-and-answer patterns themselves, which display the same civilized turn-taking that is embedded in the broader temporal economy, are not meant to prepare the assembly to take a vote, but rather to delay ready-made consensus and to defer facile solutions or decisions. In spite of all such continuities, individual scientists nevertheless find themselves in a quasi-political 'cafeteria' situation in which subjects of talk can vary widely and follow each other in rapid succession. They temporarily operate in an arena of interaction which is far more public, speedy, and permeated by power speech than the world they normally inhabit. The vast majority of academics, I suspect, would be horrified about the prospect of having to live in permanence in such a public place and to such a hectic pace.

Science in the Social Triangle

What I have pencilled in so far is a rough sketch of a spatio-temporal continuum which, rather than being centrifugally suspended between two extremes that define an essential opposition, is drawn sideways out 'from the middle' in order to proliferate smaller differences and more permeable thresholds; these, in turn, while suggesting various interminglings between scientific and political practices, nevertheless succeed in keeping them at arm's length as relatively autonomous social timescapes. In contrast to the direct liaison suggested by classical conceptions of 'scientific politics' or 'partisan science', all connections are mediated ones, dampening all efforts at reciprocal invasion by the presence of many institutional hurdles that act as spatio-temporal filters. 'Social relevance' can hence never be an instantaneous product, but is only realizable by means of successive translations and mediations across an extended buffer zone. Even though they are no longer separable in terms of their natural gravitation towards an essential 'core' or 'logic' that is philosophically defined (the search for truth, the will to power), science and politics are nevertheless identifiable in terms of a gradient of contiguous distinctions pitched at a lower operational level. The high normativity of the traditional demarcative exercise is exchanged for the minimal normativity of defending their relative autonomy on the basis of such minimal specifications of time and place. The practical logic of scientific creativity requires a systematic deceleration of communicative interaction that can only be realized through a selective privatization of social relations and institutions.

Before spelling out and further systematizing these weaker benchmarks for

What (Again) is So Special about Science?

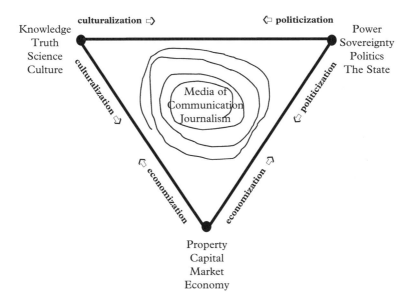

Figure 2.2: The Social Triangle

intellectual and scientific autonomy, I first intend to generalize and expand the horizontal spread of the knowledge-political continuum by marking out a third position, adding an 'economy' or 'market' pole to those of culture and the polity, and by drawing two similar continua converging diagonally upon it from the scientific and political poles. This unlocks the ontology of what can be called the *social triangle* (Fig. 2.2), which, while admittedly straitjacketing the rich mosaic of social life, has nevertheless suggested itself in one or another form to many social analysts (as different as Marx, Habermas, Bell, Cohen and Arato, or Castells) as a useful classificatory starting point (cf. also Ágh 1989; see Etzkowitz and Leydesdorff [eds] 1997 on the closely affinitive model of the 'triple helix'). The purpose of this triangular model is to find a new reconciliation between a liberal-modernist ontology of separation and demarcation and a post-liberal or postmodernist ontology of boundary-fusion and integration (Pels 1993: 208ff.; Leydesdorff 1995). I assume that social theory, in its confrontation with contemporary processes of differentiation and de-differentiation, no longer needs to choose between modernist purity and postmodern fluidity and flow, but can absorb both ontologies in a more encompassing view of how social reality is made up.

By angularly adding both a 'culture-commodity' (or 'culture-capital') continuum and a 'power-property' continuum to the 'knowledge-political' one

37

already in place, and similarly differentiating them in terms of weak boundaries, low thresholds, and a proliferation of differences (rather than the strong and singular divides that are installed by essentialist codes of Sovereignty or Property), the resultant triangle organizes a vision of a *trias societatis* that appropriately balances the mutual interpenetration of institutional principles or logics against the relative domanial autonomies that are nevertheless preserved across a great cascade of smaller demarcations. Currents of culturalization, which spill over from the cultural and scientific field, flow 'eastward' and 'southward' in order to inundate the polity and the economy (cf. the emergence of the designer economy and promotional culture; the aestheticization of consumption; the rise of an informational capitalism; the mediatization of politics; the rise of the entertainment industries and the celebrity system). Politicization drives emerge 'westward' and 'southward' from the right-hand pole to enter the domains of culture and the market (cf. the mixed or political economy of neo-corporatist institutions; forms of 'sub-politics' in science, art, sports, health services and ecological activism). Processes of economization analogously rise from the 'south pole' to flow upwards into the cultural and political domains (cf. the commodification of sports; the capitalization of art and the attenuation of status distinctions between high and low culture; corporate models in public bureaucracy; the entrepreneurial university). In this fashion, a political redefinition of science and culture (as 'nothing special') is readily combinable with a cultural, discursive, or symbolic view of politics; a political theory of property and the market may sit well together with an economic interpretation of political governance; and the culturalization (e.g. aestheticization or intellectualization) of economic life may be played against an economic view of culture and science. In all three domains and along the stretch of all three continua, the levelling effects of fragmentation, crossover and osmosis are offset and compensated for by the decoupling and buffering effects that guarantee the relative independence of culture, politics, and the economy beyond the steady flow of their triangular interweaving.

The metaphor of the social triangle hence evokes a repetitive pattern of institutional integration/differentiation that focuses a division of labour between three subsystems, action fields or social powers, which are not divided by any sharp ruptures, but retain their relative autonomy precisely because they are interconnected through broad transition zones. None of the three domains enjoys ontological primacy above any of the other; none of them is able to claim anything like a 'first instance' or 'last instance' determination; and none of them is in any sense reducible to any other. This horizontal model offers a contrast with the modernist logic of social differentiation, which is both enabled and contained by overarching factors such as a shared value consensus,

What (Again) is So Special about Science?

an infrastructural economy, or a foundational polity, which are taken to articulate and guarantee the unity of the social whole. Such domanial parity effectively excludes any strong claim for the constitutive or totalizing nature of the economy on the part of Marxist or liberal theorists (e.g. the Thatcherite advocacy of 'enterprise culture');[11] for the totalization of political sovereignty as imagined by radical right-wing theorists (e.g. 'conservative revolutionaries' such as Freyer or Schmitt); or alternatively, for the constitutive primacy of Culture, Reason, or Science (e.g. classical Enlightenment philosophy, intellectual and artistic socialism, sociologies of moral cohesion, Habermasian communication theory, or strands in cultural studies). In the economic sphere, this levelling syndrome entails a definitive rejection of the absolutist and exclusionary conception of private property and of foundationalist certainties about the productive primacy of the market. In the political sphere, it finally erodes the monolithic and centralist conception of state sovereignty, including socialist or communist variants that attempted to ordain central economic planning through state appropriation of the means of production. In the cultural sphere, one must relinquish the essentialist notions of truth, goodness and beauty that have from time immemorial underpinned a steep hierarchy between the auratic realm of the Spirit and the baser worlds of Power and Money (see Fig. 2.3).

In this fashion, all three domains are divested of their traditional ontological and foundational privileges, and are no longer thought capable of reconstituting the identity of the social whole.[12] The three regions are recast as structures without a centre whose meaning is no longer assured by any transcendental principle (cf. Rasch and Wolfe 2000). This also calls an end to the eternal seduction of the *pars pro toto* gesture, through which parts are hypostatized into

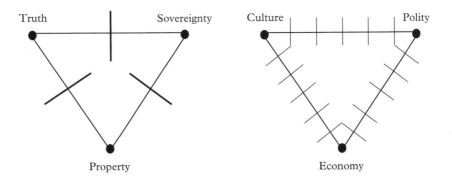

Figure 2.3: Modernist versus Postmodernist Autonomies

wholes and a particular subsystem is taken to define or represent the totality of social relationships. All totalizations remain partial and limited in scope: they never leave the particular domanial perspectives from which they are launched. Scientific accounts are no exception to this rule: instead of claiming privileged access to social reality 'as such', they never leave the narrow viewpoint and interests of a separate subsystem (Luhmann 1982: 341–44, 350ff.). The three powers and rationalities are thus similarly situated at 'ground-floor' level, co-inhabiting a flat surface that does not permit the erection of infrastructure/ superstructure hierarchies, and that enables them to nurture an agonistic co-operation on the basis of equal opportunity and equal interest. In this fashion, the idea of a balance of powers between economy, polity, and culture represents a societal generalization or 'inflation' of the liberal principle of *trias politica* (and of the Kantian tripartite division of culture into the jurisdictions of truth, goodness, and beauty), even though the equilibrium effect of these countervailing powers is achieved not so much by the splitting force of binary schematizations as by the proliferation of graded differences and passable frontiers.

The Dialectic of Differentiation

The above considerations can be further systematized by adopting some considerations on the sociology of postmodernization and the 'dialectic of differentiation' that have been proposed by Crook, Pakulski, and Waters (1992). From Spencer via Durkheim and Weber to Parsons, Luhmann and Habermas, social theorists have typically conceived of societal progress in terms of advancing differentiation and specialization between distinct but functionally interdependent roles, functions, institutions, and subsystems; while the liberal political project has traditionally been engaged in purposively distancing the societal domains in time and space and in sharpening the boundaries between them. The Kantian doctrine of the three spheres of culture (knowledge, morality, and aesthetics), for example, was generalized and sociologically modulated in Weber's articulation of autonomous value spheres (such as religion, the economy, the polity, the aesthetic, the erotic and the intellectual realms), each of which answered to its own indigenous logic and its autonomous norms and laws (cf. Brubaker 1984: 69ff.; Whimster and Lash 1987: 9–12; Crook et al. 1992: 8ff., 47–48). Whereas the Kantian threefold division of culture still singled out Reason as the privileged element that remained in charge of and unified the other spheres (also supervising the broader demarcation between culture and society), Weber's contrasting view was that the value spheres entered in irreconcilable conflict with each other (Weber 1970: 147), and that

What (Again) is So Special about Science?

there was no ultimate sphere to arbitrate between their conflicting obligations and demands. As with Kant, however, the differentiation of the spheres was dictated by immanent principles (if no longer of a cognitive but of an axiological nature), while the progress of rationalization was primarily seen as developing and protecting the plurality and integrity of the different value spheres, each of which revolved around legislative norms that were immanent and particular to it. Since norms and values were thought to reside objectively in the different life orders (e.g. the politician's 'ethic of responsibility'), which were simultaneously interpreted as incommensurable and self-legislative, the rationalization process logically entailed a progressive sharpening of boundaries and an intensification of conflicts between the spheres, which were expected to diverge due to a growing understanding of the specificity of their endogenous laws and axial principles.[13] The Modern Constitution, as Latour (1993a) has described it, is indeed a *purifying* constitution, which feeds upon policing dualisms in order to lend force and intensity to the liberal 'art of separation' (Walzer 1984).

Postmodernists such as Crook et al. on the other hand maintain that, in the contemporary period, we are witnessing extensions and accelerations of this differentiation drive that 'aggravate' into forms of *hyperdifferentiation*, which also partly reverse its direction, because they proliferate distinctions to such a massive degree that entrenched demarcations begin to leak and steady boundaries are increasingly effaced. The 'dialectic of differentiation' implies that processes of hyperdifferentiation tend to collapse the distinctions and dichotomies of cultural modernity to such an extent that they paradoxically tip over into forms of *de-differentiation*, which loosen the structural anchorage of the various spheres and reduce if not abolish the formal distances between them.[14] The differentiating tendencies are released from their underlying axial principles, reversing and involuting the modernist priorities of a progressive separation of culture or science from society, polity from economy, and economy from culture. Postmodernist hyperdifferentiation exceeds and defies any further containment by last instances; its enhanced complexity tends to create a disorganized, unsystematic and disorderly society (Urry 2000). The intensification of differentiation processes – which are accelerated by the ubiquitous advance of individualization – undermines the regional stability of the various domains, which are reconfigured as their internal boundaries multiply to the point of fragmentation (Crook et al. 1992: 225).

Cultural modernity reaches its limits when the proliferation of divisions effectively erodes the significance of distinctions between autonomous spheres; the multiplication of categorical boundaries *within* the various spheres facilitates the transgression of boundaries *between* them. In this way, postmodernization can be understood as an ironic extension-cum-reversal of the 'progress' of cultural modernity.[15] For example, the hyperdifferentiation and hyper-rationalization of science produce a pluralist branching out of approaches, methods, schools, subject areas and research fronts that breaks up the formerly unified and separate disciplines, but also tends to erase the boundaries between scientific and non-scientific forms of expertise. Hyper-rationalization corrodes the disciplinary structure of modern science, but simultaneously erodes the idea of a singular demarcation between science and society (Crook et al. 1992: 41, 209, 217). It may dialectically precipitate into forms of 'de-rationalization' if science loses its distinctive primacy and expert and lay cultures of rationality are put on a more equal footing.

In analogous fashion, the postmodernization of *political* life is reflected in a progressive decentralization and devolution of an 'overloaded' state, a tendency of power to become multifaceted and so widely distributed across society that one can no longer identify a precise set of power loci (Crook et al. 1992: 37). Expanding the analogy with science, one could similarly emphasize the demise of the rationalist or 'scientific' aura of politics ('politics is ordinary'; 'Downing Street doesn't know everything'), the fragmentation and decentring of statal bureaucratic powers and the relocation of politics towards domains such as science, technology, the economy, and everyday life.[16] While the secularization of high scientific rationality turns upon the weakening and pluralization of a previously overloaded conception of truth, the fragmentation of political powers is accompanied by the final destruction of the absolutist principle of sovereignty, as pioneered by early advocates of a pluralistic and associationist democracy such as Duguit and Hauriou (Pels 1998: 37–39, 47ff.; Hirst 1994). In this view, the state becomes one player among many in a more levelled playing field, in which 'high' political culture is brought closer to 'low' popular culture, and political knowledge is reconceived as more equally distributed between professional representatives and 'ordinary' citizens.

The core message that the triangle model is designed to broadcast is that this paradoxical balancing of differentiation and de-differentiation helps to stabilize post-liberal and postmodern forms of institutional autonomy, which no longer cluster around axial principles but profit from an intensification of the 'art of separation' (cf. Walzer 1984: 318), which replaces a few great divides by many smaller ones. This is perhaps also what Gieryn is getting at when he advises scientists who do boundary work on the science/politics continuum to 'keep politics near but out': 'to draw science near enough to politics (ideally as adjacent cultural territories) without risking spillover of one space into the other ... only good fences keep politics and science good neighbours' (Gieryn 1995: 434–36). Ray similarly emphasizes that postmodern boundary transgression is not necessarily the same as boundary collapse; one system may encroach upon another while (some of) the boundaries between them remain in place, preventing the 'chaos of unmediated complexity' (Ray 1999: 189–94, 206). For example, the contemporary redefinition of places such as hospitals, public bureaucracies, prisons, or universities as 'enterprise forms', while promoting a de-differentiation of economy, culture, and politics, does not eliminate the specificity of those places or create a 'amorphous cultural space'; difference is not necessarily flattened out but rather re-articulated according to new principles (Du Gay 1996: 178–79, 185).

In this respect, the triangle's 'third ontology' supersedes the traditional twentieth-century terms of debate in which the liberal art of fission could only be confronted by the totalitarian art of fusion (as preached and practised equally by the left and the right). For both Karl Marx and Carl Schmitt, anti-liberal institutional de-differentiation (e.g. the reunification of civil society and the state) was only imaginable under the sign of identity or homogeneity, irrespective of whether it was constituted by the economic or the political last instance (cf. Kolakowski 1977; Keane 1988: 25; Schmitt 1988; 1996; Rasch 2000b). Postmodernists sometimes appear to flirt with this identity form, not only because they come close to saying that 'everything has become cultural' (cf. Jameson 1991; Lash and Urry 1994: 142–43) while dismissing all foundational claims for the economy or polity, but also insofar as they favour a non-Euclidean topology of fluidity, undecidability and complexity which is hard pressed to recognize any stable differentiations in the seamless flow of social life (Serres 1982; Mol and Law 1994; Law and Hassard [eds] 1999; Hetherington and Lee 2000).

Absence of Haste

This 'third ontology' of flattened fences and weak autonomies may lay the groundwork for a Post-Liberal Constitution that is capable of explaining both the ordinariness and the specificity of science in more secular terms, and usefully balance it against the relative independence and integrity of other professional pursuits. For this purpose, I shall first enter a more substantive determination of the triangulated domains themselves, using a formula that immediately brings out their triadic convergence and boundary implosion, while nevertheless identifying specific 'core businesses' or key functional assignments that (weakly) define their three-sided demarcations. In a slight variation on a scheme that was proposed half a century ago by the Dutch political theorist Jacques de Kadt, the economy may be described as the field of production and organization of *things*, politics as the production and organization of *people*, and

Unhastening Science

culture (science, art) as the production and organization of *ideas* or *images* (cf. Pels 1993: 210). Constructing, managing, and deciding about ideas, people, and things (or money as the generalized equivalent of things) offers only a weak set of distinctions because, obviously, economic action also intrinsically and increasingly includes the management of ideas and people ('knowledgeable' or informational capitalism, the business firm as polity, the scientization and aestheticization of economic production), while politics increasingly covers the production and organization of ideas, images and things (scientifically based and aesthetically stylized politics, bureaucratic intelligence, state entrepreneurship), and culture increasingly encompasses the production and organization of both people and things (technology and its assessment, Big Science, university management and administration, intellectual school formation and scholarly rivalry, the public understanding of science). However, I still assume that such a broad 'division of domanial labour' defines a provisional focus for each of the domains, without specifying an endogenous logic, a normative law or a structural identity that forces them to gravitate towards conformity with a deep inner 'nature' that is naturally distinct from and incommensurable with the identifying structures of the other domains.

This preliminary categorization usefully resonates with and underpins a more systematic rerun of the phenomenological exercise in weak differentiation on which I embarked above. While their substantive 'mission statements' already tend to scale down (rather than pitch up) the distinctiveness of the three functional domains, a further pragmatic elaboration of their 'conditions of felicity' or 'styles of work' similarly identifies only minimal conditions of variation across a range of mediations. What does the production and organization of ideas minimally require, in relative contrast with the production and organization of people and things? What minimal sociological, political, economic, and psychological specifications may be developed to replace the maximum pitch of a transcendental logic of discovery? Since these are relative conditions and comparative terms, what does the production and management of ideas (or reflexivity) specifically require within other regional structures of the social triangle (e.g. research labs in industry, scientific advisory units of political parties)? Within the sphere of culture or the academy itself, or even within the narrower frameworks of individual disciplines or university departments, what are the conditions of felicity for producing, managing and deciding about ideas vs. those for managing and deciding about people or money?

In line with what has already been suggested above, I want to delineate these minimal conditions of difference in terms of variations in the structuration and management of *time and space*. Intellectual and scientific work is characterized by specific temporal and (by immediate implication) spatial settings and disciplines, which weakly but importantly distinguish it from the temporalizations and spatializations that broadly characterize the other spheres in the social triangle. Its primary condition of felicity is *absence of haste*, or a systematic and critical deceleration of thought and action, which sets science apart from the demands of urgency, immediacy, simultaneity, and publicity imposed by more speedy cultures such as those of politics or business. This gesture of unhastening the ever-increasing velocity of everyday and other professional life, which aims to unravel and reduce the astounding complexity of such forms of life, immediately goes with the pragmatic requisite of clearing a relatively bounded space, within which intellectuals and scientists can be rigorously selective about their topics and legitimately ignore the plethora of other issues that might clamour for attention. It enables them to 'freeze-frame', to go into 'repeat mode', to take things apart, zoom in on tiny details, and leisurely assess their broader significance; to slow down conversations and conflicts by means of quiet turn-takings and long communicative intervals (e.g. reading and writing rather than communicating face to face); and more generally, to postpone decisions about what the world is like and what one should do about it.[17]

In this manner, the issue of autonomy and demarcation (what makes science special?) is re-specified in terms of two minimal spatio-temporal variables. Next to a sociology of spatial distancing we also need a 'chronosociology' or 'chronopolitics' of science that identifies how different institutional orders work to different timetables and imply different rhythms of social interaction (cf. Gurvitch 1964; Young and Schuller [eds] 1988; Adam 1990). These variable tempi and loci systematically diversify between the three 'corners' or regions of the social triangle, defining their relative speed (acceleration vs. deceleration; nervous activity vs. stillness and inertia; rapid succession of images vs. 'freezing the frame') and the relative openness or closure which accommodates these various temporal regimes (publicity vs. privacy; large vs. small audiences; gregariousness vs. estrangement).[18] Calibrating both variables with the three social domains yields the following preliminary schematization (see Fig. 2.4). While culture and science are principally slow-paced and semi-private, politics 'goes public' (it is concerned with the *res publica*) and is subjected to an accelerated and intensified time regime. The market, while imposing a similar logic of hyperactivity and swift-paced busy-ness (the 'rat race'), also to some extent copies the spatial frame of science by shunning excessive publicity and preferring to operate in relative privacy or 'secrecy'.

Market liberty or free enterprise is conditioned by the same kind of structurally encoded *laissez faire* that also forbids an intensified public scrutiny or monitoring of scientific practice – an observation that holds true even if the

Unhastening Science

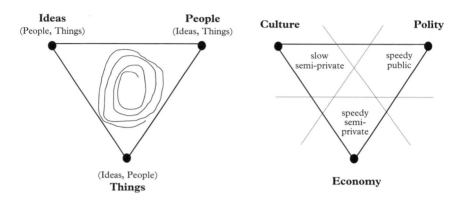

Figure 2.4: Differential Functions and Time/Space Coordinates

unconditional demarcations of absolute private property progressively cede before the more porous partitions of a mixed (political and cultural) economy. Like market liberty, the freedom of the academy can only be a bounded and relative freedom; its 'free speech' is only such in comparison with the more pressured and hasty speech of politics and business. In this respect, it is analogous to the alleged 'disinterestedness' of science which, as Bourdieu has argued, hides a specific self-interest and only appears as 'disinterestedness' in contrast to the different interests generated in political or economic settings (Bourdieu 1981; 1990b; 1993b). Given these socio-temporal variations, the idea of unhastening also provides a better account of what Habermas has rather idealistically described as the 'ideal speech situation'. While it cannot exist in a purified, power-free form, it might indicate a location on (or favour a normative extension of) the continua of knowledge-power and culture-capital that would facilitate a further deceleration of the flow of (scientific) conversation.

Image Speed

Given these terms and distinctions, it is appropriate to round off this initial phenomenological description of the social triangle by inserting a few remarks on the temporal structure of economic conduct. In this connection, it is intriguing to notice that economic and financial management resemble politics in primarily staging an *oral* or *verbal* world, a world of endless and often agonizing talk (in face-to-face situations; through incessant phoning, mobile or otherwise; and increasingly through video conferencing) which considerably outpaces the tempo of exchange among academics and scientists. Traditionally, indeed, *otium* or leisure as a condition of felicity for intellectual work has been set in contrast to *negotium*, or the speed of action and decision required equally of market traders and of rhetoricians in the public marketplace. In the new economy, especially, one can trace the emergence of 'fast' managerial styles that are geared to maximum creativity in a permanent state of emergency; time is the forcing ground of this new regime of managerial governmentality, which must adapt to a general speed-up in the conduct of business (Thrift 2002).

Reporting on research by Mintzberg, Stewart and others, Thrift has also recorded that senior managers spend between half and three quarters of their time simply talking to people, whether face to face or on the phone; and the evident purpose of their relentless travelling (long waits in airport lounges providing a ready occasion for even more mobile phoning) can only be that they need to talk face to face to people long distances away (Thrift 1999: 154; cf. Clegg and Palmer 1996: 2). In 'verbomotor' or rhetorical cultures such as these, material representations such as charts, graphs, memos or notes are primarily used to support discursive face-to-face communication; which reverses the logic and style of academic interaction, where the primary purpose of speech (if it is not directed at students in ritualized classroom situations) is to inform subsequent writing, and where the *viva voce* exchange among professionals is occasioned by and grounded in the practice of reading and writing (consuming or producing the 'dead' letter). Habits of continuous travel and continuous meeting with people generalize a performance situation that is still an exceptional one for academics (cf. the scientific conference) who, as a result of the relative estrangement or unsociability imposed by their unhastened economy of time, do not normally possess the interpersonal skills and the panache in the presentation of self that are imperative in order to sustain negotiated relationships of trust primarily by means of talk and conversation.

As indicated, the public nature of political life doubly differentiates it from the relative 'privacies' of science and of business, although it is true that the latter domains are increasingly infiltrated by the logic of politicization and mediatization. Politicians and their institutions remain uniquely swayed by the ups and downs of their public image and reputation, since they depend far more exclusively upon the regular casting of the public vote (or its speedier equivalent: the media-generated opinion poll) in order to obtain a working legitimacy; this electoral dependence forces politicians, unlike scientists and business people, to publicly account for and justify their views and decisions at regular and ever-shrinking intervals. In Bourdieu's account, the political field can therefore never become fully autonomous, since political professionals must continuously refer to their voting clientele, who in a sense have the last word in deciding the struggles between the professional inhabitants of the field

Unhastening Science

(Bourdieu 2000b: 62–63). In terms of the assimilation processes that characterize the social triangle, however, one might also follow the ways in which the logic of public discursivity and accountability slowly enters the economic sphere, effecting an incipient fusion of politics and economics that reinforces the dependence of industry on the making and breaking of reputations (of individual managers, of brand names, of the firm's solvency or market credibility) and the maintenance of public trust.[19]

Against this backdrop, it is evident that the de-differentiations of the social triangle, and particularly the logic of politicization and economization, exercise an accelerating effect on the entire constellation, and especially upon the field of science and culture. The demarcations of the social triangle are further complicated and 'weakened' by a prominent feature of modern information-dense and image-saturated societies: the 'switchboard' or 'short-circuit' function of the institutions of mass communication, which are themselves major energizers of further assimilation, and increasingly manage to impose their own fast tempo upon that of other, slower domains (see Fig. 2.2). Although not in any sense determinant 'in the last instance', they are sufficiently emergent and powerful within the social triangle to permeate and restyle the other domains to the imperatives of popular culture and mass publicity. In this configuration, the media are not simply allocated to the cultural domain but literally mediate between the different sectors, in a give-and-take that increasingly imposes an image-driven and public logic upon all of them, and that tendentially softens the residual partitions between public and private spheres (cf. Meyrowitz 1985). In the cultural field, scientists, moralists, and artists are seduced to become 'media intellectuals' and cultural celebrities who are made to speak publicly (and under rigorous time constraints) about their ideas, ethical principles, and aesthetic imaginations (cf. Bourdieu 1998b). In the political field, professional representatives are likewise progressively caught up in the maelstrom of mediatization and personalization, conniving and competing with political journalists in order to broadcast effectively their distinctive brand names and political styles (Scott and Street 2000; Thompson 2001; Corner and Pels forthcoming). In the economy, CEOs and lesser managers are similarly 'coming out' as public personae (Branson, Gates, Bezos), being sucked into the same celebrity system that, while originating from the entertainment industries, has inundated many other sectors of social life (cf. Marshall 1997). The process of mediatization hence operates as a crucial vortex that draws the other spheres into the orbit of mass visibility and public accountability, imposing upon them a form of publicity that is crucially constrained by the media's own distinctive agenda, professional interests, and specific requirements of speed.

It has also been suggested that the impact of the mass media has the

What (Again) is So Special about Science?

significant effect of rehabilitating oral and especially audiovisual culture with regard to written forms. Popular culture is often unfavourably contrasted with supposedly more 'rational' cultures (high science, high politics), precisely because it is seen as primarily oral and image-bound, and hence characterized by immediacy, embodiment, and sensation(alism), whereas more 'rational' traditions are seen as rooted in disembodied writing, information, facts, and intellectual argument (cf. Fiske 1992; Dahlgren 1992). Insofar as the parallel mediatization of science, politics, and business turns up the level of intersystemic resonance and boosts convergence from all sides of the social triangle, this process clearly entails a reinvention of orality and (tele)visuality that dramatically speeds up the tempo of interaction and alters the forms of public address in all domains. The relative slowness of textual forms of representation, which involve a reading path extending over time, sharply contrasts with the sensual immediacy of information as visual image, restoring to some extent the spontaneous gestural richness and speed of unmediated face-to-face communication (Barthes 1977; Sturken and Cartwright 2001; cf. Goody 1977: 44, 50). The universal visualization of culture in increasingly 'post-literate' societies, which is enhanced by the lightning speed of electronic communication, promotes a fast image economy and a visual literacy that undermines the prestige of the written word and reverses the century-long domination of print-based epistemology (Stafford 1994: 281ff.).

In a social universe that is thus subjected to a logic of publicity and intensifying acceleration, it is important to maintain the co-existence of differential time perspectives and temporal regimes and to defend autonomous institutional niches that permit a critical unhastening of thought and action (cf. Eriksen 2001). If time is money, then speed is power, as Virilio has remarked; which implies that acceleration needs to be confronted as a major political phenomenon (Armitage 1999: 35). In a society of increasing speed, a critical phenomenology of the 'rat race' (Virilio's 'dromology') must not only insist upon analytic distinctions between different institutional timescapes and social speed lanes, but must simultaneously nurture a chronopolitics that celebrates the relative inertia of pockets of resistance and unhastening; it must simultaneously analyse the forces of acceleration and the forces that brake or diminish speed (Kellner 1999). If science is legitimately outpaced by both politics and business, we still require a critical politics of deceleration that preserves its defining temporal rhythm and resists the stressing up of scientific work as a result of the excessive infiltration of political, enterpreneurial, or journalistic deadlines. A pragmatic (rather than transcendental) defence of scientific autonomy must focus on the preservation of this unique socio-temporal order (which itself includes a plurality of times and paces) in the face of the structural shrinkage of time that threatens to engulf it from more hasty cultures in the social triangle.

This is not to imply that standards and incentives such as user relevance, managerial efficiency, cost effectiveness, or audit accountability (or even those of political correctness and the culture of enterprise) should be expunged from the official order of science, because these criteria clearly co-determine the success of the knowledge-political hybrids that have emerged as bridgeheads between so-called pure and more applied and policy-committed forms of research. It is not a matter of objecting in principle to the infiltration of economic and political metaphors, work-styles, and criteria of excellence in science. But it is necessary to install speed limiting devices at regular intervals to brake the pace of the hasty cultures and free areas of stillness within which research and reflection may proceed more slowly and sedately. The pressures of globalization, of the morality of publish or perish, the imperatives of academic entrepreneurship and self-generated funding, the competition for promotional image and education market share, the growing salience of the intellectual celebrity system, the relentless machinery of research and teaching assessments, the endless administrative restructurings, and the resultant hypertrophy of academic leadership and management (lots of talk in endless meetings, no time for writing) together produce a threatening acceleration that undermines the weak boundaries of science and turns the tempo and habitus of the fast decision-makers into an infrastructural routine. However, as both Plato and Nietzsche were aware, we need time in order to develop 'untimely' considerations. Or as we might put it, in a variation of Konrád's (1990) title: scientific autonomy is essential for making slow observations in a fast time.

CHAPTER 3

Two Traditions in the Social Theory of Knowledge

Mannheim versus Wittgenstein

The preceding chapter has cleared a path for a more detailed inventory of the resources for developing a more pragmatically levelled and 'earthly' perspective on scientific rationality and objectivity. This chapter will sketch a field map of recent positions and divisions in the sociology of knowledge and science that have been influential in establishing such a disenchanted view. The next chapter will then enter into a sociological depth analysis of the fact/value distinction, which, like similar dualisms such as truth/power or knowledge/interest, has often acted as an epistemological placeholder and enforcer of a more ordinary and relative distinction between unhastened and institutionally secluded practices (such as science) and more hurried and gregarious ones (such as politics). The primary task of the present chapter is then to address the paradoxical fact that, while they have set fire to a whole range of other dualisms and dichotomies, science and technology studies (STS) have remained stuck in a significant residue of the dualism between fact and value. In order to account for this remarkable exceptionalism, we need to address the foreshortening of historical vision, within both STS and mainstream sociological theory, which has prompted both traditions in some way to reinvent the wheel (and to repeat some of the foundational flaws) of earlier contributions to the sociology of knowledge.

A first survey of the discursive field highlights the existence of a schism and mutual indifference between two currents of theorizing and research which I shall dub the 'Mannheimian' and the 'Wittgensteinian' traditions. This appellation is in somewhat ambiguous homage to David Bloor, who, in one of the first statements in print of the Edinburgh Strong Programme, compared the two thinkers with regard to the strategic possibility of a sociological explanation of logic, mathematics and natural science (Bloor 1973). The Mannheimian programme for the sociology of knowledge was considered 'weak' precisely for its refusal to explain cultural and natural science symmetrically and hence to extend causal sociological analysis to the hard case of the natural sciences; and for its coincident failure to demand an equally radical symmetry between the sociological explanation of true and false beliefs, thus confining the sociology of knowledge to a mere 'sociology of error'. In both respects, Wittgenstein was celebrated as offering a more attractive starting point: 'Wittgenstein solves Mannheim's problem' (Bloor 1973: 173; 1983).

The spurt of intellectual initiative that awoke the slumbering sociology of knowledge to the radical impulse of the new social studies of science in the early 1970s was therefore not initiated by Mannheimians, but largely developed without Mannheim, if not in conscious opposition to his work. Although in the discursive ferment of the 1960s and the 1970s the Mannheimian heritage was kept alive by sociologists of knowledge such as Mills, Gouldner, Coser, Shils and Wolff, their efforts did not issue in a distinct research tradition in the 1980s – with the significant exception of the grand editorial project carried off by Kettler, Meja, and Stehr (cf. Goldman 1994; Kettler and Meja 1994). For various reasons, interesting in themselves, contemporary social theorists such as Elias, Bourdieu, Foucault, Habermas and Giddens have found only restricted use for Mannheim. Insofar as they have developed distinctive sociologies of knowledge, they have also operated in virtual isolation from radical Wittgensteinian science studies.

The real 'action' and excitement in the sociology of knowledge, indeed, was not generated by mainstream sociology but emerged from the new philosophy and historiography of (natural) science. The seminal work of Kuhn, insofar as philosophical sources entered into it, took its inspiration not from the sociology of knowledge tradition but from Wittgenstein and Fleck, and initially concentrated not on 'soft' sociological, political, or historical thought but on the 'harder' sciences of nature and medicine.[1] Bloor, Barnes, and Collins, the progenitors and developers of the Strong Programme, as well as constructivists such as Knorr-Cetina, Woolgar, and Latour, followed a Wittgensteinian rather than a Mannheimian track, as did discourse analysts and ethnomethodologists such as Mulkay and Lynch.[2] Evidently, Bloor's reproach about Mannheim's 'failure of nerve' with regard to the symmetrical treatment of true knowledge and natural science was considered sufficiently damaging to turn his sociological project into a dead horse. Henceforth Mannheim was cited solely as a token predecessor (Knorr-Cetina 1983: 115, 136; Law 1986: 1).[3]

Let me at once enter some specifications that qualify this claim about a Wittgensteinian turn in science studies, in order to evade the risk of forcefully homogenizing what are in fact quite diverse streams of theorizing and research (cf. Callebaut 1993). These provisos will simultaneously elaborate significant reservations about Bloor's opposition of a 'strong' Wittgensteinian to a 'weak' Mannheimian programme in the social theory of knowledge; in fact, the legacies of both Mannheim and Wittgenstein are much more interpretively flexible and interpenetrable than is suggested by such sweeping categorical gestures. Indeed, Bloor's critical reading of Mannheim has been plausibly described as actually more of a correction and expansion of the classical sociology of knowledge than an across-the-board attack on it (Lynch 1993: 42ff.; Kim 1994: 391); while his recruitment of Wittgenstein in support of such a 'strengthened' naturalistic sociology has similarly come under spirited attack (Sharrock and Anderson 1984; Hacking 1984; Coulter 1989; Lynch 1992a; 1993). Appropriations of Mannheim have diverged as sharply as have those of the later Wittgenstein. While pleading an anti-causalist, anti-epistemological and praxeological version of Wittgenstein, for example, Lynch uncritically supports the same naturalistic reading of Mannheim that is still taken for granted by Bloor. Other critics of the Strong Programme, such as Hekman, have dismissed such a reading as a 'serious misunderstanding', and opposed it with an anti-epistemological and anti-foundationalist reading that brings the Mannheimian project close to that of Gadamerian hermeneutics. The intriguing thing here is that Hekman symmetrically reads Wittgenstein as still caught up in the epistemological concerns that dominated Enlightenment thought, and as failing to make the crucial move from epistemology to (Gadamerian) ontology (Hekman 1986: 120, 128).

If, as a result, neither 'Mannheim' nor 'Wittgenstein' identifies a stable corpus of intellectual claims, and if each may easily take the epistemological role of the other, it may appear that the tension and the rift are less between firmly entrenched ancestral traditions than between conflicting naturalistic and anti-naturalistic interpretations of *both* thinkers. Nevertheless I still presume that talk about a Wittgensteinian turn plausibly draws together many divergent strands in the science studies movement into a relatively coherent picture. I also presume that a more attentive reading of Mannheim, which more focally acknowledges the productive indecisions and complexities traversing his project, will help to specify an intellectual position that may counter some of the deficiencies that have meanwhile evolved within this Wittgensteinian project. The purpose of this chapter is therefore to inquire whether, if perhaps Wittgenstein has to some extent solved Mannheim's problem, the reverse might not also apply; in other words, whether the Wittgensteinian programme in the social studies of science, which has so successfully reinvigorated science and technology studies over the past decades, has not meanwhile run into difficulties that might to some extent be resolved by reverting to a broader and more traditional

Mannheimian agenda of research. The gap that has evolved between the two rival perspectives can and should be bridged, perhaps by a 'relative synthesis' of Mannheimian inspiration. This first of all requires some preliminary recovery work. As I pointed out, it has not been solely the Wittgensteinian tradition of science studies that has promoted forgetfulness about Mannheim; the mainstream sociological tradition itself has similarly found itself reinventing Mannheimian wheels. Hence this chapter makes an effort to recover something like an original Mannheimian inspiration from beneath the brushwork of *both* modern social studies of science *and* modern sociologies of knowledge.

Value-Free Relativism and its Discontents

Before sketching a summary programmatic outline for such a synthesis between the two traditions, I shall first attempt to identify the relative coherence of the Wittgensteinian project and locate its current impasse. I take it to indicate a comprehensive style of thinking and research that may be succinctly characterized as 'value-free relativism'. This project integrates a strong anti-epistemological and anti-normativistic temper with a radically contextualist and descriptivist methodology. Suspicious of all theoretical generalization and normative legislation, it advocates the close ethnographic analysis of concrete examples of everyday and scientific language use and practices (the case studies method). Language is preferably conceived as (speech) *action*, challenging traditional views that emphasize reference, correspondence, and mimetic representation. It thus stresses the *performative* nature of language-in-use: words are also deeds, which help to bring into being what they describe. The meaning of words resides in their indexical usage, and remains confined to the language games or forms of life in which they are put to practical effect (Pitkin 1972: 39, 289; Phillips 1977: 27–30). The relativity of scientific truth claims is thus merely a particular instance of a more widespread scepticism concerning the rule-bound character of practical activity. Far from compelling particular courses of action, traditional notions about truth, reason, and logic offer post hoc rationalizations that account for the orderliness of practices and the shared nature of conventions (Collins 1991a; Woolgar [ed.] 1988; Lynch 1993).

Wittgensteinian 'value-free relativism' thus insists on the substitution of a naturalistic for a normativistic conception of rationality, according to which discrepancies in belief become 'simply a matter of cultural variation, so that all beliefs in all cultures become equivalent for the sociologist' (Barnes 1976: 125; Bloor 1991 [1976]). Naturalism and relativism in the theory of knowledge therefore imply *symmetry* of explanation between truth and error, and *impartiality* or moral indifference towards nature and society – the two core tenets of the Strong Programme that have been progressively extended and radicalized in subsequent waves of constructivist STS (see more extensively Chapter 6). Constructivist naturalism neutralizes the question of legitimacy or illegitimacy of knowledge in favour of ethnographic redescriptions of scientific discourses and practices which, in line with Wittgenstein's own analyses of ordinary language use, nearly 'leave everything as it is' (Sharrock and Anderson 1984: 377; Collins and Yearley 1992: 308–309). *Nearly*, since the critical edge of such dispassionate investigations (conducted from Wittgenstein's 'Martian' point of view), is not to harvest 'denunciations' of erroneous native beliefs or irrational native actions but to 'display' the historicity and pragmatic contingency of all scientific practices and claims (cf. Latour 1993a: 43–46).[4] On this interpretation, the label 'value-free relativism' appears sufficiently comprehensive to include the drift from the causalist 'politics of explanation' of Bloor (who is more hesitant to follow Wittgenstein on this issue) towards the 'politics of description' deployed by Knorr-Cetina, Callon, Latour, and Law; while it also encompasses sibling rivalries such as those between Collins and Latour and between Bloor and Lynch, all of whom operate a more or less radical methodology of disinterested analysis.[5]

There are two major weaknesses in this Wittgensteinian agenda, which predictably mirror its major strengths (cf. Fuller 1988; Fuchs 1992; Lynch 1993; Radder 1996). First, its ethnographic descriptivism easily degenerates into a species of empiricism that conflicts with its own precept of reflexivity, since it ignores the normative and political constitution of its own knowledge claims. While undermining quite a number of entrenched epistemological binaries (cognitive vs. social explanations, science vs. politics, culture vs. nature), 'value-free relativism' remains mired in the dualism of facts vs. values, and is unable to deal successfully with the problem of critique (Pels 1991a; 1995a; 2000: 10–18; Proctor 1991: 224ff.; Radder 1996; Fuller 2000b). Second, its initial focus on the micro-dynamics of laboratory settings and scientific controversies tended to privilege an actor-centred bias that occluded broader macro-institutional constraints on scientific production. Despite the Strong Programme's lingering macrosociological sensitivity, and major recent efforts to reincorporate larger societal contexts in the study of local scientific work (Haraway 1991; Knorr-Cetina 1982; 1999; Shapin and Schaffer 1985; Cozzens and Gieryn [eds] 1990; Fuchs 1992; Latour 1993a; Jasanoff 1995), constructivist studies after the ethnographic turn equally tilted towards the micro-side of the micro/macro dualism as to the factual side of the fact/value dichotomy. There is still a challenge to 'get constructivism out of the lab' (Gieryn 1995: 440).

My own gamble is that such programmatic shortcomings may be alleviated by promoting a (partial and guarded) shift towards a more comprehensive and normatively sensitive Mannheimian research agenda. This requires a return from a doubly restricted *sociology of science* to a more broadly conceived *social theory of knowledge* or *social epistemology* (Fuller 1988; 1992; 1993; Harding 1991; Pels 1991a; 1998; 2000; Fuchs 1992; Roth 1994).[6] It implies a considered shift from a micro-oriented ethnography of laboratory life and scientific controversies towards a macrosocial theory of knowledge as classically outlined in the works of Marx, Durkheim, Mannheim, and Berger and Luckmann, which reinserts science studies into the more general concerns of cultural studies and social and political theory. Such a comprehensive socio-political theory of knowledge and culture (cf. Bourdieu and Wacquant 1992; Beck 1992; 1997; Beck, Giddens and Lash 1994; Webster 1995; Castells 1996; Robins and Webster 1999) extends its scope of analysis more emphatically from academic-scientific beliefs and thought styles towards non-academic (ideological, political, journalistic, everyday) ones and their multiform interrelationships. It also favours the recuperation of the problem of ideology critique (the Mannheimian problem of ideology and utopia), which, as I have indicated, has been unjustly cashiered from the roster of symmetrical science studies and that of post-modern philosophy and anti-foundationalist social theory more generally (Simons and Billig [eds] 1994; W.T. Lynch 1994; Žižek [ed.] 1994). It also includes an attempt to reconnect the social theory of knowledge to a social theory of intellectuals, experts, and professionals, or more broadly, to a macro-theory of the knowledge society and its emerging strata of epistemocrats or cultural capitalists (e.g. Bauman 1987; 1992b; Eyerman and Jamison 1991; Stehr 1994; Webster 1995; Pels 2000).[7] Last but not least, it includes an attempt to restore the breadth and depth of what Mannheimians such as Stehr and Meja call the 'magic triangle' of symmetrical interdependency between epistemology, sociology and ethics, thus counteracting the tendency of an imperialistic sociology (and of postmodernist philosophy in general) to abrogate epistemology and ethics (cf. Stehr 1981; Stehr and Meja 1982; Radder 1996; Winner 1993; Squires [ed.] 1993; Bauman 1993; Fuller 2000b).

All of this, however, should be undertaken while preserving the crucial epistemological gains that have been harvested in the course of two decades of research and reflection in the radical microstudies of scientific knowledge: their ethnographic precision, their symmetrical inclusion of natural, practical, and technological knowledge as objects of analysis, their anti-realist and anti-universalist approach to issues of representation, their radical reflexivity, and their 'strong' thesis about the essential inseparability of cognitive and social dimensions in knowledge formation, which is also echoed by the Foucaldian slogan *pouvoir/savoir*. Precisely where this synthetic effort will take us is at present difficult to predict.[8] Nevertheless, the primary task of a social epistemology, as proposed here, would consist in an attempt to restyle or reinvent some ideas from the classical tradition by reimporting the gains of contextualism, constructivism, and reflexivity, and thus to reinvest the critical results of science and technology studies into social and political theory itself (cf. Barnes 1988; Latour 1991; 1993a; Law 1991).

Reinventing the Wheel

I have already noted the curious fact that, since the early 1970s, the mainstream sociological tradition has been just as silent about Mannheim as has the alternative tradition inspired by Wittgenstein, Fleck, and Kuhn. Whereas one can still trace conscious lines of descent connecting Mannheim to American sociologists of knowledge such as Merton, Mills, Coser, Shils, Berger, and Gouldner, major contemporary European thinkers such as Elias and Bourdieu, who found themselves much closer to the sources of the sociology of knowledge tradition both geographically and intellectually, have somehow managed to 'skip over' Mannheim in order to reinvent many of his crucial insights without due awareness or acknowledgment. Norbert Elias's socio-genetic and processual theory of knowledge, despite his sometimes condescending tone about the work of his former principal, is substantively informed by unacknowledged terminological and substantive borrowings from Mannheim's early work (e.g. Mannheim 1982 [1922]; for a different view see Mennell 1989 and Kilminster 1993). Pierre Bourdieu's investigation of intellectual, scientific, and cultural fields, which originally developed in remarkable isolation from both German and Anglo-American sociological lineages, is likewise marred by a virtual absence of references to Mannheim's work, even though core ideas of his field theory of science are clearly prefigured there (Pels 1998: 225ff.). Mannheim's celebrated 1928 lecture on cultural competition, which Elias at the time enthusiastically hailed as a 'spiritual revolution' but subsequently ignored,[9] also anticipates the basic lineaments of Bourdieu's quasi-economic model of science, including the central idea of the selection (*Auslese*) of truth as a product of the criss-crossing censure induced by interested intellectual competition (Mannheim 1952: 196–97; Meja and Stehr [eds] 1982: 326).[10]

This abridgment of historical consciousness may also explain why Karin Knorr-Cetina's early work (1977; 1981), while sympathetically building upon and engaging with Bourdieu's quasi-economic theory of scientific competition, largely credits Bourdieu with a model that can already be found in outline in Mannheim. A similar substitution is evident in some early 'economistic' articles by Latour (e.g. Latour and Fabbri 1977; Latour 1993b: 100ff.) and in Latour and Woolgar's benchmark study *Laboratory Life* (1979). Knorr-

Unhastening Science

Cetina's subsequent more critical discussion of the economic model of science repeatedly identifies Merton as one of its progenitors, while once again failing to mention the obvious ancestorship of Mannheim (Knorr-Cetina 1982: 104; 1983: 129). Bourdieu confuses Merton and Mannheim in similar fashion. In critical homage to the former, and in defence of important features of the Mertonian approach against the 'levelling' and 'nihilistic' impulse of radical science studies, Bourdieu sympathetically cites Merton as follows, embracing this position as entirely his own:

> In the cognitive domain as in others, there is competition among groups or collectivities to capture what Heidegger called the 'public interpretation of reality'. With varying degrees of intent, groups in conflict want to make their interpretation the prevailing one of how things were and are and will be. (Bourdieu 1990b: 297–98)

Since it was Merton, Bourdieu explains, who established that science must be analysed sociologically through and through, the Strong Programme did little else but crash through a wide open door. This Mertonian piece of wisdom, however, is an immediate paraphrase of a crucial passage in Mannheim's 1928 essay (1952: 196–97). Merton himself provides the reference in a footnote, in explicit mention of Heidegger's *Sein und Zeit*, 'as cited and discussed by Karl Mannheim' in 1928 (Merton 1973: 100–101).[11]

While the longitudinal relationship between Bourdieu and Mannheim thus raises a few questions, the lateral relationship between Bourdieu and radical constructivists appears to have become one of sincere mutual annoyance. In true Parisian fashion, Bourdieu and Latour have succeeded in putting an ocean of intellectual distance between them (cf. Callebaut 1993: 107, 473). Bourdieu has summarily thrown together the Strong Programme (which is not distinguished from Latourian actor-network theory or any other variant of science studies) with modern currents in reflexive anthropology and literary deconstruction as opening the door towards a narrowly disguised 'nihilistic relativism', which stands squarely opposed to a truly reflexive social science (Bourdieu and Wacquant 1992; Bourdieu 1993c; 2001: 41–66). The constructivist notion about a seamless web connecting science and society is said to introduce a monistic reductionism, a 'short-circuiting' of explanations that fatally regresses behind Merton. If the Strong Programme would take seriously its own tenet of reflexivity, Bourdieu contends, it would recognize that its grandiose and self-certain statements only perform a distinctive break, while serving little else but the initial accumulation of scientific capital.[12]

In his turn, Latour has increasingly toned down his early enthusiasm for Bourdieu's field theory of science, while the latter now evidently feels that his ideas have been plagiarized. Since their early work, which is still heavily impregnated by the quasi-economic model, both Knorr-Cetina and Latour have increasingly veered towards a quasi-*political* or *power* model of science, and have grown increasingly critical of the 'economistic' metaphors of credit, profit, and cultural capital (e.g. Latour 1986). In addition, Latour has summarily incorporated Bourdieu in his sweeping critique of the 'poverty of sociology', which allegedly remains stuck in the Kantian divorce between culture and nature, and is unable to account for the quasi-objects and the new hybrid powers created in laboratories (1993a: 5–6, 51, 54). Bourdieu is thus offered as a typical victim of what Latour calls the 'Modern Constitution', which continues to separate the representation of facts in nature from that of citizens in the social world, while the networks, hybrids, and quasi-objects are presently proliferating so massively as to undermine this separation from all sides (1993a: 5). Knorr-Cetina, for her part, somewhat downplays this critique of Bourdieu's 'forgetfulness of the object' but still pontifically agrees with Latour that Bourdieu's sociology of science 'is orthogonal to, and beside the point of, the most interesting developments in recent sociology of science' (Callebaut 1993: 473–74).

Even so, it is not Bourdieu or Latour but Mannheim who should be identified as the *Urheber* of the agonistic model of science in either of its currently popular quasi-economic and quasi-political framings (respectively, the idea of intellectual competition and profit-seeking, and the consonant idea of science as a 'continuation of politics by other means'). If perhaps the Starnberger School actually coined the term 'science dynamics', it was Mannheim who came very close when referring to the 'dynamics' of intellectual currents, their 'inner rhythm', or their 'rhythm of development' (1952: 329–30, 326). Nor did Gouldner and Friedrich invent the reflexive slogan of a 'sociology of sociology' or even that of a 'Marxism of Marxism': Mannheim did, as early as 1922 (1982 [1922]; 1968 [1936]: 111). Foucault's *epistēme*, Althusser's *problématique*, and Kuhn's concept of paradigm are all foreshadowed in Mannheim's early work on *Weltanschauung* (1952: 33ff.). Many ideas of the Kuhnian turn, which reputedly launched the success story of the modern social studies of science, were prefigured in Mannheim's sociology of knowledge; including the critique of a priori rationalism, the tendential blurring of the division between facts, theories, and values, the idea of scientific revolution as a *Gestalt*-switch (e.g. Mannheim 1982 [1922]: 127–28), of the rootedness of scientific viewpoints in underlying worldviews, and of the constitutive role of scientific communities. Curiously, it took an iconoclastic historian of natural science, following a Wittgensteinian track, to reawaken the social sciences to much of their own Mannheimian heritage (Phillips 1977).

Unhastening Science

Thought Styles

I shall elaborate another example a little more extensively, since it so deftly illustrates the lack of interaction between the two traditions that already manifested itself during the 1970s. It concerns the concept of 'style of thought', which social students of science routinely ascribe to Ludwik Fleck's 1935 constructivist classic *Entstehung und Entwicklung einer wissenschaftlichen Tatsache*, but which is found as early as 1921 in Mannheim. Fleck's work was rediscovered through Kuhn, who acknowledged that the former had anticipated many of his ideas (1970a: vi–vii; 1979). Although the concept of *Denkstil* makes its appearance in 1929 in Fleck's rejoinder to a 1928 article by Kurt Ziegler (Schäfer and Schnelle 1980: xxii; 1983: 16), Mannheim explicitly traces it to a 1910 Hungarian article by Lukács (Frisby 1991: 124).[13] Fleck himself, however, and his modern editors after him, consistently fail to refer to Mannheim. This is quite remarkable in view of the fact that Fleck does mention sociologists of knowledge such as Durkheim, Lévy-Bruhl, Gumplowicz, Jerusalem, Scheler, and Simmel, and hence comes extremely close to Mannheim's contemporary work (Fleck 1980 [1935]: 62–70, 145). It is almost inconceivable that he could have missed Mannheim's well-published contributions.[14]

Mannheim, as indicated, liberally employs the sociological concept of *Denkstil* as early as 1921 in his article on *Weltanschauung* (1952: 34–35; 1982 [1922]: 86–88, 124–28). The concept is reapplied in his study of historicism (1924) and in his early programmatic article on the sociology of knowledge (1925) (1952: 84ff., 134ff.). His *Habilitationsschrift* 'Conservative Thought' (1927) begins with a more systematic exposition of the style concept, which exemplifies its strategically central position for Mannheim's sociology of knowledge as a whole (1953: 74–84). The widely acclaimed and discussed lecture on 'Cultural Competition' of the following year once again gives it critical prominence, as is evidenced by Mannheim's singular remark that 'theories of knowledge are really only advance posts in the struggle between thought-styles' (1952: 228, 210). In the course of the debate following Mannheim's lecture, the concept was used by Alfred Weber as if it were already common currency (Meja and Stehr [eds] 1982: 374–75). In Mannheim's *Ideologie und Utopie* of 1929 the concept of *Denkstil* is once again omnipresent.

In both Mannheim and Fleck, the concept is critically poised against the mimetic objectivism of dominant natural-scientific discourses. Fleck's version inaugurates an extensive analysis of the construction of a medical fact (the Wasserman reaction) in its specific social and intellectual context. Mannheim's concept precisely captures the specific difference that is said to obtain between cultural or human studies and the natural sciences, emphasizing the interpretive advantage that the former enjoy over the latter. The very fact that natural scientific thought is interpreted as a particular *style* (and hence as non-exclusive, historically contingent, positionally determined and perspective-bound) marks his critical distance from the universalist claims of rationalist epistemology (cf. Barnes 1994). 'Thought style' closely fraternizes with concepts such as *Weltanschauung*, *Gestalt* or 'totality', which have a similarly synoptic intent (e.g. Mannheim 1968 [1936]: 111). A style of thought expresses a fundamental volitional impulse or 'world postulate' (*Weltwollen*) that resides beneath and structurally organizes more articulate cognitive contents. Ultimately rooted in pre-theoretical, 'irrational' motives, it provides the connecting link between discursive contents and prediscursive experiental group strivings and rivalries. As such, it is merely a different name for the 'existential attachment' of thinking itself (Mannheim 1952: 210). Like the concept of *Weltanschauung*, the concept of *Denkstil* therefore provides an anti-reductionist intermediary term that highlights the indirect connections between intellectual contents and social class interests, which are often mistakenly correlated in direct fashion: 'We cannot relate an intellectual standpoint directly to social class; what we can do is find out the correlation between the "style of thought" underlying a given standpoint, and the "intellectual motivation" of a certain social group' (Mannheim 1952: 184; 1953: 74–84).[15]

An intriguing short-circuit connection between these various traditions is suggested by Mary Douglas's (1986) attempt to stretch not Mannheim's, but Durkheim's sociology of knowledge to (almost) include the Fleckian notion of thought style. While Durkheim's 'sociological epistemology' remained undeveloped and Fleck moved far beyond it, they nevertheless share an emphasis on the social basis of cognition that is also conveyed by the idea of a thought style. Fleck's reflexive gesture was to apply this idea to modern society and even to science – a move that would have horrified Durkheim and his followers, who have traditionally exhibited 'an excessive respect, bordering on pious reverence, for scientific facts' (Douglas 1986: 14, cf. 46–47). Evidently, Durkheim's sociological view of religion was not supposed to apply to secular beliefs, and least of all to science and sociology. 'Fighting as allies', however, Douglas supposes, the strength of each can supplement the weakness of the other. A combined Durkheim-Fleck approach to epistemology would prevent either science or religion from being accorded too much privilege: 'Both science and religion are equally joint products of a thought world; both are improbable achievements unless we can explain how individual thinkers combine to create a collective good' (1986: 37).[16]

Unhastening Science

Failures of Nerve

As noted, the first statements of the Edinburgh Strong Programme still include critical discussions of Mannheim, whereas subsequent generations of sociologists of science tend to treat him as largely irrelevant. If only for this reason, it is important to reinspect the 'failure of nerve' argument, which is encountered most curtly and polemically in Bloor 1991 (1976), Barnes 1977, and Woolgar 1988a: 23, 48, and occurs in a more sophisticated and quiet version in Bloor 1973, Barnes 1974 and Mulkay 1979: 10–17, 60–62.[17] As indicated before, this critical argument suggests that Mannheim failed to carry through the activist, relativist, and symmetrical impulses of his own sociological programme, and conceded too much to the contemplative model of a realist and rationalist epistemology. This led him to exclude mathematics, logic, and the natural sciences from a comprehensively social and contextual analysis. In 1973, to be sure, Bloor still generously admitted that Mannheim's conception of the sociology of knowledge was a 'close approximation' to the Strong Programme, although he 'faltered' with respect to the all-important symmetry principle (1973: 175). Somewhat later, however, he accused sociologists such as Mannheim, who continued to privilege natural scientific knowledge as an exceptional case, of a betrayal of their essentially naturalistic disciplinary standpoint and a 'lack of nerve and will': 'Despite his determination to set up causal and symmetrical canons of explanation, his nerve failed him when it came to such apparently autonomous subjects as mathematics and natural science' (1991 [1976]: 4–5, 11, 157).[18]

There is no question of refuting this type of critique or of showing its lack of textual foundation. The methodological demarcation between cultural or human studies and natural science is a well-documented position in Mannheim from his earliest writings (e.g. 1982 [1922]: 75–76, 98–99; 1952: 35–36, 44, 101–102, 130, 135). A representative example occurs in the 1928 lecture on intellectual competition, where Mannheim emphasized that the postulate about the *Seinsrelativität* or positional and perspectival determination of thought did not touch the exact natural sciences, but remained confined to historical, political, cultural and social, and ordinary everyday thought. In the latter type of 'existentially determined' knowledge, the results of thought processes were in part determined by the nature of the thinking subject. While in the natural sciences thinking was carried on by an abstract 'consciousness as such', in existentially determined thought it was the 'whole man' who did the thinking (1952: 193–94).

However, the attempt to bring this demarcation under the psychologically debunking category of a 'failure of nerve' appears to miss and to misrepresent

Mannheim's own critical epistemological intentions. First of all, the 'failure of nerve' argument is distinctly anachronistic. The neo-Kantian distinction between culture and nature and between cultural and natural sciences, which Mannheim adopted as his starting point, was primarily intended to *liberate* or *emancipate* the sciences of culture and society from the epistemological primacy of natural-scientific method, and hence to relax and delegitimize the exemplary function of natural-scientific rationality (cf. Frisby 1991: 194; Mannheim 1952: 37, 67, 70–71). It was Mannheim's explicit aim to recover the autonomous logic of cultural and social knowledge, which had been discredited by the unwarranted extension of natural-scientific methodology (1952: 76, 82). Both the positivistic and the Kantian conception of science mistakenly conceived of exact natural science as the only ideal prototype to which all sciences should aspire. However, since each sphere of reality had its own kind of 'givenness', each type of thinking should be understood in 'its own innermost nature' (1952: 70, 126, 194). Natural-scientific knowledge invoked only one of the capacities for creating knowledge possessed by the human spirit, but tended to hypostatize this one form of knowledge as knowledge *per se* (1982 [1922]: 76, 185, 195). It would be altogether wrong, proceeding from such an overly narrow conception of knowledge, to deny existential determination to knowledge in general or to treat this condition as an eliminable defect (186).[19]

In consequence, the knowledge-political motive that underlies Mannheim's *distinction* between cultural and natural science is largely identical to the one that subsequently informs the Strong Programme's attempt to *erase* it. Evidently, Mannheim did not suffer from a failure of nerve, but courageously attempted to break the spell of dominant epistemological conceptions of truth and rationality, marching against the same 'static philosophy of Reason' that subsequently constituted the point of attack of the more radical Wittgensteinians. Mannheim's stated objective was to undermine scientism by freeing 'half of the world' and half of the sciences from its suffocating embrace. Liberation from epistemological rationalism is precisely also what constitutes the 'Kuhn effect', which breaks the spell of unitary scientific methodology by offering a novel and disrespectful interpretation of the development of natural science itself. Barnes (1977: 3) therefore misreads Mannheim when he concludes that the latter took the disinterested knowledge of the natural sciences as preferable to sociology or history or political thought. On the contrary: Mannheim regularly insisted upon the distinctive superiority of the cultural, historical, and social sciences over against the natural sciences precisely because the former were existentially bound and hence shared unique cognitive profits, offering a 'deeper penetration into its object than is ever possible in the exact sciences' (1982 [1922]: 252; 1968 [1936]: 150–52; cf. Hekman 1986: 74).[20]

Unhastening Science

There are some additional ironies in play. As suggested before, Bloor's symmetrical extension of causal sociological method towards logic, mathematics, and natural science operates a form of sociological naturalism that imitates features of the same objectivistic scientism from which Mannheim took care to distance himself (cf. Hekman 1986: 39ff.; Lynch 1993: 50–52, 71). The Strong Programme, Bloor argues, simply extends and articulates the naturalistic, value-free stance of sociology, which is simultaneously 'the basic standpoint of the social sciences'. Bloor thus cavalierly identifies his own naturalistic standpoint with that of 'most contemporary science', which is primarily 'causal, theoretical, value-neutral, often reductionist, to an extent empiricist, and ultimately materialistic' (1991 [1976]: 157). Mannheim's intention to save value-bound interpretation and understanding from the regime of empiricist causal explanation (e.g. 1952: 81–82) is thereby simply reversed. In addition, Mannheim's productive hesitations with regard to the epistemological separability between facts and values are countered by a Weberian distinction between evaluation and explanation.[21] Although Bloor admits that the Strong Programme is itself value-based, its core underlying value entails a conception of the natural world as 'morally empty and neutral', affirming the same moral neutrality that we have learned to associate with all the other sciences (1991 [1976]: 13). Mannheim, on the other hand, discusses something close to Bloorean symmetry in his treatment of the 'non-evaluative general total' conception of ideology (1968 [1936]: 71–72), only in order to plead the inevitable transition towards a more evaluative point of view.[22]

In view of such misrepresentations and difficulties, it must not be reckoned a failure or a weakness of Mannheim's programme that natural science was considered a special case and provisionally immunized from sociological scrutiny. In its own historical context, Mannheimian asymmetry was as 'strong' as Bloorian symmetry was in the context of the early 1970s, and there is a tinge of contextual unfairness in accusing Mannheim of having failed to apply his perspectivist and relationist method to the natural sciences themselves. If Mannheim could still write that, while the development of natural science was propelled by an 'immanent logic of things', each phase of the social sciences was connected 'in both content and method with the total process' (1982 [1922]: 277), the Kuhnian turn in the historiography and sociology of natural science has convincingly proven otherwise. It has revealed the workings of a cultural and social logic that has softened the 'hard' sciences beyond Mannheimian imagination, conclusively disproving his contention that the exact sciences make statements 'into whose content the historical and local setting of the knowing subject and his value orientation do not enter' (1952: 101). However, if *all* the sciences now stand revealed as in some sense 'existentially

Two Traditions in the Social Theory of Knowledge

determined', strong symmetry is less a critique of the Mannheimian programme than a contemporary extension of it (Lynch 1993: 42). If *all* knowledge is now seen as 'a continuation of politics by other means' or as subject to a logic of interested competition, it should be acknowledged that it was Karl Mannheim who first liberated this idea for the cultural and social sciences.

Productive Indecisions

I have begun my argument by drawing attention to the intellectual rift between a Mannheimian and a Wittgensteinian tradition in the study of knowledge and science, pleading a partial return to Mannheimian macrosociological, epistemological, and normative concerns in order to help resolve some internal Wittgensteinian difficulties. This synthetic move towards a social epistemology is expected to repair what some commentators deplore as the 'unsplendid isolation' of radical science studies, which, it is said, have lost touch with the main body of sociological work and have so far failed to produce a truly general theory of knowledge and science (see e.g. Fuchs 1992). But this move also bridges the divide from the opposite side, realigning some major recent macro-sociologies of knowledge, such as those of Elias and Bourdieu, more closely with the core concerns and insights of micro-oriented constructivism. In this fashion, the revitalization of social and political theory, which has been rather unilaterally proclaimed by actor-network theorists such as Latour (1991) and Woolgar (1997), may be reciprocated through a sociological and political broadening of the research agenda of science studies themselves.

Let me conclude this review by specifying some of the themes of a social epistemology that is able to renegotiate some of the (in)differences between the two traditions. However paradoxically, the heritage of Mannheim is inspirational here precisely because his work is enlivened by the presence of a number of unresolved ambiguities.[23] As noted, Mannheim has repeatedly been accused of indecision on crucial points of epistemology, of hovering between the old and the new, and of 'lacking the courage of his convictions'. My contention is that these hesitations are to some extent productive, insofar as they provide useful correctives to particular Wittgensteinian exaggerations, and hence offer a more ecumenical point of departure where analytic differences are not split but surpassed. More particularly, they help us to evade the logic of inversion that still constitutionally ties radical science studies to the a priori assertions of traditional epistemology that they have polemically set out to reject (cf. Darmon 1986). In this sense, we may generalize Kettler, Meja and Stehr's observation that Mannheim's 'productively unresolved' position (in their case, between Hegelian Marxism and Weberian sociology) makes his

Unhastening Science

work 'a timely, heuristically valuable starting point for fresh study' (1990: 1470).

The following remarks do not so much focus upon the larger 'social' in social epistemology, i.e. upon sociological macro-processes of intellectualization, scientization, or professionalization (where a Mannheimian programme builds upon acknowledged strengths), but retrieve and clarify a few points of socialized epistemology itself. Fundamental to the project of a social epistemology, first of all, is a spirited resistance against the eliminativist tendency present in naturalistic science studies, which, as we have seen, regularly vote for a sovereign dismissal of normative epistemology in favour of a rigorously empirical study of 'what actually happens' in science. Mannheim's first productive indecision is located precisely at the point of intersection of this classical disciplinary contest. His initial polemical separation between normative epistemology and empirical sociology appears to anticipate the abolitionist temper of modern science studies, but is subsequently revoked in favour of a sociological *reconstitution* of questions of truth, rationality, objectivity, and value (cf. 1952: 192, 226–27; 1968 [1936]: 1ff., 256ff.). Clearly, Mannheim's purpose was 'not so much to criticize epistemology as to rescue its project in the light of historical developments which rendered it redundant' (Scott 1987: 42). It was to replace an academic epistemology, which remained tied to justificatory and foundationalist concerns and a static, individualistic polarity between subject and object, by a new relational and positional epistemology, which would effectively 'reckon with the facts brought to light by the sociology of knowledge' (Mannheim 1968 [1936]: 70–71, 264) – more specifically, the fact of 'the essential perspectivism of human knowledge' (cit. Wolff 1959: 571). It was thus not to discard the problem of truth and objectivity altogether, but to bring it 'one notch lower, closer to the level of concrete research' (1952: 103).[24]

Another closely related indecision on Mannheim's part engages the long-standing controversy about cognitive vs. social explanations of the development of knowledge, where his views are mobilizable as an antidote against remnants of both cognitivistic and sociologistic reductionism. Even though (*pace* Bourdieu 1990b: 298) SSK's conception of social determination has never been of the unilateral short-circuit variety, the logic of inversion has regularly lured its theorists in this abolitionist direction; as is exemplified by Latour's Seventh Rule of Method, which advised a moratorium on cognitive explanations of science and technology originally extending for a ten-year period (1987: 247, 258).[25] Mannheim's ground-breaking analysis of the role of competition in cultural and intellectual life, while departing from the radical assertion that this 'external' social factor 'entered as a constituent element into the form and content of every cultural product or movement', was carefully demarcated against both cognitive internalism and unbridled sociologism (1952: 191–92). The metaphors of a politics and an economics of knowledge, which alternatively express this internal socio-cognitive link, but which also sometimes serve as vehicles of intellectual rivalry between contemporary schools of sociology of science (cf. Bourdieu's economic field theory vs. Latour's political actor-network theory) are still liberally mixed in some of the most central passages of Mannheim's original 1928 essay (1952: 196–98, 210–14; Pels 1997).

Anticipating Bourdieu's anti-economistic economy of practices, Mannheim defended himself against the criticism of 'projecting specifically economic categories into the mental sphere'. The original discovery of the role of competition by the Physiocrats and Adam Smith only demonstrated the impact of this general social relationship in the particular context of the economic system. The ultimate aim was to strip our categorial apparatus of anything specifically economic in order to grasp the social fact *sui generis* (Mannheim 1952: 195). In close analogue, Mannheim strove to avoid the alternative pitfall of political reductionism by emphasizing that the explanation of intellectual movements in political terms was 'not intended to give the impression that mental life as a whole is a purely political matter, any more than earlier we wished to make of it a mere segment of economic life'; it is simply the case that the vital and volitional element in thought is easiest to grasp in the political sphere (1952: 212). Whether expressed in an economic or political metaphor, it was evident that the activist, volitional and interested determination of thought should not be taken as a simple reflex of various social locations, but presupposed a specific freedom of the intellectual sphere (1952: 195, 228–29). Mannheim accordingly distinguished between material *interestedness* and *committedness* to styles of thought or intellectual forms, and between *social* and *intellectual* stratification (1952: 183–87). While it is perhaps too much to say that claims such as these anticipate Bourdieu's field theory of science, they do help to tone down extremist suggestions about a seamless web connecting science and society, and about the imperative dissolution of the modernist divide between scientific and political representation (cf. Latour 1993a).

Normative Complexities

This anti-reductionist impulse also helps to specify Mannheim's intriguing intermediate stance in the familiar contest between partisanship and value-freedom – one that is doubly defensive against the Marxian danger of direct politicization and propagandistic thought and against the Weberian danger of uncommitted, disengaged intellectualism. Thinking, for Mannheim, is inevitably positionally determined and hence committed, but not *politically* determined or evaluative in any *immediate* sense. His criticism of the liberal-modernist divorce between facts and values (e.g. 1952: 216–17) – which notably agrees with Carl Schmitt's view – and his contrary assertion of the inevitable value-groundedness of thought, do not issue in a total renunciation of the ideal of political neutrality, but rather in an attempt to revitalize its original impulse (Meja and Stehr [eds] 1982: 401). Thought does not progress by completely restraining its evaluative and pragmatic orientation, but by retaining its original *élan politique* while simultaneously subjecting it to reflexive intellectual control (1982 [1922]: 42, 89n). Value-freedom, Mannheim argues in a turn of phrase that he would repeat many times, 'is possible in sociology and social knowledge in the sense that one ought to refrain from any valuation. But at a much deeper level, valuation cannot be excluded; namely at the level of the perspectivity that has entered into the formation of concepts' (1982: [1922]: 247). In this manner, the rejection of a 'superficial' amalgamation of facts and values and of science and politics is balanced against the admission of the inevitable presence of a mixture of facts and values and a politics of knowledge in the deep structure of theoretical definition and empirical analysis.

Since I take this issue of facts vs. values or normativism vs. naturalism as strategically crucial for a social epistemology as here proposed, I am interested in detailing it a little more closely. I have already pointed out that STS's scramble to consign various inherited philosophical dichotomies to the intellectual scrap-heap has so far piously halted before the deconstruction of this particular dualism. Once again, Mannheim is interesting because his agenda reaches beyond the straightforward replacement of normative enquiry by a self-satisfied sociological naturalism. His purpose is not to reject but to reformulate a concept of 'noological' rationality that follows through the implications of the historicity and positionality of both epistemological and normative beliefs (1968 [1936]: 265–66). Although questions of fact and right are provisionally divorced for methodological purposes, they are reconnected in order to afford a necessary transition towards an evaluative conception of the sociology of knowledge, and towards an explicit defence of the utopian element in sociological thought (1968 [1936]: 78ff., 235–36).

Lynch has argued, as we saw, that the Strong Programme's core postulates of symmetry and impartiality were actually less an attack on Mannheim's programme than a more radical extension of it. In his view, Bloor, Barnes, Collins, and even reflexivists such as Woolgar, further radicalized what Mannheim called the 'non-evaluative general total conception of ideology', which advocated a provisional suspension of all judgment as to the correctness of the ideas under analysis, and hence can be easily harmonized with a naturalistic and agnosticist reading of the later Wittgenstein (Lynch 1993: 42ff.). What this analysis ignores, however, is that Mannheim explicitly distinguished *two* types of approach to ideological inquiry that arose on the level of the general-total conception of ideology: an approach characterized by freedom from value-judgments ('non-evaluative relationism'), and an epistemological and metaphysically oriented *normative* approach that he called 'dynamic relationism' (1968 [1936]: 71, 76, 88). According to Mannheim, what was originally simply a methodological technique ultimately disclosed itself as a *Weltanschauung*, which encompassed metaphysical-ontological value-judgments of which we were previously unaware. The very possibility of being completely emancipated from ontological, metaphysical, and ethical presuppositions was considered a 'positivistic prejudice' (1968 [1936]: 78–79). Although we might have begun non-evaluatively, we were forced eventually to assume an evaluative position that explicitly avowed the metaphysical presuppositions undergirding empirical knowledge. This normative shift was seen as 'typical of the whole development of contemporary thought' (1968 [1936]: 80–86).

Lynch's own reflexive praxeology, for all its legitimate objections to Bloorian SSK and its scientization of the later Wittgenstein, continues to share much of this empiricist prejudice by refusing to entertain a normative epistemology of whatever kind (cf. Lynch and Fuhrman 1991; 1992; Lynch 1992c; 2000). Indeed, insofar as ontological concerns inform more radical versions of STS, such as actor-network theory, they do not implicate articulated normative commitments; the constitutive postulates of symmetry and agnosticism ('following the actants') continue strongly to work against this.[26] De Vries' construction of an anti-Kantian and ontological (more precisely: Latourian) Wittgenstein, for example, maintains a distance from normative concerns, recommending that we merely 'expose' the way in which forms of life actually work as heterogeneous ensembles of words, actions, and things (1992: 30–31). From a different viewpoint and tradition, Hekman, who makes such a laudable effort to rescue Mannheim from Bloorian scientism by highlighting his transition from neo-Kantian epistemology to ontology (a move she thinks Wittgenstein was unable to make), fails to specify in what sense such an anti-foundationalist and hermeneutical ontology integrates a value perspective (Hekman 1986).

It cannot be my purpose here to speculate about the general relationship between epistemology and ontology, even though a social epistemology would provisionally refuse any strict separation between the two concerns, as it has before refused the forced choice between normative epistemology and empirical sociology. In this respect it would remain opposed to STS's epistemological

69

Unhastening Science

'indifference', including the various praxeological, ontological, or anthropological radicalizations of its central postulates of symmetry and impartiality. Indeed, it would prefer to consider the contest between sociological explanation and ontological description as a fraternal rivalry among fellow Wittgensteinians, and as less significant than the broader contest between naturalism (causalist *and* descriptivist) and normativism. In this range of dispute – and probably going beyond what Mannheim himself had in mind – it would explore the possibility of an intentional *confusion* of facts and values, as so stringently forbidden by Barnes, Bloor, and many other proponents of 'value-free relativism' inside and outside of STS.[27] Following Mannheim's suggestion about the proximity between evaluative and ontological judgments, it would favour a much closer fit between the concerns of epistemology, ontology, and ethics.

It is *this* Mannheim who today may still be fruitfully opposed to the symmetrism and agnosticism of Wittgensteinian ethnography. By highlighting the normative complexity of his approach, we are at least reminded that the contest between normativism and naturalism is still open. So are the other contests that Mannheim failed to decide.[28] They can only be brought to a close if the well-entrenched polarization between what I have called the Mannheimian and the Wittgensteinian traditions is considered deeply problematic from both sides, and if each side recognizes that it stands in need of the other in order to facilitate the 'next step' in the development of a social theory of knowledge.

Indifference or Critical Difference?

A closely affiliated debate on the possibility of a critical normativity has recently swept ethnomethodology – which has long served as a grounding inspiration and a *compagnon de route* for Wittgensteinian science studies. This debate shows that, far from being settled or superseded, the fact/value issue continues to be a highly contentious one; as is the normative significance of the Wittgensteinian pragmatics of language that forms its interpretive backdrop. McHoul (1988) initially proposed a reading of Coulter's work that sought to extend the latter's avowed descriptivism into a 'critical pragmatics' with intrinsic evaluative and political intent, opposing the latter's conviction that sociology remained free of internally generated normative commitments. Bogen and Lynch (1990) branded this attempt to press ethnomethodology into the service of social critique an ill-fated piece of interventionism, opting for a more limited 'therapeutic' critique of prevailing explanatory and transcendental programmes in philosophy and the social sciences (cf. also Bogen 1996; Lynch 2000: 26). In reply, McHoul (1990, cf. 1994) restated his interest in a 'critical descriptivism' that would more definitely acknowledge the power-charged and politically committed nature of all social-scientific inquiry. Jayyusi's respecification of the fact/value issue similarly combined the Wittgensteinian idea of an internal connection between normative, conceptual, and practical judgments with a more politically sensitive pragmatics; in her estimate, the agnostic position held by Bogen and Lynch retained a residue of the classical fact/value dichotomy (Jayyusi 1991: 249, 251n).

The presumption of an internal connection between naturalistic description and normative appraisal, and hence of the antiquarian status of conventional fact/value distinctions, therefore opens up two different epistemological pathways. On one side beckons a tradition that has been described as 'naturalist', which is loosely traceable to the later Wittgenstein. It intends to 'redescribe', 'display', 'reveal', 'lay out for view', or 'just follow' what practical speakers and ordinary actors actually say and do, with the purpose of disestablishing transcendental and a priori conceptions of rational action. It is naturalist in the sense of remaining suspicious of all normative legislation of an epistemological or socio-political kind, and in the sense of offering 'disinterested', 'non-evaluative', or 'neutral' analyses of language-in-use and practical conduct. This tradition incorporates both Bloor's and Collins' more traditional Weberian versions of value-free sociology and the more novel forms of ethnographic descriptivism practised by, for example, Knorr-Cetina, Woolgar, Callon, Latour, and Lynch. Not wedded specifically to positionings on the explanation-description axis, to 'rule-scepticist' vs. 'anti-scepticist' interpretations of Wittgenstein, or to strong Weberian vs. weaker conceptions of scientific neutrality, its family resemblances and fraternal rivalries appear sufficiently articulate to warrant inclusion in the composite picture I have drawn.

The alternative view finds the issue of the normative legitimacy of knowledge claims irrepressible, and hence attempts to preserve some kind of normative and critical epistemology. This standpoint may be loosely associated with a more normatively sensitive Mannheimian view – that is, if Mannheim's sociology of knowledge can be successfully quarantined from the 'non-evaluative' interpretation that is still shared by contenders such as Bloor and Lynch. Rather than distancing description from critique (which is quickly assimilated to 'denunciation'), this older project reflexively reintegrates them also on the second-order level of sociological analysis. Although the agnostic tradition allows the 'laminated' nature of facts and values on the level of ordinary language and practical action, it hesitates to duplicate this insight for the professional practice of ethnographic inquiry itself. The upshot of this reflexive duplication is once again to accentuate the *critical distance* between 'outside' observer's and 'inside' actor's standpoints or, in ethnomethodologese, between the different interests and perspectives of analysts and members.

Unhastening Science

According to Bogen, Lynch, and Coulter, partisanships and normative commitments are only operative in ordinary settings of practical action whereas, in the analytical setting, the critical impulse remains confined to a therapeutic indifference towards transcendentalist or foundationalist postures in (rival) philosophies and sociologies. In a more normatively reflexive social-epistemological conception, however, the scope of critique is widened to encompass ordinary life itself (within science and outside). The point about ethnographic neutrality is that it faces two ways in order to change the balance of forces: while *de*valuing foundationalist theorizing, it simultaneously *re*values ordinary reasoning and action. Its stated impartiality is therefore doubly partial: critically poised against the former, it is sympathetically disposed towards the latter, even to the extent of critically identifying with it. By recommending an end to all explanation, generalization and theorizing (cf. Button 1991), and by abstaining from all judgments of adequacy, relevance, or value of the actor's own practical reasonings, the policy of ethnomethodological *indifference* therefore runs the risk of erasing all *critical difference* between analytical and practical accounts, and hence of becoming ensnared in what it seeks 'disinterestedly' to describe. Its refusal to transcend the logic of everyday practice, or to perform any kind of epistemological break with mundane rationality, envelops it in a condition quite similar to that which it so compulsively resists where competing theories are concerned. But precisely to view language as a game in which 'words are also deeds', and to view social practice as a performance in which facticity is recursively and circularly constructed, is already to bring a critical outsider's perspective to bear upon knowledges, both 'ordinary' and scientific, insofar as these are framed by the 'natural attitude' and stabilized by means of reifications.

The ethnographic attitude misunderstands itself as merely descriptive insofar as it is already critically disposed by the very fact of topicalizing central features of practical conduct (its lamination of facts and values, its essential performativity) that are not so immediately transparent to the actors themselves, and may even be actively resisted in their mundane reasoning about a taken-for-granted world 'out there' (cf. Pollner 1987). Its nervous disavowal of transcendent theorizing is exaggerated into a constructivist idealization of the language and practice of ordinary members – which is little else but sociological populism in disguise (cf. Fuller 2000b: 27, 29). There is a peculiar irony in such a linear inversion of the objectivist and transcendentalist position. On the one hand, objectivist sociology, while officially proclaiming an epistemological break with the mundane world, is derided for inadequately extracting itself from it and for uncritically reiterating and even aggravating common-sense assumptions about realism, facticity, and social order. On the other hand, ethnomethodology itself, while officially refusing such an epistemological break, surreptitiously performs it; misrecognizing its own critical distance from the mundane world, it subtly colonizes it by reading its own performative epistemology into actors' practices. Objectivist social theory claims to transcend common sense, only to repeat the natural attitude; ethnomethodology claims to use nothing but ordinary members' resources, only to redescribe the ordinary world in such a manner that every actor becomes a practising ethnomethodologist. One one side we have a break that occludes a continuity; on the other we have a continuity that occludes a break. The critical project that I defend in this book puts an end to this mirror-play. It rehabilitates the classical intuition about the need for an epistemological rupture without counting upon any kind of transcendental or foundational legitimacy. The demarcation is twofold: against objectivistic sciences and against objectivistic forms of common sense. To both, it critically opposes a constructivist, anti-reificatory epistemology. It cannot rest satisfied with 'leaving everything as it is', or 'just following' what actors say and do. It needs to follow them *critically*.

CHAPTER 4

The Natural Proximity of Facts and Values

The Fact/Value Split

Values, the cultural pessimist laments, are no longer what they used to be. The language and practice of contemporary morality, critics such as MacIntyre, Bloom, Scruton and Finkielkraut agree, suffers from a deplorable state of confusion: ethical judgments are permanently contested and seem no longer capable of being rationally justified. But Facts also, fellow pessimists hasten to add, are only a slim shadow of their former self. Ever since Popper exchanged the inductivist metaphor of the rock-bottom of knowledge for the critical-rationalist one of the swamp, empirical science appears to have lost all solid footing, rendering Kant's first question 'What can we know?' just as precarious and unanswerable as his second: 'What ought we to do?' To compound this twin complaint, there is a third, which completes the doleful *a capella*. The venerable art of *distinguishing* judgments of value from judgments of fact, or of avoiding logical contaminations between them, appears similarly lost to the contemporary mind. Traditionally hailed as one of the foremost accomplishments of Western rational culture, the categorical separation between *Sein* and *Sollen*, already undermined by neo-Marxist politicization drives since the late 1960s, is today unceremoniously swept into the intellectual dustbin by an entire generation of pragmatist, constructivist, and postmodern thinkers. The principle of value-freedom is routinely debunked as an antiquarian dogma that nobody in his or her right philosophical mind would care to reanimate.

In this chapter, I undertake to replace this threefold pessimism with a three-fold optimism. The failure of the classical Enlightenment project to secure a rational foundation for moral values and empirical explanations is not a twin intellectual distaster, but might be celebrated as a double awakening from an impossible and perilous dream. The 'unfoundedness' of both genres of

The Natural Proximity of Facts and Values

judgment is not a double trap-door leading towards cognitive irregularity and moral licence, but a sign of epistemological and axiological modesty, signalling the absence of the symbolic violence that so naturally accompanies the (universal) temptation to universalize our set preferences and pet ideas. However, this chapter will not be focally concerned with such currently over-rehearsed critiques of foundationalism. It will instead encircle the problem of the *mutual connection* or *interaction* between values and facts, in order to defend a view that is squarely opposed to the dualistic standard conception found in Moore (1993 [1903]), Weber (1970), Kelsen (1967 [1934]), Popper (1962), Dahrendorf (1968), or Kolakowski (1972; 1978), and which is summed up by Kelsen in characteristically apodictic terms:

> The difference between *is* and *ought* cannot be explained further. We are immediately aware of the difference. Nobody can deny that the statement: 'something is' – that is, the statement by which an existent fact is described – is fundamentally different from the statement: 'something ought to be' – which is the statement by which a norm is described. Nobody can assert that from the statement that something is, follows a statement that something ought to be, and vice versa. (Kelsen 1967 [1934]: 5–6)

This view has become so widely adopted in analytic philosophy, logical positivism, ethical emotivism, behaviourism, and mainstream social and political theory during the past century that it has acquired an air of naturalness and self-evidence, virtually assuming the status of a cultural institution (H. Putnam 1981: 127; Hennis 1994). To counter it, I want to venture a claim about the 'natural proximity' of facts and values, which is just as optimistic as my previous claims (Pels 1990; 1991a). The idea that empirical-cognitive and evaluative statements *are* deeply intermeshed and *should* naturally go together must not be taken as a symptom of philosophical incoherence, signalling a regression to naïve and dangerous forms of monism, but as a promising springboard for theorizing and research. Althought it has held its ground somewhat longer than other binary oppositions in the modernist canon, the fact/value distinction is presently over-ripe for deconstruction if not for a final implosion.

This 'natural proximity' of facts and values is only logically conceivable when it is not itself offered as another factual observation, but as a theorem in which constative and normative modalities are themselves immediately interlaced. As such, the new principle meets an elementary demand of reflexivity, since it is not permitted to escape its self-imposed methodological prescription. The theorem about the inseparability of facts and values is just as performative and circular as the dualistic opposite that it it reverses and rejects; the crucial difference being that the former reflexively affirms what the latter chooses to camouflage under a claim of ontological objectivity.[1] This is why an expression such as 'natural proximity' fits my purpose better than, say, 'logical' or 'ontological' proximity; not merely because we need to distance ourselves from the policing powers surreptitiously exercised by argumentative logic and ontological essentialism since their 'invention' in the Socratic dialogues,[2] but also because the adjective ironically exploits the native ambiguity of conceptions of Nature and the natural, which have always been uneasily harnessed between the modes of facticity and validity.

The standard view of scientific rationality carves out a profound schism and a decisive asymmetry between the judgmental spheres and validity domains of *Is* and *Ought*, explanation and evaluation, empirical knowledge and morality, cognitive and ethical theory. These realms of being and knowing are supposedly governed by philosophical principles that are generically asymmetrical and mutually irreducible, and hence open up an unbridgeable gap between the two realms. This asymmetry may be interpreted ontologically, such that values and facts are assigned to incompatible worlds or existential realms, or epistemologically, such that facts, conceived as objectively knowable things-in-the-world, are brought into contrast with the essentially precarious, contingent, and subjective status of values. The standard argument also routinely adopts a logical-linguistic form, ascertaining that prescriptive claims cannot be legitimately derived from or reduced to empirically descriptive ones, and that to do so is to commit what G.E. Moore has familiarly dubbed the 'naturalistic fallacy'. It is precisely this anti-reductionist purpose that has charged the fact/value dualism with much of its traditional resilience and fighting force.

This axiom of (onto)logical asymmetry is supported by a historical narrative that follows a brazenly Whiggish style. The facticity/validity split is typically granted a finalistic prehistory that traces a gradual unfolding from monism towards dualism (see Fig. 4.1), proceeding from a primitive stage of osmosis towards rational insight into the necessity of dissociation and heterogeneity (cf. Weber 1970; Popper 1962; Radnitzky 1973). Hence the dichotomy is simultaneously profiled against a historical point of departure and against monistic residues in current lay reasoning and ordinary language, which still allow for a continuum of interactions between value-judgments and judgments of reality. Indeed, it is a critical mainstay of the traditional view that everyday discourse is replete with ill-monitored transitions between description and prescription, and that their strict methodical divorce is a primary criterion of superiority demarcating science from common sense.

This separatist conception, however, seriously underrepresents the extent to which both 'facthood' and 'valuehood' are themselves historical inventions

The Natural Proximity of Facts and Values

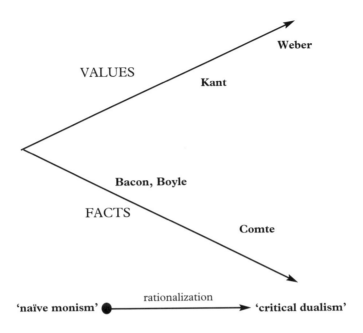

Figure 4.1: The Standard View of the Fact/Value Split

and constructions, which have gradually evolved as complicitous counter-concepts in a two-pronged process of generalization and abstraction (Edel 1980: xix, 339ff.; Graham 1981: 1–32; Doeser 1986). The modern concepts of fact and value were crafted in the course of an antinomic polarization that gradually 'emptied out the middle', circularly defining 'value' as the receptacle of everything that was not covered by the emerging notion of 'fact' and vice versa, ultimately establishing them as axial concepts in separate domains of ethical judgment and empirical-causal science. As MacIntyre has emphasized, this historical enactment of (what he critically refers to as) the *apparent* division between fact and value required a parallel and simultaneous purification of the concepts of morality and factuality: '"Fact" becomes value-free, "is" becomes a stranger to "ought", and explanation, as well as evaluation, changes its character as a result of this divorce between "is" and "ought"'(1984: 84).

By taking this reciprocal articulation for granted and projecting it backwards into history, mainstream epistemology, ethics, and social science have honoured the dualism with an accolade of timeless truth, tracing its genesis as an irrevocable advance of rational enlightenment. But this view not only neglects the sheer protean complexity that the fact/value nexus has exhibited over time (cf. Proctor 1991), but also tends to ignore the socio-logic of intellectual rivalry (such as the deepening institutional separation between the domains of science and the state; see further Chapter 8), which has governed the dynamics of the fissure and the broader social forces which it articulates and in which it is embedded. The interest in a sociologically richer and more differentiated historical account has been reinforced by a mounting scepticism about the logical tenability of the fact/value dichotomy itself.[3] It has hastened the recovery of two symmetrically interdependent 'creation stories', which relate how a gradual process of *reification* or 'hardening' of facts has been mirrored and reinforced by a simultaneous process of *idealization* or 'mollification' of moral values. In this interpretation, the twin concepts have increasingly profiled themselves 'at each other's expense', in a synchronous movement of diffusion and disentanglement that has progressively unravelled their primordial symbiosis. Emerging as diametrical antipodes, facticity and validity finally arrive at a cognitive-logical back-to-back position, which has been frozen at the moment of greatest amplitude, when facts are exclusively made to refer to objective things-in-the-world (or things-in-themselves) and values to subjective states of the mind.

Objectification and Subjectification

If the point zero of this dual articulation can be located anywhere in historical time, it might be found in the Platonic and Aristotelian blending of the true and the good, which was preserved more or less seamlessly in the medieval scholastic doctrine that *ens et bonum convertuntur*. However, as the institutional-intellectual conflicts that pitted medieval theology against philosophy and religion against science acquired a sharper edge, the continents of normative discourse and the discourse of truth were slowly set adrift. The Renaissance revolution in the natural sciences oriented the new experimental sciences to an understanding of 'the language of nature' as it spoke in its purest voice, liberated from premeditated ideas, prejudices and ideologies. Pioneered in Galilei's *nova scientia*, this new naturalistic perspective received a fully articulated philosophical expression in both Descartes' rationalist and Bacon's empiricist programmes. Descartes' principled divorce between spirit and matter directly reflected Bacon's distinction between ideas (opinions, hypotheses, prejudices, ideals) and 'things'. Boyle's novel conception of scientific 'matters of fact' similarly identified them as referring to objective states of affairs in a value-free and theory-less external world, and hence as promising the absolute certainty and universal assent that theories and values were no longer able to offer (Shapin 1984). The accompanying institutional demarcations were already clearly implicated by the Royal Society's famous resolve 'not to meddle with

The Natural Proximity of Facts and Values

divinity, morality and politics' but to restrict its concerns to the improvement of natural knowledge (cf. Ornstein 1983: 108–109).

The epistemological properties that positioned the concept of facticity at a widening remove from that of validity ('nakedness' in the sense of both value-freedom and theory-freedom) hence emerged from a prolonged process of semantic divergence, purification, and intensification. Hume still drew a somewhat inconclusive boundary between Is and Ought, even though the relevant passage in his *Treatise* has been welcomed by analytic philosophers as clearly anticipating their conception of the naturalistic fallacy (Hume 1978 [1739–40]: 469; MacIntyre 1969). A further deepening of the demarcation is traceable in Kant's critical distinction between pure and practical reasoning, which split the Socratic question about the nature of the good life into two separate inquiries into the conditions of possibility of theoretical and those of ethical rationality. The fledgling social sciences further widened the rift by demarcating themselves from moral and political philosophy and embracing the epistemological paradigm of the natural sciences. Scottish-English political economy, the French Physiocratic school, Montesquieu's proto-sociology and the historical philosophy of Condorcet all invoked the Cartesian-Baconian concept of facticity, and described socio-historical reality as being governed by value-free empirical laws.

In the German intellectual universe, Hegelian absolute idealism, which temporarily reintegrated the normative and the natural in quasi-Platonic fashion, provoked an empiricist reaction that found its most forceful expression in Ranke's alleged dictum that historiography should represent past reality 'wie es eigentlich gewesen ist' (cf. Vierhaus 1977). Early sociology, as it emerged in the writings of Ideologues such as Destutt de Tracy and positivists such as Saint-Simon and Comte, aligned itself closely with this empiricist perspective. Durkheim's primary methodological rule to conceive of social facts as 'things', and his principled separation between *choses* and *idées* (a category that included both ideas and ideals) resonated with both the Cartesian and the Baconian epistemological programmes. This reified notion of facticity, which also governed the positivist philosophies of Mill and Mach, was channelled far into the twentieth century through the logical positivism of the Vienna Circle and the analytics of Hempel and Nagel.

On the opposite side of the fissure one may follow a similar articulation of the concepts of value and morality, which, in their perfected 'emotivist' form, were finally positioned at right angles and at maximum distance from this objectivist paradigm of factuality. Originally, Greek and Latin ethical conceptions still blended facts and values by identifying morality with a person's good character; the Renaissance conception of *virtù* still typically mingled the description of character with an identification of characteristics of excellence (Skinner 1978: 120ff.). (Neo)-Aristotelian teleology contrasted man as an empirical creature with the potentialities that slumbered in his essential nature; ethics and practical reason were seen as providing reliable indicators for proceeding from man's factual to his potential state. This teleological pattern remained largely intact when it was absorbed into a theistic framework, irrespective of whether it adopted a Christian form as in Thomas Aquinas, a Jewish one as in Maimonides, or an Islamic one as in Ibn Roschd. Geared either to the discovery of 'true' human nature or to the substance of divine law, this entire tradition still endorsed a view of moral utterances as permitting judgments of truth and falsity (cf. Graham 1981: 15–16).

This intellectual schema underwent substantive alterations in early Protestantism and Jansenist Catholicism, which began to limit and differentiate the conception of reason: in the new theological vision reason stood powerless before the passions, and could no longer plot the transition from objective potentiality to human action. While still considered capable of discovering factual and mathematical relations in the realm of nature, rationality was gradually restricted to the definition of means rather than ends in the realm of practice. As we noticed before, a further distancing between moral precepts and natural facts was observable in the writings of Enlightenment thinkers such as Hume, Diderot, and Kant. The endeavour of eighteenth-century philosophy was still to ground a rational ethics upon universal characteristics of human nature. There was a rising awareness, however, that natural facts might be logically irrelevant to such value-statements, and that it was illegitimate to proceed from strictly factual premises to moral or evaluative conclusions. While Hume, as we saw, had only vaguely intimated this principle, Kant was more categorical in establishing that the moral law could never be derived from statements of a cognitive nature – even though he maintained that both types of judgment could be universally and rationally grounded.

The development of English utilitarianism from Bentham through Mill to Sidgwick similarly emphasized the essential elusiveness of a teleological framework for morality. The value concept in its truly modern sense, however, emerged only in the mid-nineteenth century in German philosophy, following the 'culturalization' of the narrower technical definition of value that was first developed in classical political economy. Neo-Kantians such as Lotze and Windelband began to identify philosophy with a general value theory, pushing the concept towards an axial location in the cultural sciences and turning it into a prime marker of the nature/culture dichotomy (Lemaire 1976: 134ff.; Schnädelbach 1983: 197). Neo-Kantianism further distanced the realm of Being from that of the Good by ontologically uprooting the latter: while still laying claim to objectivity, values were no longer seen as factually existent, while *Sein* was reduced to a pure facticity that was no longer capable of grounding the *Sollen* ('Seiendes ist, Werte gelten') (Schnädelbach 1983: 198–200). From Windelband and Rickert this subjectivized value concept trickled into the second *Werturteilsstreit*, in the course of which Max Weber famously defended the principle of value-freedom over against Schmoller's and Wagner's normative science of society (Von Ferber 1965; Turner and Factor 1984; Proctor 1991; Hennis 1991). Nevertheless, as is suggested by his metaphorics of the 'absolute polytheism' of values and the 'eternal struggle of the gods', Weber still hesitated and compromised between the objectivism of universal cultural values and the subjectivism of individual value orientations (cf. Lemaire 1976: 139, 143; Brubaker 1984: 62ff.; Oakes 1988; Parsons 1967: 90). A further 'mellowing' and subjectification of value emerged in the ethical emotivism of G.E. Moore (*Principia Ethica*, 1903), who more exclusively argued that ethical judgments could not claim an objective status but merely expressed subjective preferences, attitudes or feelings, and that normative conflicts could therefore never be resolved on rational grounds (cf. Mackie 1977).

As I have indicated, the standard 'Whig' narrative brings this double process to a standstill and naturalizes the dualism as ontologically given. However, there is no intrinsic reason why the extreme pitch of the dualism marks a philosophical finality and is logically immune to further historization. My brief historical narrative has already emphasized the semantic fragility of the fact/value split and the 'contrived' and contested nature of the epistemological demarcation, suggesting its close implication in new styles of philosophical and scientific professionalism and in the legitimation of a broad institutional separation between science and society. Such a critical awareness tends to reverse the standard representation of the relationship between the quest for scientific autonomy and the philosophical principles that are conventionally adduced to support it. It indicates that the fact/value distinction, instead of providing an unquestionable logical foundation that dictates the institutional differentiation between science and society, should rather be seen as a performative distinction that co-emerges with and co-produces this differentiation, precisely by describing the rift as ontologically pre-given. The ontological fixture functions as a form of power speech (a form of intellectual 'divide and rule') which empowers the institutional demarcation by failing to acknowledge the force of its essentialist form.

Unhastening Science

The Diamond Pattern

However, this genealogical strategy of reversal does not initially take us very far beyond the idea that there is a more lively interaction between the domains of fact and value than the standard view finds permissible (Graham 1981; Doeser 1986). In addition, it easily invites a form of historicism that is suspicious of all normative epistemology, and accordingly repeats the fact/value dualism on its own second-order level of observation. In this respect, it joins the broad current of 'value-free relativism', which, as I noted in the previous chapter, provides the epistemological underpinning of much recent work in language pragmatics, discourse analysis, ethnomethodology, and constructivist studies of science. Although it has radically weakened the traditional rationalistic and empiricist foundations of scientific work, this broadly Wittgensteinian tradition continues to embrace an ethos of methodical distanciation and ethnographic descriptivism that throws cold water upon the ambition to develop an alternative normative sensibility (Fuller 1988; 1989; 1992; 2000b; Laudan 1984; 1987; 1990; Radder 1996).

In the following, my objective is to radicalize the critical impulse that slumbers in this type of historicism and articulate it into an explicitly normative project. Working my way towards this alternative, I shall first redescribe the antinomy of facts and values in the more radical and specific sense of a quasi-Hegelian *contradiction*, which progresses in a familiar three-step sequence from an originary union towards dualism and divorce, and henceforward to a reconciliation and reunion of its former opposites. Deepening the claim that the fact/value dualism is not a timeless axiom, an ontological datum, or a transhistorical epistemological truth, it is reconceived as the transitory 'middle term' in a sequential movement of alienation, contradiction, and reconciliation. In its second phase, this movement *reverses* the parallel processes of the objectification of facts and the subjectification of values in order to restore the 'natural proximity' that tied them together in the original situation. The weak postulate about the *interaction* between facts and values is strengthened into an anticipatory claim about their progressive *fusion* or *osmosis*. This movement transcends the historical antinomy and draws its two 'secessions' together in a novel synthesis that resembles (but also reveals important differences from) its monistic, undifferentiated point of departure. The diverging movement flips over at its point of greatest amplitude into a countermovement of dual approximation and convergence, resulting in a diamond-like pattern (see Fig. 4.2).[4]

In this quasi-dialectical scenario of divergence, antinomy, and convergence, the point of *arrival* simultaneously constitutes a utopian, logico-normative point of *departure* from whence the entire movement is projected backwards in time.[5] The diagram thus does not merely reconstruct a secular pattern of fission-fusion, which claims to offer an adequate representation of intellectual-historical realities, but simultaneously clarifies and justifies a critical position; it deliberately 'acts out' and advocates what it seemingly sets out merely to describe. In this manner, it reveals an awareness of the reflexive paradox that would arise as soon as it presented itself as an exception to its self-made rule (cf. Pels 1998: 257–59). This paradox inevitably threatens all (onto)logical assertion of an alleged gap between judgments of fact and value (their distribution over different worlds, spheres of being, or logical domains) because such arguments inevitably contravene their own grounding principle by immediately mixing description with evaluation. All ontologies describe 'normative facts' and hence function as performatives that partly create what they state; but few ontologies are prepared to acknowledge this inherent normativity and performativity. This reflexive problem already arises in Hume, the (contested) father of the critique of the naturalistic fallacy.[6] It recurs in trenchant and dramatic form in the Weberian postulate of the 'absolute heterogeneity' of the judgmental domains of *Sollen* and *Sein* – which indeed offers a sharply paradoxical example of its own opposite, i.e. of their natural proximity.[7] As I shall explain further below, even Habermas's recent attempt to diminish the divide while retaining the idea of different ontological reaches of the normative and the cognitive is troubled by this paradox of recursive self-implication.[8]

On the factual side of the diagram we may trace a gradual reversal of previous drives for the scientific reification of reality. The beginnings of this weakening of 'hard' facts may be found in classical American pragmatism, in

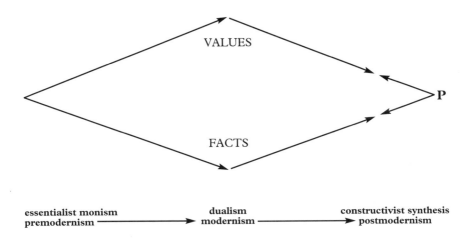

Figure 4.2: The Diamond Pattern

late-Wittgensteinian and Austinian speech act theory,[9] and in Popper's critique of inductivism; but this meltdown is considerably accelerated as a result of the postmodern and constructivist turns in philosophy, cultural studies, discourse analysis, and the social studies of science. Sociologists such as Gouldner and Bourdieu have expressed various forms of scepticism about the feasibility and desirability of a value-neutral social science – a mood that is shared by political and ethical theorists such as Taylor, Lukes, MacIntyre and Unger, historians such as Duby, Pocock, Ginzburg, Skinner, Greenblatt and Schama, and philosophers of history such as Rüsen, White, and Ankersmit. Further dilutions of the traditional notion of facticity are encountered in Foucault's genealogies, in Putnam's 'internal realism', in Rorty's neo-pragmatism, and in the postmodernism of Derrida and Lyotard. Discourse analysts such as Fairclough, Potter, and Wetherell emphasize the inextricable linkage between factual and evaluative features in the performativity of natural speech (cf. Potter and Wetherell 1988; Fairclough 1995; Potter 1996). Constructivists such as Knorr-Cetina, Latour, and Woolgar entirely reverse conventional conceptions of reality by interpreting facts as subjective claims that have succeeded in wiping out all traces of their own knowledge-political construction (Knorr-Cetina 1981; Latour and Woolgar 1986 [1979]; Latour 1987; Woolgar 1988a).

On the value side of the diagram, conversely, one may trace a partial de-subjectification of the epistemological status of normative judgments, where neo-Aristotelian and cognitivist approaches (MacIntyre, Nussbaum, Kohlberg, Habermas) and feminist reorientations of the ethics of care (e.g. Gilligan) have rebridged the gap from the opposite side, reanimating elements of fact/value 'contamination' and teleological purpose that characterized an earlier ethics of virtue. However, insofar as these approaches have tended to reimport quasi-essentialist and quasi-foundationalist arguments, they have 'held up' rather than accelerated the movement towards a fuller synthesis. This synthetic state is only reached if the 'value-relativist' position is generalized to include statements of a cognitive and descriptive nature, which blurs the strong demarcation between fact and value that still dominates the Weberian and Moorean versions of the 'subjectivist' argument. Some postmodernist arguments about the contingency of value positions imply such a radical mixture (Herrnstein Smith 1988; Bauman 1993; Squires [ed.] 1993). It is this particular brand of methodological relativism that finally homogenizes their epistemological status and nudges them towards the diamond's point of osmosis.

English empiricism originally advocated a divorce between facts and values precisely in order to distance itself from absolutist grand theory and find a secure (though fallibilist) foundation in empirical fact. The Kantian and neo-Kantian demarcations expressed an even stronger awareness of the elusiveness of foundational securities for moral values and of the ineradicable perspectivism of theory – but the demarcation was also there to protect what was still conceived as the rational basis of empirical knowledge. More radical forms of epistemological doubt, however, once again demand a closure of the gap and a fuller recognition both of the value-infected character of facts and of the reverse condition: the deep impregnation of values with factually descriptive information (cf. H. Putnam 1981; 1990; MacIntyre 1984; Doeser 1986). While a limited critique of essentialism has historically unfolded within a dualistic framework, a full-blown critique requires going beyond it and working from a position of amalgamation and osmosis. The diamond hence simultaneously follows a *linear* movement from absolutism through modernist critique towards constructivism, and a *circular* movement that seemingly regresses towards its own point of departure. But obviously, the natural proximity of values and facts, which initially dons the iron suit of philosophical essentialism, returns at the end of the day in far lighter relativist attire.[10]

As already suggested, this presentation highlights some striking resemblances between the fact/value diamond and the peculiar cadence of the Hegelian dialectic. Both are seen to pass through a tension-ridden state of bifurcation (*Entzweiung des Geistes*) towards a final stage where, after an intermezzo of alienative disruption, the original unity is recovered on a higher metaphysical plane. The primitive holism that initially fuses Being and Goodness in the prism of the Absolute is replaced by a more sophisticated synthesis that restores facts and values to something like their original proximity. Like the Hegelian dialectic, the diamond scheme simultaneously claims 'realistic' adequacy and follows an 'idealistic', future-oriented teleology. But in the present case, there is of course no finalistic and necessitarian logic that chronologically unfolds 'from left to right'. Its true origin and source of intellectual energy is not found at the Far Western tip of the diamond but at its Far Eastern one, in a spot that is only stabilized by extrapolating the dual trend towards its point of intersection. Its retrograde (and deliberately circular) logic hence swims against the current of historical time; and the locomotive engine of the process is not found in something like the necessary self-development of the Idea but in social processes of intellectual competition, social classification, and institutional demarcation.

While Hegelians permit themselves to be 'carried' by an Absolute Spirit that attains its reflexive apotheosis in Hegelian philosophy itself, the diamond scheme must instead be read as an interested, context-specific attempt at knowledge-political reconstruction or retrojection from a presentist position (P), which emphatically conceives of itself as contingent, value-committed, and performative. Similarly to the Hegelian dialectic, it does not claim to supersede its self-exemplifying proposition about the natural proximity of facts and values. Unlike the Hegelian triad, however, it refuses the arrogant absolutism that projects the end position as the only rationally sustainable one. Although the diamond scheme claims adequacy to reality, it is not exclusively assembled from historical data. As a form of reality-oriented 'fiction' that emerges from the hermeneutic mobilization of historical 'facts' by a knowledge-political perspective, it has a rather precarious suspension point. It does not rest on solid ground but hangs (quite comfortably, though) by the slenderest of intellectual threads.

Naturalistic and Normativistic Fallacies

A further intriguing aspect of the diamond scenario is its capacity to accommodate and reorganize a number of important historical *reduction debates*. The gradual opening of semantic space that results from the structural divergence between facts and values also offers the opportunity of shifting or exchanging philosophical priorities, which may result in the promotion of either facts or values to the status of ultimate grounding of knowledge, science and morality. In the newly emerging conceptual space, norms can either be reduced to facts or facts can be derived from or grounded in norms. If the former reduction (of which Marxism stands routinely accused) is equal to committing the 'naturalistic fallacy', the latter (which is attributable to both neo-Kantian sociologists such as Weber and neo-Kantian socialists such as Bernstein and De Man) appears to drag us into the opposite 'normativistic fallacy'. In the Far Western corner of the diamond this double reduction and double fallacy still hardly manifest themselves, while they once again tendentially lose their meaning in its Far Eastern corner. Fact–value or value–fact reductions only get moving as soon as one discerns an ontological breach between their different jurisdictions; and they only acquire the acute form of 'reduction dilemmas' as this breach freezes into a supposedly timeless distinction. In this manner, the historical differentiation between the realms of fact and value – and between the great rival traditions of naturalism (or positivism) and normativism (or idealism) – is inseparable from the socio-cognitive mechanism of reversal and reduction that gradually shaped their contours and locked them together in a centuries-long embrace.

The diamond scheme hence simultaneously charts a horizontal dimension that plots a linear, diachronic pattern of fission and fusion, and a vertical one that permits the positioning and comparison of various theoretical positions in terms of a logic of primacy or reduction. This bi-dimensionality is a further distinctive feature that renders the diamond preferable to the standard view, which is only capable of differentiating along the horizontal dimension and fails to chart the vertical one adequately. In Weber (1970: 129ff.), Popper (1962: 61–73; 369ff.), Parsons (1967: 79–101), Dahrendorf (1968: 1–18) and Kolakowski (1972; 1978), all theoretical positions tend to be grouped around the same linear track of rationalization and disenchantment that extends from 'naïve monism' towards 'critical dualism', erasing important nuances in an across-the-board critique of the 'confusion' between facts and values. The diamond scheme, on the other hand, differentiates more easily between the (symmetrical) modality of *osmosis* and the (asymmetrical) modality of *reduction*, and more carefully traces the various directions that attempts at reduction or assimilation may take.

Such reductionary dilemmas are already encountered in germinal form in the late medieval rivalry between theological and philosophico-scientific truth, as it was played out, for example, during the Galilei trial. Either the religious value sphere could be treated as dogma and science be turned into the handmaiden of theology or, following the early Enlightenment thinkers, the sphere of morality could be subordinated to the discipline of natural science. The first attempts to establish a 'natural science of society' in Condorcet's social mathematics, Destutt de Tracy's 'ideology', and Saint-Simon's and Comte's social physics, clearly articulated this reductive ambition. On the other side of the divide, Kantian idealism not merely demarcated theoretical from practical reason, but also postulated the epistemological priority of the latter over the former. Kant's overriding purpose was to 'defend moral consciousness, which expresses an unconditional Ought, in its sovereign irreducibility', and hence to bring out the moral commandment 'in its native and original sublimity' (Banning 1958: 92). Lotze and other neo-Kantians such as Windelband, Rickert, and Weber similarly sought to establish the primacy of cultural values (and of the cultural sciences) in an attempt to rescue them from inappropriate naturalization and scientization. According to Gouldner and others, Weber's demarcation between facts and values was primarily urged by his concern to preserve the autonomy of a realm in which personal will and moral conviction could be unreservedly acted out in the face of the universal advance of science-based rationalization. The order of value-committed belief and that of scientific knowledge, though separate, were not accorded equal partnership, for in Weber 'reason only consults conscience and perhaps even cross-examines it. But conscience has the last word, and passion and will the last deed' (Gouldner 1973: 23; cf. Parsons 1967: 82, 96; Mommsen 1989: 8–10).[11]

Weber's dualistic axiom thus cannot be divorced from knowledge-political motives, which in his case were primarily directed against the positivistic scientization of morality that he found operative in Schmoller's and Wagner's

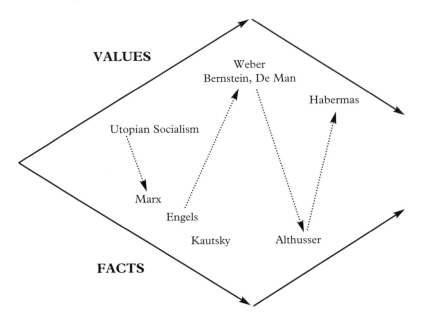

Figure 4.3: Socialist Reductions

normative economics and in Marxism's materialist reduction of ethical and political judgment. Weber's reinstatement of the dignity of the normative over against the scientism of political economy fitted into a broader pattern of reversal and reduction, which included ideological disputes beyond the academic universe. Marx, for example, had already turned the objective theory of justice that characterized utopian socialism 'back upon its feet' in order to assert the primacy of scientific materialism and the ultimately derivative nature of morality and justice. Neo-Kantian and ethical socialists such as Sombart, Bernstein, Adler, Cohen, De Man and Banning, however, were anxious to turn Marxism 'back upon its head' again. Their principled separation between the neutral-scientific account of inexorable social-economic laws and socialism's normative political project was precisely meant to assert the primacy of the ethical commands of justice over the dictates of scientific understanding (cf. Lukes 1985: 5ff.; Proctor 1991: 121–33). The subsequent development of Marxist socialism can be seen to continue this oscillatory movement, veering back via the Althusserian critique of socialist 'humanism' towards the recovery of the salience of the normative dimension in Habermas's dialectic of system and lifeworld (Fig. 4.3).

As a result of this relative disjuncture between its horizontal and vertical dimensions, the diamond permits an intriguing shift in the point of departure and field of fire of epistemological critique. This critique is now diverted from the fact/value osmosis itself and refocused to identify the different reductionary movements that proceed to establish the primacy of one genre of judgment over another. Since the point of osmosis or 'confusion' now indicates the diagram's quasi-teleological point of arrival, the critical risk is no longer generated by the integration of facts and values *per se*, but by attempts to *derive* one type of judgment asymmetrically from the other.[12] In cheerful anticipation of their ultimate (con)fusion, it is of course meaningless to polemicize about logical or ontological priority. In this respect, the diamond scheme traces an intriguing pattern of complementarity and collusion between the 'naturalistic' and 'normativistic' fallacies, which are locked in a repetitive pattern of mutual reversal. This does not imply that the idea of either of these reductionary fallacies has become entirely pointless. In their classical formulations, however, they remain strongly parasitic upon the dualistic logic of value/fact separation, and hence mistakenly identify the naturalistic reduction of values to facts or the normativistic prioritization of values with the value/fact mixture as such. The issue of value/fact *differentiation* and the issue of epistemological *priority* are far more loosely coupled than the anti-naturalist or anti-normativist critiques normally recognize. While it still holds true that normative claims cannot be grounded upon empirical ones or vice versa, this is the case not so much *despite*, but precisely *because* facts and values are (deemed to be) naturally inseparable.[13]

This initially surprising claim may be clarified by briefly considering the conventional epistemological critique of Marxism's naturalization of morality and ethics. The young Marx of course provides an eloquent example of a reversal of the normativistic fallacy which immediately slips into the naturalistic one. Both the 'ethical' revisionists of neo-Kantian stripe and contemporary critics such as Kolakowski (1978) and Lukes (1985) closely identify Marxism's objectivism with its supposedly inadequate separation between scientific and ethico-political concerns. Marx's theory, it is repeated, erroneously dresses up norms as facts, and is therefore ridden with camouflaged value-judgments. But such well-rehearsed criticisms of its crypto-normative infrastructure, which are certainly justified in themselves, do not hinge upon the prior validity of a neo-Kantian epistemological framework. Indeed, it could well be argued that the latter's dualistic premise precisely obfuscates and misrepresents the distinctive *asymmetry* that characterizes Marxist materialist reductionism. Rather than superseding the Kantian divorce between Is and Ought, Marx more nearly *inverts* it, retaining significant residues of this distinction due to the logic of this inversion. In Marx, the ethical dimension is not so much *merged with* that of factual and lawlike explanation, but *absorbed in* and *subordinated to* it. A critique that departs from the principle of natural proximity would not target this assimilation *per se*, but would rather focus upon the distinctively asymmetrical or 'unequal' manner in which it is executed; which in this case forces values to merge with facts under the sovereign dictate of the latter. Undoubtedly, it is a core epistemological weakness of Marxism to have insufficiently recognized the moral embeddedness of its analytical and empirical propositions. But the issue is not so much *that* it has mixed up norms and facts, but rather that it has not mixed them up in a principled enough manner.[14]

Rivalry Effects: Philosophy versus Sociology of Science

One strategic illustration of the cartographic advantages and the reflexive looping back of the diamond scheme can be found in what might be called a 'quasi-Whiggish' reconstruction of the debates between philosophers and sociologists/historians of science that also occupied us in the previous chapter. By setting these debates within the larger framework of divergence–convergence between normativist and naturalist approaches, the diamond pattern reveals the same gap that Laudan, for example, has attempted to bridge in his proposals for a 'normative naturalism' (Laudan 1987; 1990). Fuller has constructed a similar bridge in his efforts to help science studies recover from their 'normative anemia' (Fuller 1988; 1989; 1992; 2000b; cf. Radder 1996: 177–83). Let me once again insist that this presentation, in deliberately mixing descriptive and normative reconstructions, is simultaneously presentist and reflexive, because it does not claim exemption from its self-embedded principle of the natural proximity of facts and values (Fig. 4.4).

Evidently, this abbreviated field map of debates and rivalries in science studies, while conscientiously reaching for descriptive adequacy, simultaneously supports a normative argument in favour of what it purports to describe: the gradual intertwinement of rationalist philosophies of science and historicist social studies of science, if these are identifiable as discourses that prioritize the normative or view themselves as primarily descriptive. On the one hand, the diamond scheme traces intellectual lines of descent and rivalry constellations that are, I hope, descriptively acceptable to many students of the field to whom my normative concerns are foreign. On the other hand, it advocates a more radical integration along the fact/value axis, critically rejecting both the uncontrolled projection of idealizations into 'what actually happens' in science, and the lingering crypto-normativism of much current work in the history and sociology of science. It anticipates a point at which the naturalization of the normative merges with the normativization of the natural, and at which offering the description is continuous with positing the norm.

The Natural Proximity of Facts and Values

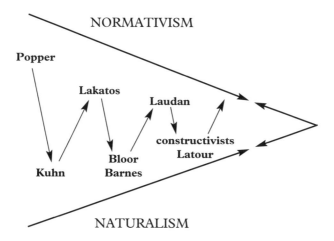

Figure 4.4: Philosophies versus Sociologies of Science

Let me fill in this synoptic picture of closure between the ideal and the real (between science's sanctimonious and ugly faces?) with some very broad strokes. We may begin with the famous Gestalt switch that occurred when Kuhn accused Popper of overgeneralizing the normative picture of science as essentially dedicated to 'permanent revolution', which in his view tended to obscure the significance of the more mundane realities of 'normal' science (Kuhn 1970b). From Popper through Lakatos towards Laudan one may trace an ever closer integration of naturalistic historio- and sociographic findings in the normativist mindset. Lakatos adopted the main contours of Kuhn's picture of normal science, while still upholding the priority of normative methodology over a descriptive psychology and sociology of science (Lakatos 1970). Laudan dropped even this Lakatosian requirement of rational reconstructibility, underwriting the 'Whig History' accusation levelled by Bloor against Lakatos, cutting the link between rationality and scientific progress, and allowing a key role for the history of science in the process of evaluating methodological rules. However, although methodology and epistemology needed to be conceived 'far more than they normally are, as empirical disciplines', such naturalization did not empty them of normative force and preserved an important prescriptive role for the philosopher of science (Laudan 1984: 39–41). 'Normative naturalism' should not be squeamish about committing the naturalistic fallacy, since normative and descriptive concerns are interlaced in virtually every form of human inquiry (Laudan 1990: 56).[15]

On the opposite side of the divide, Bloor's familiar critique of Lakatosian rational reconstruction as a Whiggish mobilization of history cleared the path for the Strong Programme's counter-image of the natural world as 'morally empty and neutral', which separated evaluation from explanation in a value-free, causal, and symmetrical approach to both error and truth (Bloor 1981; 1991: 6ff., 9–10; Barnes 1974). As Barnes specified, sociology was concerned with 'the naturalistic understanding of what people take to be knowledge and not with the evaluative assessment of what deserves to be so taken' (1976: 1). This type of Wittgensteinian 'value-free relativism' was further articulated in Collins' relativist empiricism, Mulkay's discourse analysis, Woolgar's reflexive constructivism, and the actor-network theory proposed by Latour, Callon, and Law. The empiricist heuristic of 'following the actors', which was already implicit in Kuhn (and explicit in Garfinkel), acquired an increasingly radical methodological profile. Introducing a representative collection, Callon, Law, and Rip for example declared that we should not be 'blinded by morality' in the study of how a society took shape, but should instead attempt to 'describe, with neither fear nor favour, what it is the actors do'. It was not a matter of taking sides, of applauding or rejecting; we should instead try to 'understand, through careful study, the roots of scientific power, and where this is taking us'. Their professed aim was not to pass judgment, but rather 'to reveal the forces that are at work' (Callon et al. 1986: 5–6, 223).

In terms of the diamond scheme, the constructivist studies of science therefore offer a vivid mirror image of their 'best enemy's' position. The crypto-normativism of STS has been fully matched by the traditional 'crypto-descriptivism' of rationalist philosophies of science. While historians and sociologists of science have accused philosophers of surreptitiously reifying their values, philosophers have routinely slighted the sociological naturalists for concealing their values in their facts. This adversarial complicity also suggests that the shorthand description of their mutual rivalry as one conducted between 'pure' normativists and descriptivists may be misleading, insofar as it merely adds up their contrasting self-images instead of going beyond them. Indeed, the fact/value dualism offers a framework of polarization that is readily resorted to in conditions of disciplinary or paradigmatic competition. To some extent, it is an artefact of contrary but complicitous attempts to secure cognitive distinction, which flattens out the ambiguities by which the interacting positions are subterraneously linked. The orthodoxy of normativism vs. naturalism, in other words, precisely obscures the fact that both approaches immediately feature various *mixtures* of description and prescription.[16] Whereas the ethnographic redescription of 'what really goes on' in scientific laboratories is always close to a normative celebration of the ordinary practicality of scientific 'ethnomethods', the Popperian emphasis upon 'best practice' is an immediate

The Natural Proximity of Facts and Values

normative transcription of the empirical strategies of those who are selected as the greatest philosophers and scientists (Popper 1970).

My objection is therefore not that both sides have mixed what should be kept asunder on logical grounds, but that they have done so in an *uncontrolled* and *asymmetrical* fashion, and have failed to carry the mixture to its 'natural' conclusion. As indicated above, my quarrel is not with the reductionist fallacies themselves but with the lack of deliberate control with which they are committed as a matter of everyday and unremarkable routine. In this perspective, the weakness of the philosophy of science is not that it contains too much value and too little fact, but that it slides between the two registers without being quite aware what it is doing. The issue is not so much that a primarily normative philosophy of science does not stand up well in empirical terms (although that is of course the case), but that it insufficiently monitors the natural proximity between evaluation and factual description that it itself embodies and enacts. This deeper symbiosis permeates (for example) Lakatos's conclusion to his celebrated article on the methodology of scientific research programmes: 'The history of science has been and should be a history of competing research programmes [...], but it has not been and must not become a succession of periods of normal science: the sooner competition starts, the better for progress' (1970: 155). But similar aporias arise on the other side of the intellectual fence. Both Feyerabend and Bourdieu have rightly complained about Kuhn's rather ambiguous handling of the logical relationship between description and prescription (Feyerabend 1970: 198–99; Bourdieu 1981: 280).

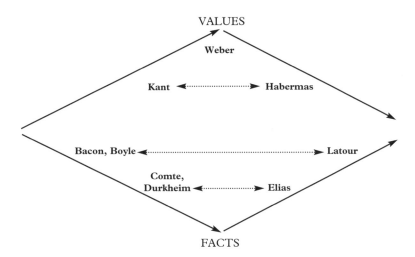

Figure 4.5: Homologous Positions within the Fact/Value Diamond

Bloor's naturalistic programme is explicitly framed by normative principles such as impartiality and symmetry, while its 'value base' incorporates the idea of the world as 'morally empty and neutral'. The Strong Programme mimics its philosophical competitors in its resolve 'to represent as natural what it takes for granted', thus repeating a rather deliberate version of the naturalistic fallacy (Bloor 1991: 9–10).

In the following sections, I shall further illustrate the organizing capacities of the diamond scheme by means of brief excursions into three contemporary intellectual positions: those of Elias, Habermas, and Latour. Their different locations in the Eastern sector in some respects mirror a triad of structurally similar locations in the Western sector: those of Durkheim, Kant, and Boyle. This paired elucidation may further clarify the diamond's cognitive field and its horizontal and vertical differentiations, while enabling us to spell out some of its critical functions in the face of influential conceptual alternatives (Fig. 4.5).

Elias: Involved Detachment

The first of these vantage points is offered by Elias's neo-positivist sociology of knowledge, which intends to supersede the static opposition between subjectivist (or involved) and objectivist (or detached) social relationships and sociological methods in the horizontal dimension, but, along the vertical axis, strongly maintains the priority of an objectifying empirical science that subsumes normative issues under factual ones (Elias 1978a: 223–25, 233–35, 243–45, 309–10; Layder 1986; Pels 1991b; 1996). On the one hand, Elias's processual methodology 'liquefies' the static polarity between subject and object and emphasizes their ontological interconnectedness. The stark opposition between commitment and value-freedom is replaced by a continuum of interactions, balances, and mixtures, where 'involvement' refers to the relative dominance of the properties of the observer in obervation statements and 'detachment' to the relative dominance of the properties of the object itself (Elias 1987; 1991; Mennell 1989: 159ff.). Although all forms of observation retain a residual subjectivity, Elias insists in Durkheimian fashion that it is the researcher's duty to advance as much as possible towards the pole of detachment, i.e. to distance oneself maximally from one's interests, sentiments, opinions and ideals in order to draw closer to 'reality itself'. The descriptive and processual continuum hence also incorporates a normative threshold where the balance of relative involvement tips over in favour of relative detachment. It is a *normative continuum* that features an in-built gradualistic criterion of truth, which self-evidently links involvement with affect-ladenness, interest, and prejudice, while detachment and self-control are identified with objectivity and 'adequacy to reality' (Elias 1978a: 254–56; 1978b: 22ff.; 1991: 109–12).

Reflexively and critically speaking, therefore, Elias's continuum harbours the same intrinsic mixture of facts and values for which alternative approaches are rather severely taken to task (cf. Elias 1978b: 22–25, 152–57). This natural proximity is not openly accredited but takes the form of a historicist *reduction* that absorbs all normative energy and hides it from view. Moreover – and in tension with Elias's own dismissal of the ontological reification of norms – this crypto-normative continuum is projected back into history in order to frame a comprehensive developmental theory of knowledge that closely resembles the 'historical positivism' of Comte and Durkheim (cf. Heilbron 1995; Pels 1996; Mennell 1989; Kilminster 2000).[17] In primitive societies, human beings are still tightly constrained by their fear of and identification with natural phenomena; their everyday attitudes still display a great deal of involvement and only a small measure of detachment. This lack of emotional control keeps them cloistered in magical-mythical conceptions. Gradually, however, they manage to untie this double bind, breaking the vicious circle of insufficient self-control and lack of control over nature, and decreasing the fantasy content of knowledge in favour of its reality content. The stock of human knowledge becomes increasingly objective, i.e. more in agreement with actual facts and events than with wishes and desires and their accompanying fantasy impressions. The scientific revolution of the sixteenth and seventeenth centuries forces an epistemological break that decisively shifts the balance of involvement and detachment in the direction of the latter. This take-off creates a deepening discrepancy between the conceptualization of natural phenomena and that of human and social relationships since, in Elias's view, the social sciences have so far been unable to perform such a definitive breakthrough. Nevertheless, the promise is held out of a similar revolutionary acceleration in the social sciences, if these succeed in disentangling the double bind, in improving the detachment ratio of their statements by means of greater self-control, and hence by producing knowledge that offers a greater degree of 'process control'.

In spite of the initial distance taken from the categorical or 'static' distinction between values and facts, Elias's historicist *démarche* therefore never escapes the classical parameters of sociological positivism (cf. Mennell 1989; Pels 1991b; 1996). The first of these is a taken-for-granted opposition between factual knowledge and socio-political ideals and a summary identification of the latter with preconceived fantasies and ideologies – a contemporary reprise of the Baconian dichotomy that also inspired Saint-Simon, Comte and Durkheim, and that routinely equates commitment, passion, partisanship, and personal presence with limitation, bias, subjectivism, and epistemological pollution. A

Unhastening Science

second dogma is a self-certifying theory of truth and method that, in spite of its processual reformulation (Elias 1991: 115), retains a conventional trust in the notion of correspondence with the facts or 'adequacy to reality'. The two dogmas converge in Elias's favourite view of the scientist as a destroyer of myths, who, through factual observation 'endeavours to replace myths, religious ideas, metaphysical speculations and all unproven images of natural processes with theories – testable, verifiable and correctable by factual observation' (1978b: 52). They also define his (Comtean and Durkheimian) derivation of the relative autonomy of social science from the prior discovery of 'society' as an irreducible set of social facts (Elias 1978: 45) – which tends to divorce the scientific object from the knowledge-political project and reverses their order of constitution (Tenbruck 1981; Pels 1996: 115–16; 2000: 51).

A third positivistic plank is the exemplary function of 'natural-scientific' detachment and self-control and the attendant optimism about scientific and social progress. Although Elias warns against any mechanical transposition or excessive generalization of natural-scientific models of rationality, his reservations focus upon the alleged superiority of processual explanations over logico-deductive ones, upon the interaction between inductive and deductive stages in the research process, and upon the issue of the 'subject character' of the social-scientific object – leaving intact the shared moral that both social and natural scientists need to distance themselves from their own roles as participants in order to reveal reality 'as it is' (Elias 1987: 51–52, 58, 65, 100–102). This approach favours the retention of a traditional ethic of value-freedom, according to which all fact/value mixtures are considered ill-judged and contrary to scientific propriety. The great pioneers of sociology, Elias declares, focused upon the inherent structure of social development but simultaneously remained under the spell of their hopes and fears, of 'the enmities and beliefs resulting from their role as immediate participants'; hence their writings inextricably interweave involvement and detachment. The obvious lesson is that mixture equals confusion and that the two components must be seen as logically distinct (Elias 1987: 13, 94). Science must be ruled as much as possible by autonomous values that are internal to it (such as the quest for truth) and as little as possible by heteronomous or external ones (Elias 1987: 34–35, 39; 1978b: 17–18).

In this manner, Elias's balancing act of involvement and detachment remains caught in an unreflected contradiction between descriptive gradualism and normative polarization. His continuum of mixtures embodies an unacknowledged normative tension between its two polar extremes, and hence displays a far deeper proximity between involvement and detachment than the model of balances officially recognizes. It requires a more reflexive awareness of this natural proximity to argue that all facts are indeed *normative* facts, and that factual descriptions do not so much reflect reality as help to create what they state. This performative awareness changes the direction of our thinking about the relationship between detachment and scientific autonomy. In the spirit of the general argument developed in this book, the critical issue can no longer be how we can prepare a definitive breakthrough from involvement to detachment in the sense of epistemological neutrality. We can only defend the autonomy of science by means of a *politics* of institutional distanciation that permits a systematic freezing of interactional frames and a critical slowdown of observations and reflections. This 'temporizing' or 'waiting' fails to guarantee access to 'objective reality'; but it does generate perspectives and interests that differ interestingly from more hurried ones in the social universe.

Habermas in Performative Contradiction

Within the logic of the diamond scheme, Habermas's social philosophy may be regarded as the most heroic contemporary attempt to straighten out the moving front of the 'double decadence' of facts and values and to stabilize it at a specific point in philosophical space. On the one hand, Habermas distances himself from the dualistic Weberian position that projects a principled asymmetry between normative and cognitive validity, suggesting that the truth of empirical knowledge and the justice of intersubjectively valid norms should be treated as epistemologically equal. Simultaneously, however, he erects a defensive parapet against further surges of relativism by establishing a rational argumentative foundation for both evaluations and cognitions. The neo-Kantian chasm is thereby diminished but not entirely closed (Habermas 1971; 1986; 1996). This epistemological equilibration is accomplished by means of a critique of logico-deductive explanation and the substitution of the correspondence theory by a consensus theory of truth, both of which weaken the traditional objectivist foundation of analytical propositions. In compensation, values regain a partially cognitive status and the validity basis of normative judgments is 'rationalized' over against the alleged arbitrariness of Moorean emotivism, Weberian value-subjectivism and Schmittian decisionism. Habermas thus advocates a partial intermeshing of *Sollen* and *Sein* without contemplating their integral fusion, relegating normative discourse and the discourse of truth to autonomous ontological realms that can only be rationally reintegrated upon due recognition of their prior independence.

This procedural levelling of the epistemological status of facts and values has a definite polemical edge, which may be clarified by invoking the nuclear distinction between labour and interaction (Habermas 1971). Initially designed as a critical interpretation of the Marxian distinction between forces and relations of production, it was merged with the Weberian distinction between means-oriented and ends-oriented rationality to form a comprehensive opposition between goal-oriented (strategic or instrumental) action and communicative action or, more succinctly, between 'system' and 'lifeworld'. Strategic action presupposes an objectifying attitude of the subject towards the world of things and is geared towards the pursuit of individual and collective interests, while communicative action is fundamentally oriented towards other subjects and the accomplishment of intersubjectively shared understandings. In this aspect, Habermas's basic typology of action runs parallel to the distinction between the factual and the normative, insofar as the objectifying intercourse with 'nature' is located in the realm of technical-empirical facts (the 'system' is defined as a norm-free domain), whereas the domain of interaction is organized by 'normative structures'; communicative action is essentially framed by intersubjectively valid norms (Habermas 1971: 91–92; 1979: 95ff.; Giddens 1982: 111).[18]

Over against both Marx and Weber, Habermas is concerned to demonstrate that processes of rationalization on the level of communicative action are independent from and equivalent to those that unfold in the realm of goal-oriented action. The resounding theme of his early critique of Marx, and of his parallel critique of the positivist 'halving' of rationality, is that interaction cannot be reduced to the instrumental logic of labour. In a social-ontological version of the critique of the naturalistic fallacy, Habermas protests the effacing of this distinction in order to rush to the defence of the ontological autonomy of normative structures. In his subsequent work, however, this anti-reductionist campaign gradually strengthens the implicit tendency to privilege and secure the ontological *primacy* of communicative over strategic rationality. Rather than discarding the dualistic framework that is imposed by dichotomies such as 'subject–subject' vs. 'subject–object' relations or 'rational' vs. 'empirical' action coordination, Habermas thus increasingly tends to *reverse* the priorities of classical Marxism and positivism. Mutual understanding is the 'originary' mode of language use and, in the final analysis, society must be integrated through communicative action (Habermas 1986: 288, 293–95; 1996: 27).

The core pathology of the colonization of the lifeworld is cast in similarly dualistic terms. System and lifeworld each possess their own legitimate remit of action, but the threat of absorption or corruption of the latter by the former demands a reassertion of the primacy of normative rationalization, which anchors the systems in the lifeworld and normatively domesticates them. This 'last instance' priority of normative rationalization is also historically retrojected as a developmental logic that gradually uncouples the systems of state and

The Natural Proximity of Facts and Values

market from the originary totality of the lifeworld, while the latter remains effective as the ontological grounding and generative force of social constitution (Habermas 1987: 153ff.). In tribal-archaic and politically stratified class societies, the realms of symbolic and material reproduction are still immediately intertwined; but the subsequent rationalization of the lifeworld affords the autonomization of material reproduction in self-steering political and economic domains, where action coordination proceeds by means of 'empirical' media such as money and power. The emergence of these autonomous realms, in which actors confront each other strategically, is ultimately predicated upon rational learning processes on the level of lifeworld communication, and more particularly, upon the gradual differentiation of claims to analytical truth and normative justice and the concomitant institutionalization of science, morality, and law. In this fashion, the articulation of the reified, norm-free systems of money and power requires continual 'confirmation' by the normative institutionalization processes in the lifeworld, which remains the subsystem that uniquely guarantees the integration and complexity of the differentiated whole (Habermas 1987: 154–55, 173–79; cf. Rasch 2000a: 29ff.).

As in the case of Elias, my sketch of Habermas's position cannot be other than abbreviated and schematic. But it suffices to note that the (onto)logical distinction between lifeworld and system represents far more than an empirical-historical hypothesis, since it is immediately charged with a volatile normative energy. This profound entanglement of the factual and the normative is more evident as soon as we examine some of its collatoral extensions. While the opposition between symbolic and material reproduction or that between linguistic and non-linguistic media of action coordination can be read as relatively neutral and descriptive, adjacent distinctions between rational and empirical motivation, between good reasons and (bad?) incentives, or between action that is oriented towards mutual understanding and action that is directed towards the attainment of private goals, evidently carry a normative surplus charge. This suggests that the multi-layered differentiation between system and lifeworld itself betrays a 'natural' proximity of facts and values that is much more intimate, constitutive, and 'nuclear' to it than is accounted for in Habermas's official ontology. This ontology, which separates normative and cognitive validity claims in order to be able to rejoin them in a 'rational' manner, features a subliminal fact/value mix that undercuts its own premises by immediately transgressing its grounding proposition. As an ontological decision, the paramount distinction between strategic and communicative action constitutes a performative *amalgam* of strategic and communicative action elements – which precipitates Habermas into a performative contradiction as deep as the one that he finds embedded in all forms of postmodernist relativism.[19]

Unhastening Science

A similar paradox emerges from Habermas's familiar thesis about the communicative *telos* that inhabits all natural language use (e.g. 1979: 1ff.; 1986: 287). Curiously, his argument appears to repeat the naturalistic fallacy by projecting a string of supposedly 'unavoidable' idealizations into the pragmatic conditions or 'presuppositions' of natural everyday speech, which comes close to presenting a consensualist value preference as an empirical condition of ordinary communication. In Habermas's account of linguistically mediated communication, validity and facticity are indeed interlinked, but only insofar as such counterfactual idealizations (which include the criticizability of claims, the sincerity of communicative intent, and the universality of the anticipated audience) are read into the 'really existing' facticity of the lifeworld. As Habermas has recently summarized his position:

> In seeking to reach an understanding, natural-language users must assume, among other things, that the participants pursue their illocutionary goals without reservations, that they tie their agreement to the intersubjective recognition of criticizable validity claims, and that they are ready to take on the obligations resulting from consensus and relevant for further interaction ... A set of unavoidable idealizations forms the counterfactual basis of an actual practice of reaching understanding, a practice that can critically turn against its own results and thus *transcend* itself. Thus the tension between idea and reality breaks into the very facticity of linguistically structured forms of life. (Habermas 1996: 4)

In this respect, the theory of communicative action is able to absorb both the proximity and the 'essential tension' between facticity and validity into its fundamental concepts (cf. Habermas 1996: 8). But the prior specification of these infrastructural conditions demands a much closer performative alignment between communicative and strategic dimensions in a politics of knowledge that reflexively subverts this primordial distinction. If, on the other hand, language is viewed as a form of action that *inherently* mixes communicative with strategic intent and values with facts (a proposition that can itself only be stated as a performative mixture), we had better drop any residual urge to assert the ontological primacy of one form of speech action over another. Rather than 'naturally' emerging from the transcendental-empirical structures of the lifeworld, the ideal of an 'unconstrained' speech situation rather suggests a blueprint for a specific *knowledge-political* practice that finds itself at home in a different, 'unhastened' timescape. Only by this temporal-institutional privilege is it able to fulfil its promise of rationality and criticizability, allowing for the *time* and *space* in which speakers can unfold their capacity to deepen and defend their claims by adducing further arguments.[20]

The Natural Proximity of Facts and Values

Latour's Political Epistemology

Latourian actor-network theory may appropriately localize the third intellectual position examined here, that of 'agnostic constructivism' as it has emerged subsequent to the Kuhnian turn in science studies. The critical anthropology that Latour and Woolgar originally deployed against established philosophical explanations of scientific success was quick to expose the discovery of scientific facts as objective things-in-the-world as an interested tribal myth. Facts, they contended, only appeared at the end of a long construction chain, in the course of which controversies were progressively closed and subjective claims hardened into self-evident and routinized knowledge, ultimately producing a dehumanized, asocial, ahistorical 'nature'. A fact was therefore a statement that had successfully erased its contingent social roots and historical trajectory; it was a subjective claim that had been solidified by a successful process of reification. In this fashion, the traditional normative dilemma of subjectivism vs. objectivism or constructivism vs. realism was defused by spreading it across a 'microhistorical' or 'psychogenetic' continuum that followed a trajectory of reification, at the end of which facts duly emerged as 'mirrors of reality' (Latour and Woolgar 1986 [1979]; Latour 1987, 1988a; Woolgar 1988a).[21]

Evidently, this analysis of the reification of facts was not intended critically, at least not in the conventional sense of contrasting a laudable epistemological relativism with the less acceptable realist practices of working scientists. As Latour's subsequent work further clarified, he never saw it as his purpose to 'unmask' objective facts as naturalistically inflated or ontologically secured subjective claims, or to privilege the outsider's analytical view over that of the practitioners themselves. Rather, his analysis appeared to suggest that working scientists were perfectly aware of the quasi-political ruses and tricks with which they operated this naturalistic stabilization, switching opportunistically between relativistic and reificatory modes of representation as circumstances and audiences required; and that, accordingly, ethnographers of scientific practice could modestly limit themselves to 'following the actors' rather than imposing their allegedly more sophisticated epistemological theories upon their subjects of study. Since scientists were duly aware of both their fabrication and their fetishization of the facts, analysts of their practices and beliefs merely had to reveal and articulate what they were actually doing, without the need to improve upon them (Latour and Woolgar 1986 [1979]: 257; Latour 1988a: 148).

In Latour's *Science in Action* (1987), for example, science was depicted as a Janus *bifrons* who continually alternated between the two faces of ready-made, official science and of unofficial, pragmatic and political science 'in the making'.

101

Unhastening Science

As soon as one turned from black-boxed facts to their process of construction it became evident that, contrary to the outwardly polished image, content and context or knowledge and interest continually interpenetrated and intermingled; science in many respects resembled a form of politics 'continued by other means'. For Latour, however, this classical topos of the discrepancy between 'saying' and 'doing', between shiny façade and sordid reality, did not function on a critical level. The face of science-in-the-making was not used as a critical foil against and a preferable substitute for the face of ready-made-science: the two images were arranged in descriptive juxtaposition and taken equally seriously in epistemological terms. There was hence no normative schism between participant's image and observer's image; both were deemed part of the empirical reality of the research process. In this fashion, science could be pictured as a living contradiction between 'science in action' and 'ready-made science', in which two voices permanently and simultaneously talked at cross-purposes (Latour 1987).

This model has been subsequently generalized to explain the peculiar 'duplicity' of what Latour has called the Modern Constitution, which officially separates science from politics, objects from subjects, facts from values, and natures from cultures (the work of purification), even while it surreptitiously mixes all of these categories and proliferates their hybrid combinations (the work of translation or mediation). The two founding fathers of this Constitution, Hobbes and Boyle, are appropriately interpreted as simultaneously adhering to realism and constructivism. While Boyle states that the facts speak for themselves, he simultaneously undertakes to fabricate them in his laboratory. While Hobbes states that people make their society through the social contract, he also reifies the Leviathan and mobilizes material things to add to its natural strength (Latour 1993a). Earlier and subsequent analyses of Pasteur's politics of knowledge identified the founder of microbiology as a duplicitous agent who self-consciously juxtaposed and shuttled between these seemingly incompatible philosophies of science. In Latour's view, Pasteur's experiments invariably created two epistemological planes: a constructivist one in which he acted as the spokesperson who accepted responsibility for his fabrications, and a realist one in which he delegated his action to nonhuman entities in a transcendent 'nature out there'. Once again we should not impose our critical categories but merely deploy those of the actor himself, who knowingly reified his entities, and redefined and incorporated other actors into his own programme, while simultaneously professing merely to read the Book of Nature (Latour 1983; 1988a, 1999a; cf. Pels 2000: 10–15).

Modernity is therefore not a plot that must be unveiled, a web of illusions that must be torn away in order to reveal the real practices or interests that

The Natural Proximity of Facts and Values

underlie all sanctimonious legitimation. The relation between the work of reification and purification and that of mediation and construction is not that between illusion and reality, or between false consciousness and truth. Ethnographers and historians of science simply have to acknowledge that 'we have never been modern', since we have always surreptitiously mixed what we have officially separated: politics and science, humans and nonhumans, values and facts, social construction and natural reality. Following Boltanski and Thévenot (1991), Latour demands the end of 'denunciation' and of critical 'debunking', since social science can no longer claim the privilege of rising above the actor and criticizing his or her false consciousness in the name of a 'deeper' reality. This requires us to shift from a critical sociology towards a sociology of criticism that is no longer interested in the ritual of unmasking but simply describes the various modalities with which we all accuse and criticize one another, without being able to claim a transcendent analytical position (Latour 1993a: 40–45). In his recent work on political ecology, Latour's objective is similarly 'not to reverse the established conceptual order' but to describe the actual state of affairs, since 'political ecology already does in practice everything that we affirm she should do'; we merely have to explicate its practice as against its official philosophy (1999b: 16).[22]

From the perspective adopted here, however, the Latourian account of modernist duplicity carries a number of empirical difficulties and invites some unattractive normative consequences, which require it to be confronted simultaneously on both planes. The agnosticist slant of the argument hinders a more reflexive acknowledgment of the actual intensity with which it hybridizes factual and value propositions, and hence understates the performative effect that it exercises upon any account of 'real-world' (scientific) practices. The reiterated characterization of modernist representational practices as 'duplicitous' appears empirically contestable, resulting in no small measure from an unacknowledged and anachronistic 'shifting out' of the analyst's interpretation of science to the 'actual practices' of working scientists – an exercise that comes close to repeating the naturalistic fallacy in a less palatable form. Judging from the textual evidence, for example, it seems just as risky to ascribe any strong form of constructivism to Pasteur as to read it into Boyle's experimental naturalism or the absolutist political philosophy of Hobbes (cf. Pels 1995a; see more extensively Chapter 6 below). The advocacy of an a-critical 'sociology of criticism' is similarly weakened as soon as we discover that 'common sense' accusations and denunciations often follow a reifying and foundationalist pattern, and that the adoption of 'transcendentalist' critical positions is a routine feature of what might be called 'everyday essentialism' (Pels 2002). The refusal to distinguish or hierarchize the perspectives of actors and observers or to exploit the discrepancy between what they officially say and unofficially do falls to the ground as soon as one realizes that it remains the apparent privilege of the Latourian observer to self-consciously 'add up' and 'explicate' the two conflicting tasks of reification and construction/mediation about which the moderns are thought to be 'explicit'. Such presumed discrepancies between 'sayings' and 'doings' or between 'theory' and 'practice' reimport a familiar ideology-critical figure, which, in Latour's case as in more traditional ones, invariably boils down to a literal 'contra-diction' between *two sayings* or *two theorizations* (those of the actors and those of more detached analysts), one of which appropriates the category of 'practice' in order to (dis)simulate critical distance from it.

The upshot of this is that Latour fails to develop a critical theory that successfully challenges the reification of both natural and social facts. The Latourian ethnographer never undertakes to *confront* the scientist but – in true ethnomethodological fashion – tends to *admire* her for the cleverness and panache with which she pulls it all off. The constructivist face of the Janus *bifrons* never actively frowns upon the realist face, because reification and black-boxing are not critical targets but stated and legitimate *objectives* of all practices of translation and mediation. There is thus no epistemological room for improving upon the strategic and opportunistic plying back-and-forth between reification (or speaking for facts in such a way that they speak for themselves) and translation (or acknowledging the performative and political nature of representation).[23] The a-critical, agnostic description of 'what we all do' and 'what we have been doing all along' ends up by absolving both 'commoners' and scientists from hardening their claims with any means they see fit to use.[24] But this is little more than a considered 'political' reversal of the traditional presumption of normative philosophies of science that truth is sufficiently energetic and strong to spread of its own accord. What is insufficiently challenged is the very pursuit of hardness, solidity, and certitude that also drives the former, even though cognitive guarantees are now traded for the hard political work of forging networks of human and non-human allies.

In terms of the diamond matrix, therefore, the Latourian approach to knowledge politics, while successfully surpassing the dualism of reason and force, still insufficiently hybridizes and mixes the categories of fact and value. It is quite remarkable, in this respect, that some of Latour's recent writings offer a close graphical approximation of the diamond scheme, in tracing the progression of a broad 'front of modernization' that distinguishes ever more clearly and definitely between 'the efficiency and objectivity of the laws of nature' and 'the values, rights, ethical requirements, subjectivity, and politics of the human realm'. While the modernist narrative of progress proclaims that 'what was confused will become distinct' and that subjectivity and objectivity will no longer be mixed up, science studies have reversed this arrow of time, enmeshing humans and nonhumans 'at an ever greater level of intimacy and on an ever greater scale ... The confusion of humans and nonhumans is not only our past *but our future as well*' (Latour 1999a: 199–201; cf. 1999b: 250–58).[25] Abandoning the obsolete boundary between values and facts, Latour now attempts to delineate a new 'separation of powers' that purports to rescue the important differences contained in the former dualism, in order to redistribute them across a different set of conceptual oppositions (1999b: 149ff.).

While an exposition of this new (and complicated) conceptual strategy would carry us too far afield, it should at least be noted that Latour, when laying out the different 'competences of the collective', does now incorporate an unambiguous place for morality, and that moral competence is precisely defined as introducing an *uncertainty* about the proper relationship between means and ends, a concern with the *quality* of the means, and a *caring* disposition not to treat entities (whether human or non-human) simply as means. Morality is less preoccupied with values or rights than with *slowing down* the process of conjugating means to ends (1999b: 211–12; cf. 2002). In other words, the explicit restaging of morality coincides with the introduction of an ethos that appears to preach the virtual opposite of the quest for reificatory certainty and indifference about the means of stabilization that has so massively dominated his previous works. In line with this, and in contrast with the 'fast and hard' image of the political craft of gathering allies and solidifying one's networks, Latour now also emphasizes the need for political ecology and political epistemology to 'take their time', advancing at the pace of a tortoise, in brief to *decelerate* in the face of the urgencies of action; political ecology must glide from certainty to uncertainty and is better off preferring weakness over strength (1999b: 11–16, 41).

Scientific and 'Common' Sense

Summing up, the three positions discussed appear to 'suffer' from the same 'infirmity' which, in the context of the reflexive diamond pattern, must instead be considered a sign of rather good health. They all favour a greater intensity of interaction between factual and value claims than is accountable in the traditional dualism. But they also display a deeper and more 'natural' proximity between facts and values than is foreseen by their own methodological self-conceptions. Indeed, all three positions appear to slide into one or another version of the reductionist fallacy, by insufficiently acknowledging the value-pitched character of their various descriptions: in Elias's case, of the changing historical balance of involvement and detachment; in that of Habermas, of the transcendental presuppositions of ordinary language use; and in that of Latour, of the supposedly agnostic description of 'what we all do' when we construct knowledge in scientific laboratories. Within the diamond matrix, this deeper symbiosis between facts and values is revealed in the form of a historical dialectic of divergence, antinomy, and convergence. The postulate of natural proximity is itself reflexively incorporated into this scheme, which is self-consciously 'retrojected' under the guidance of a performative hermeneutic and a reverse teleology that posits its effective beginning at the end of the entire oscillatory movement.

In conventional accounts of scientific rationality, the genesis of the fact/value distinction is normally depicted as an irreversible process of intellectual enlightenment. The diamond-shaped framework on the other hand suggests that the rationalization process of modernity, although it may have historically profited from this epistemological bifurcation, also runs objectivistic risks which, if seriously faced, force the abandonment of the dualistic position. The three positions reviewed above may therefore also be valued more positively, as providing new intellectual tools that spur the advance of the rationalization process beyond the limits of the conventional fact/value split. Elias sharpens our awareness of the need to historicize and render fluid the traditional dualism of subject and object and of the attitudes of involvement and detachment. Habermas crucially emphasizes the need to view the theory of rationality in a broader compass and to strike a new equilibrium between facticity and validity. And Latour strengthens our conviction about the performative nature of facticity and thereby radicalizes the critique of epistemological foundationalism beyond Elias and Habermas.

Finally, and following the main thrust of my analysis in this book, the diamond scheme harbours crucial implications for the tension-ridden and contested relationship between scientific and 'common-sense' rationalities. On the one hand, the idea of a natural proximity between facts and values implies that science loses much of its traditional superiority over what Habermas calls 'the communicative praxis of everyday life', where descriptive and prescriptive claims supposedly mix in unrestrained fashion. The sublime profile of science, insofar as it is still shored up by the fact/value dichotomy and the ethos of value-freedom, is deservedly 'vulgarized' and levelled down to more mundane dimensions. On another level, however, critical theory should be able to *dis*engage itself more clearly from both science and 'common sense', insofar as both everyday and scientific representations continue to commit the reductionist fallacies in their asymmetric form, and performatively underwrite the reification of reality and the absolutization of norms. Over against such solidifications, the diamond pattern upholds a critical irreductionism. A critique of unmonitored reductions of values to facts remains crucially important, because it is precisely by means of the evaluative and performative infrastructure of constative claims that a multitude of compulsory 'realities' are proliferated and reproduced. Social science often merely repeats such everyday reifications in more complicated jargon and surrounds them with a protective legitimatory shield. A truly critical theory of knowledge will be most successful in combating such unreflective duplications by self-consciously throwing values and facts together.

CHAPTER 5

Knowledge Politics and Anti-Politics: Bourdieu on Science and Intellectuals

Autonomy and Duality

In this chapter, my aim is to engage more directly with Bourdieu's field theory of science and his notion of the intellectual 'politics of reason', which supply intriguing thought material for a further elaboration of a pragmatic, knowledge-political conception of intellectual autonomy. Similarly to Latour's theory of translation, Bourdieu's field theory relativizes the science/politics divide, but does not satisfactorily resolve the conundrum of fact and value. However, while Latour defends the view that the purpose and strength of scientific laboratories is precisely to erode the differences between science and society, Bourdieu continues to profess a principled loyalty to the idea of scientific and intellectual autonomy. Indeed, his approach positions the concept of autonomy so centrally that it virtually performs as a sociological deputy of more traditional criteria of truth and rationality: professionalism and autonomy go far towards defining the specificity of the intellectual and scientific fields in their most authentic state. In this respect, and in critical abnegation of the de-differentiation as proclaimed by radical constructivists and 'nothing-specialists', Bourdieu maintains a principled distinction between an 'external reading', which thematizes the incorporation of science in the wider social cosmos, and an 'internal reading', which is oriented towards the concatenation of knowledge and (symbolic) power inside the scientific microcosmos itself (e.g. 1990b: 298).

Before entering upon a discussion of Bourdieu's perspective, however, I first need to clarify my usage of the protean concept of 'intellectual' and its relation to adjacent concepts such as 'professional', 'expert', and 'academic scientist'. First, it makes little sense to hope for objective definitions of any of these categories by means of enumerative or 'finger-pointing' exercises. All definitional portraits of intellectuals, professionals, experts, and scientists are simultaneously self-portraits, often attempts to delineate a cherished identity, and hence liable to entail claims that are essentially knowledge-political in nature. For Bourdieu, the very definition of intellectuals, scientists, and professionals is itself continually at risk and at stake in struggles within intellectual, professional, and scientific fields.[1] Secondly, following the lead of Feyerabend, Foucault, and Bourdieu himself, I shall refrain from drawing any hard-and-fast demarcation lines between intellectuals and professionals or scientific experts. The reported demise of the 'general' and the concomitant rise of the 'specialist' intellectual has erased much of the sociological and normative distance between properly intellectual and professional work, which can no longer be usefully counterposed in the manner of a universalist, critical dedication to Culture or to the moral community as a whole vs. a particularist devotion to self-serving corporative interests. Cultural pessimists such as Hofstadter, Bloom or Jacoby consistently underrate both the elitist hazards of emphasizing the universalist calling of intellectuals as well as the technical and social necessity of specialization, ignoring the extent to which professionalism has become an inescapable foundation for the credible performance of intellectuals in the public world.

Nevertheless, Feyerabend, Foucault, and Bourdieu do portray intellectuals as somewhat *more* than experts and scientists, insofar as the former tend to take on wider issues that open up the professions to each other and to the political field. In contrast to the deceased grand intellectual, however, the new professional intellectual remains more closely involved with partial fields of competence and their local politics of knowledge, and refuses to enter the arena of big politics as an alter ego and potential substitute for the professional politician. As an 'anti-political politician', she first of all engages with political issues within the profession, and typically meddles with big politics where domestic professional issues border on or erupt into larger public debates. Meanwhile, she does so as a part-time but sophisticated amateur, anxious both to guard her own professional autonomy and to respect that of the politicians she challenges and criticizes. As we shall trace below, the problem of the public role of intellectuals is thereby recast as a problem of the 'two faces' of intellectual professionalism, irrespective of whether intellectuals speak for people or for things, or whether they speak in public places or in the more restricted spaces where they primarily encounter their fellow professionals.

In this respect, the academic and intellectual profession, while perhaps representing 'nothing special' in philosophical terms, still retains a particularity of its own, a specific difference that it is important to emphasize and maintain. The postmodernist disenchanters of the intellect and the constructivist normalizers of science have perhaps been too zealous in relativizing and levelling down distinctions. Feyerabend's tarnishing of intellectual experts as 'one special and rather greedy group' already implied a rather drastic idealization of the layperson and her capacity for democratically supervising science (1978: 85–86, 96–98). A similar anarchist streak is encountered in Foucault, who, in his effort to defrock all intellectual prophets, lawgivers and problem-solvers, at one point exclaimed: 'Down with spokespersons!' (1985: 81–82). Bourdieu's suggestion that every man should become his own true spokesperson, 'speaking instead of being spoken', likewise looked forward to this anarchist utopia (1993a: 7). In all of these postures, however, the initial radicalism was eventually tempered. Feyerabend wisely refused to root out the experts entirely, advising to use but never fully to trust them (1978: 97). Foucault was sensible enough not to preach the annihilation of all those who spoke '*for* and *above* the others', spelling out more modest tasks for what he called the 'specific' intellectual (1980: 126ff.). As we shall see below, Bourdieu even advocates a *Realpolitik* of reason, according to which a self-consciously corporatist defence of the autonomous interests of intellectuals providentially coincides with a 'politics of the universal'.

As suggested earlier, more mundane and practical arguments for intellectual autonomy can be based upon a new consideration of the differential timescape of science, or its specific technique of *unhastening* the tempo of thinking and research. If this is a plausible idea, the role of intellectuals might be reidentified in terms of differential interactional speeds: as go-betweens who temporarily enter the fast domain of publicity and politics, not in order to remain there and turn into professional spokespersons, but to 'do their thing' and then return to the slower routines of academic research and writing. Intellectuals can be seen as mediators who simultaneously serve the publicizing (i.e. the quicker circulation) of slow academic expertise, and the deceleration or halting of the 'fast thinking' of political professionals. They fill in some of the intermediate spaces that connect science and politics along what has been called the knowledge-political continuum, and in this respect rub shoulders with journalists, policy advisers and professional spokespersons, who occupy such intermediary positions in a more permanent fashion. In this minimal sense, we might still want to distinguish academic from intellectual, and intellectual from fully political roles, even though the knowledge-political continuum admits only weak boundaries and low thresholds between these different professional functions.

Although no longer protected by a traditional logic of justification, the professionalized production of knowledge does therefore include a minimal craftsmanlike claim, if not to cognitive and methodical superiority, then at least to the legitimacy and value of the special perspective and particular skills generated by a systematic deceleration of thought and action. In other words, because scientific expertise is manufactured in relatively insulated and distanced institutions that feature a peculiar temporality or time-frame, it is simultaneously superior and inferior to the knowledge generated in faster-moving realms such as professional politics and business or the unspecialized maelstrom of ordinary life. Within the framework of the social triangle that Chapter 2 has mapped out, scientific institutions obey a distinctive, more relaxed time economy than other specialized occupations such as journalism, politics, and business; they are screened off to a larger extent from the pressures of publicity, and produce for a comparatively more esoteric collegial audience. The defence of intellectual autonomy therefore entails the protection of this institutional capacity for slow, systematic, and methodical gathering of knowledge, in a minimal definition that does not entail any commitment to a particular scientific methodology.[2]

'Autonomy' is such a felicitous knowledge-political term in this context because it not only specifies a crucial precondition for sustained professional competence, but simultaneously sensitizes us to the corporatist and conspiratorial dark side inherent in all forms of professionalism. In this dual role, it also defines the axis of recent sociological debates about the productive and repressive functions of modern higher knowledge professions. The classical functionalist view disseminated by Durkheim, Spencer, Tawney, and Parsons emphasized positive characteristics such as institutionalized expertise, democratic control over knowledge and technology, and a collective ethos of disinterested public service. The critical approach elaborated from the mid-1960s by authors such as Johnson, Freidson, Larson, and Illich chose to highlight darker traits such as social closure, self-interest, privilege, control over clients and their needs, and the ideological legitimation thereof (Gyarmati 1975; Freidson 1986; Derber, Schwartz and Magrass 1990). The former approach, which rubbed shoulders with the vested complimentary self-image of professionals, encoded the quest for autonomy in the bid for rational self-regulation, which was supposedly required by the production and application of superior knowledge and complex skills in the service of the general interest. In the critical approach, autonomy was instead connected to the private interest of professional elites who wished to extend their monopolistic control over a specific market of expertise and the material and immaterial profits generated by it (Sarfatti-Larson 1977; Ruschemeyer 1983). As a result of such contradictory determinations, the concept of professional autonomy gradually adopted a Janus face: it came to display an intrinsic duplicity or *duality* in which good and evil, functional necessity and dysfunctional domination, appeared to conspire closely.[3]

In the following I shall focus more intently upon this 'simultaneity' or intrinsic coalition of the two faces of modern intellectual professionalism. I shall proceed from the hypothesis that its enabling and disabling dimensions are interconnected in a much more immediate and constitutive sense than can be grasped from the dualistic opposition in which both traditions hold each other prisoner. 'Duality' does not mean that the productive and exploitative dimensions, functions and dysfunctions, can be joined together through simple addition. It rather entails something like a generative connection or symbiotic coincidence of light and dark sides – polarities that appear to presume and precondition one another in a sense that falls out of range for one-dimensional 'optimistic' or 'pessimistic' approaches. Applied to the issue of intellectuals and scientists, one implication is that their social indispensability is organically tied to the social threat generated by their collective presence in the division of labour. What makes intellectual work socially risky inheres in the same structural properties that define its social utility. Intellectuals are socially 'freed' to exercise the office of professional thinking, researching, speaking and writing – activities that require a specific slowing down of the pace of communication and interaction. Because they have been educated and trained in such unhastening skills and exercise them as workaday routines, they are in many respects better at them than ordinary people. But the same professionalism and speciality turns intellectual spokespersons into a social threat, since by definition they speak *on behalf of* and *in the place of* others; and in extreme cases strive permanently to exclude others from entering the arenas where one may speak and write and from the means by which one may learn to do so (see extensively Pels 2000). In this conception of intellectual autonomy, the optimistic and pessimistic connotations of 'disinterestedness' and 'interest' intermingle, dismissing traditional dualisms between truth and power or involvement and detachment in the duality of what I have called 'self-interested science'.[4]

Protection of relative autonomy towards the *outside* (the celebrated 'academic freedom') is therefore not a normative principle that self-evidently emerges from the 'nature' of intellectual and scientific work, but an entrenchment that requires incessant knowledge-political efforts towards the formation and upkeep of corporatist interest coalitions. *Inside* the relatively autonomous and hence collectively distanced field, knowledge-political competition rages and once again erases all traces of disinterestedness. Initially, therefore, detachment is little more than a secondary effect of the collective unhastening and estrangement that an autonomous intellectual industry is capable of imposing upon its industrialists. Secondly, within this relatively slow-paced economy, detachment is produced by the interested objectifications to which intellectual competitors mutually force each other in their agonistic exchanges of criticism and counter-criticism. Both externally and internally such objectifying mechanisms have everything to do with the politics of knowledge and almost nothing to do with the classical ethos of value-freedom.

The Scientific Field

It is precisely this tense coalition between knowledge-political duality and professional autonomy that is at the heart of Bourdieu's view of the social logic of science. Indeed, Bourdieu's field theory of science balances a *generic* definition, which has a *levelling* or disenchanting effect, with a *specification*, which produces a contrary *demarcating* effect. The generic definition undercuts the traditional celebration of science as an exception to the rule of a general theory of fields or a generalized 'economy of practices'. The operation of the scientific field presupposes the existence of and itself generates a specific type of interest; scientific practices only appear disinterested when compared with alternative interests produced in different fields. The competitive game of science is oriented towards the specific stake of gaining a monopoly over scientific authority, in which technical competence and symbolic power are inextricably intertwined. Hence the scientific struggle is characterized by an 'essential duality' in which intrinsic intellectual and extrinsic material interests, intellectual and political strategies, epistemological conflicts and power conflicts are indissolubly joined (Bourdieu 1981 [1975]).

In the specifying operation, the emphasis is displaced from the interested, political-strategic, 'capitalizing' character of scientific practice towards the 'other means' with which this quasi-political or quasi-economic rivalry is conducted. Here we are not so much concerned with knowledge *politics* but rather with *knowledge* politics; not with the accumulation of economic or political capital but rather with that of *cultural* or *informational* capital. The axiom about 'intrinsic duality' is now read in reverse, since the analytic focus is redirected towards the specific manner in which rational-scientific knowledge and technical competences turn into vehicles for the accumulation of symbolic power or symbolic capital. Not 'ordinary' profit-seeking or 'naked' power-grabbing (insofar as these exist at all) are operative here, but the urge for recognition, for a brilliant reputation, for a distinctive name; a form of interest that is simultaneously constituted and mystified by the overt dismissal of 'ordinary' and 'vulgar' objects of interest such as money and power.

In this manner Bourdieu's model does justice both to the internal cohabitation of knowledge and politics and to the external divorce between the professional manufacture of knowledge and 'outside worlds' such as politics, business or journalism. Both in the interface between science and the outside world and in the agonistic world inside, the traditional opposition between disinterestedness and interest evaporates, severing the time-honoured connection between intellectual distanciation and the positivist or Weberian conception of neutrality. Indeed, if the truth about the social world is not constituted *in spite of* but precisely *as a result of* the knowledge-political interests of the sociologist,[5] the political logic of knowledge can only fully unfold as soon as science has emancipated itself sufficiently from external political or economic interests. On balance, the very science that nervously advertises itself as neutral and detached turns out to be pulled by political strings, while truly autonomous science can live without this posture of objectivity and neutrality.

Thus distanciation, in both its external and internal dimensions, is immediately derived from specific forms of engagement. With respect to the broader relationship between theory and practice or between the intellectual world and that of everyday life, Bourdieu draws our attention to the social determinations that condition the intellectual attitude as such, the 'scholarly' or 'contemplative' gaze that is directed towards social reality. As soon as we begin to observe, we effect an epistemological break which is simultaneously a social break, since we withdraw more or less completely from the social world. This posture of the 'impartial spectator' is not only socially exceptional but is also supported by concrete social privileges. Detachment is not a product of methodological morality but of the 'scholarly situation', the specific habitus of professional 'schoolmen', and presupposes the *scholē* or *otium* or the specific 'idleness' that marks out the contemplative life. Distanciation is a practical relation to practice. It is this practical engagement that the 'ethnocentrism of the *savants*' functions to conceal.[6]

A similarly intrinsic rapport between involvement and detachment is manifest in the internal analysis. The scientific field constitutes the theatre of a more or less unequal struggle between (usually) two parties who are endowed with unequal quantities of specific capital and are therefore unequally equipped to appropriate the products and profits of scientific labour: the dominant or established elite and the dominated groups of outsiders and newcomers. The established group normally opts for a strategy of conservation of the scientific status quo with which its interests are immediately linked. New entrants are capable of 'choosing' a comparatively tranquil strategy of succession, or they can adopt the risk strategy of subversion of vested scientific authority, entailing novel redefinitions of the stakes and confines of the game. The struggle in which every actor has to engage in order to enforce his own authority as a legitimate producer is invariably also a struggle to impose the 'mode of production' and the definition of scientificity that maximally concur with his own specific interests (Bourdieu 1981 [1975]: 263, 269–72). It is this objective logic of rivalry between established and newcomers and the resultant cross-checking of their mutual products which unintentionally turns the pursuit of self-interest into an engine of scientific progress. The anarchic antagonism of private interests is transformed into a scientific dialectic, as every actor is forced, when challenging and resisting his opponents, to adopt an instrumentarium that lends his polemical purposes the universal scope of a methodical critique. This is how social mechanisms enforce the realization of universal norms of rationality. The *libido dominandi* can only be satisfied by submitting to the specific censorship of the field, where it is sublimated into a *libido sciendi* that can only triumph over its adversaries by opposing one theorem, demonstration, or set of facts to another (Bourdieu 1990b: 300; Bourdieu and Wacquant 1992: 115–16, 178).

Facts, Values, and Performative Effects

Although 'duality' and 'autonomy' thus figure prominently in Bourdieu's sociology of science, the epistemological status of these key terms remains somewhat ambiguous. 'Duality' is of course first of all an escape word, a polemical foil that twice delimits one's position over against free-floating intellectualism on the one hand and all forms of (economic and political) interest-reductionism on the other. This is enough to transform it into a term of embarrassment that only provisionally liberates us from the Procrustean alternative imposed by such well-established conceptual dichotomies. In addition, adjectives such as 'intrinsic' or 'essential' may be epistemologically perplexing, as they suggest the presence of an essentialist dialectic that claims to recover objectively determined contradictions in the heart of things. Bourdieu himself chooses to present the principle of duality as a theoretical summation of a string of empirical-analytical observations (1981 [1975]: 260; 1991: 180–83, 197; 1993b: 74). The same empirical rigour is invested in his concept of relative autonomy and his view of the historically structured specificity of the intellectual, scientific, and other cultural fields. However, concepts such as 'duality' and 'autonomy' also appear to effectuate a subterranean *normativity*, which is nowhere accounted for or recognized in so many words. Clearly, Bourdieu's Nietzschean (or rather, Spinozist) impatience with all forms of utopian moralism is so pervasive, and his trust in the capabilities of a 'rigorously' empirical science so overwhelming, that his field theory regularly slides towards objectivism and the risk of reification.

Although a score of classical dualisms are convincingly put to rest, Bourdieu therefore appears to underestimate the continuing hold upon his work of the obstinate antinomy of normativism vs. naturalism or that of facts vs. values. On this point, his reflexive critique of objectivism is insufficiently radical. Even though he intends to think beyond the old antinomy and reaches for a form of ethics that purports to supersede it, it remains unclear which version of the classical antinomy is put to the test (the neo-Kantian one? The positivist one?) and how precisely the rift between *Sein* and *Sollen* should be mended. Bourdieu also retains a distinctly Durkheimian order of epistemological priority, according to which the ethical quality of social science is ultimately derivable from its explanatory power. It is the knowledge of objective determinations and objective necessities that circumscribes a form of freedom which in turn may condition a modest practical ethics. Reflexive sociology is an ethic *because* it is a science (Bourdieu and Wacquant 1992: 171–72). Elsewhere Bourdieu frames his critique of the naïve utopianism of the classical philosophy and sociology of science in such a manner as to virtually presuppose the fact/value distinction. Merton is indicted for incorrectly offering the normative rules that figure in scientists' self-images as descriptions of the positive laws of the scientific mode of production itself, whereas 'the market in scientific goods has its laws, and they have nothing to do with ethics'. This is enough to suggest that these 'immanent' laws of the field can be determined according to a canon of empirical exactitude that is exempt from all normativity (Bourdieu 1981 [1975]: 266).

Both the principle of duality and that of autonomy, however, can only be sensibly defended if one acknowledges that they represent more than descriptive generalizations, and circularly knot together normative and empirical propositions in a seamless epistemological web. In both cases the description immediately posits the norm. 'Inescapable' or 'intrinsic' duality, of course, is one way of saying that the good or functional and the bad or dysfunctional aspects of intellectual practices are tied together much more closely than is normally accounted for.[7] Due to such rapprochements, the polarity changes its normative hue. The good shades into something less good, but what used to be considered an unmitigated evil becomes a little less threatening in return. The *démasqué* of the mandarins and of their narcissistic ideology of disinterestedness is balanced by the epistemological rehabilitation of previously dark and marginal phenomena such as competition, self-interest, and the accumulation of symbolic capital. For example, the definition of the intellectual and scientific world as an agonistic arena or as a capital market in which unequally endowed parties compete for mutual distinction constitutes much more than a cool observation of the impersonal operation of objective field laws. The description includes some pertinent normative suggestions: that scientific rationality is not a singular and uniform phenomenon, that it is not definitionally constrained by the quest for consensus, and that conflict, competition, and struggle fulfil a positive, generative function in the production of knowledge (cf. Bourdieu and Wacquant 1992: 178–79, 188–89). A similar observation applies to the definition of the stake of the intellectual game as itself one of the stakes in the intellectual game, and the concomitant thesis that there are no arbiters in the field who are not themselves interested parties to the dispute (Bourdieu 1981 [1975]: 264).

If it is true that the bulk of analytical concepts and propositions in Bourdieu's sociological lexicon carry a normative surplus and mingle facts with values, they can more fruitfully be read as *performative* definitions that (re)describe their object in such a manner that the description simultaneously (re)creates what it purports to describe. As I have argued before, performativity is a useful denotation of the knowledge-political construction effect of representations that naturally mix normative and empirical judgments. Notwithstanding Bourdieu's extraordinary sensitivity to the ordering function of language and to the constructivist truth that words can make and break things, his approach once again appears to incur a reflexive deficit. His failure to appreciate the full performative significance of his propositions tends to transform all supposedly 'realistic' statements about objective structures, relationships and positions, or about the incontrovertible primacy of objective laws of the field, into so many reifications. In apparent violation of his stated principles, Bourdieu adopts a rhetoric of totality and transcendence according to which the 'complete system of strategies', the 'game as such' may now be reviewed from an allegedly neutral position outside and above the game: a singular and privileged point of view able 'to take (in thought) all possible points of view' (1981 [1975]: 283; 1988a: xvi, 47–48; 1996b: 34; Bourdieu and Wacquant 1992: 259–60).

If we refocus the problem of specificity or relative autonomy, we encounter a similar objectivist slant. Despite a deep sensitivity to the performative effect of social and social-scientific classifications, which definitely sets him apart from the crowd of his fellow sociologists, Bourdieu still underestimates the extent to which every codifying representation of social likenesses and differences, of frontiers, domains, articulations and instances, functions to co-produce the same social universe that it claims objectively or realistically to mirror. The defining sociologist engages just like any ordinary actor in the struggle for the imposition of legitimate classifications, and classifies others in order 'to tell them what they are and what they have to be' (cf. Bourdieu 1991: 243). In similar fashion, sociological claims about specificity or relative autonomy, according to which different social fields are expected to 'obey' different objective logics or different immanent necessities, are not *ostensive* but *performative* definitions by means of which the sociologist forces the social facts that he defines into obedience and submission (cf. Latour 1986; Barnes 1988: 52–54).[8] Like everybody else, Bourdieu participates in the social struggle for classification; like everybody else, he 'works upon' the field rather than divining its innermost logic.

In this interpretation, the relative autonomy of the scientific field not only constitutes an empirical-historical variable but also erects a normative yardstick. The historical-sociological analysis of science immediately (and circularly) suggests its own criterion of scientificity. The autonomization of the scientific field secures the development of specific field laws that in turn guarantee the advance of rationality. As the accumulated scientific stock multiplies, the struggle for social recognition is increasingly sublimated into a struggle for scientific recognition, i.e. a type of recognition that can only be conferred in a field of 'restricted production', i.e. a system of peer review by colleagues who are simultaneously competitors. The performative implication is that scientists *should* actually recognize no other clients than their competitors, and *should* in fact exclude as illegitimate all exchanges that are not considered legal tender in the field. The same performative tonality defines Bourdieu's conception of the 'inaugural revolution', according to which the development of the relative autonomy of a particular field takes a qualitative jump that fundamentally alters its domestic logic of competition. Because the costs of entry rise steeply and continued investment in the field becomes increasingly drawn-out and costly, homogeneity among the competitors increases, the contrast between conservative and subversive strategies diminishes, and great periodic revolutions are replaced by numerous small permanent revolutions that are liberated from external political effects and 'social arbitrariness' (1981: 273–74). An 'authentic' scientific field is accordingly defined as one in which polemical disputes are encased in a more fundamental consensus about the stakes of the game and the means with which to resolve disagreements; a dissensus that derives its productivity precisely from the objective agreement which undergirds it.

A Politics of the Universal?

In my view, therefore, the principles of knowledge-political duality and relative autonomy must be read as performative claims that together lay the foundations for a new normative-empirical epistemology.[9] The double operation of Bourdieuan field theory, which evens out the deconstruction of the traditional concept of truth with a *social* demarcation between science and other social fields, appears rigorously to exchange normative propositions for empirical ones, but at closer range turns out to be unable to divest itself of a residual normativity. Truth and rationality are conceived as compelling effects of a social mechanism of nonviolent (but not interest-free) rivalry, which can only be installed in a self-regulating and autarchic scientific field. Thus, knowledge-political rivalry and relative field autonomy take the place formerly reserved for classical criteria of rationality. They constitute two sides of a new political criterion of scientificity. As indicated, however, Bourdieu's work also bears traces of an objectivistic misrecognition of these basic principles. His claim to take in the entire complex of strategies in the field still underwrites the possibility of a sovereign, totalizing point of view and a neutral outsider's gaze, which are doubly incompatible with the radical perspectivism of a dual theory of interests. This 'field objectivism' invites reifying definitions of the limits and specific stakes of particular fields, which in turn conflict with the neighbouring conviction that both, and hence also the identity of the legitimate players in the field, are permanently and irreversibly contested and contestable.[10]

In this way, the idea of a neutral foundation for objective truth and for universal conditions of scientific rationality is readmitted through the sociological back door. Bourdieu tends to advance his field theory as a sociological underpinning of the 'universal validity of scientific reason', for example when he crucifies the 'nihilistic subjectivism' of the Strong Programme and of Latour and Woolgar, or when he takes on postmodernist philosophy more generally (Bourdieu 1981 [1975]: 276; 1990b: 299; 1993c: 365–74). In conscious paradox, this brand of universalism is consistently grounded in a particularistic theory of interests. Rather than quarrying in a Habermasian vein for universal preconditions of communicative rationality, Bourdieu insists upon specifying the social and historical conditions that force actors to *take an interest* in the universal, i.e. that constrain them to contribute to the production of universal truth through the push and pull of their strategic self-interests. That is why he counteracts the idea of a universalist pragmatics of language with his own alternative of a 'politics of the universal' or a '*Realpolitik* of Reason' (1989; 1990a: 31–32; 1996a: xvii, 337ff.; Bourdieu and Wacquant 1992: 175ff.).[11]

In this Smithian defence of universalism by particularistic means ('private vices, public virtues'), the idea of relative autonomy once again occupies pride of place. This fact is clearly corroborated by Bourdieu's conception of the 'anti-political' calling of intellectuals. Intellectuals are 'paradoxical beings' who enter upon the historical stage by superseding the opposition between pure culture and social engagement. They are 'bi-dimensional' beings who belong to an intellectually autonomous field and hence must respect its indigenous laws, while simultaneously deploying their specific expertise and authority in a political activity outside of it. They intervene in politics, but without political weapons, and without turning into full-time politicians. Contrary to the traditional antinomy between autonomy and engagement, it is therefore perfectly possible to exercise both simultaneously (Bourdieu 1989: 99–101; 1996a: 129–31, 340–48). This portrayal of the identity of intellectuals is once again predominantly introduced as a *realistic*, empirical-genetic definition. Ethical-political interventions by intellectuals should be grounded in a rigorous understanding of the operation and historical constitution of the intellectual field. However, both formulation and context suggest that this definition also harbours performative elements that give the figure of the intellectual a distinctly critical profile. A non-polemical, non-perspectival characterization of his 'essential' identity, such as Bourdieu still seems to advocate here, also runs against his conviction that the definition of what makes a true intellectual is always and invariably at stake in the intellectual field itself (Bourdieu and Wacquant 1992: 100).

Leaving aside this epistemological conundrum, we may notice that Bourdieu's conception of an 'anti-political politics' of the intellectuals simultaneously strengthens their autonomy and guards against the temptations of the ivory tower. This is realizable if institutions are created that enable them to intervene in politics under their own authority, especially with the aim of gaining control over the material means of cultural production and intellectual legitimation. Intellectuals are therefore under an obligation to value the defence of their autonomous interests as their primary task: defence of the autonomy of the field must be accorded top priority (Bourdieu 1989: 103).[12] For too long intellectuals have sheltered behind the interests of other groups and classes in order to celebrate these as universal interests; now it is time to speak up for their own. Intellectuals must no longer feel remorse in defending their own corporatist privileges, since 'by defending themselves as a whole, they defend the universal'. Protection of their corporate autonomy effectively coincides with the defence of the social conditions of the possibility of rational thought, as they are paradigmatically exemplified in the scientific field, where no one is able to succeed over anyone else without arming himself with better arguments, reasonings, and demonstrations (Bourdieu 1989: 103–104).

What is most striking in this 'corporatism of the universal' is its almost dialectical configuration of opposites. On the one hand, Bourdieu does not hesitate to picture the defence of intellectuals' professional autonomy as a matter of collective interest bargaining. But nor does he shrink from sublimating this corporative interest into a universal interest, turning universalism into an immediate corollary of field autonomy. Consequently, his *politique de la raison* is still tainted with a residue of rationalistic idealism and by a concomitant underestimation of the dark sides of intellectual professionalism – including his own. The corporatism of the intellectuals is euphemistically wrapped in a residually Hegelian notion about the universal class. Unlike Gouldner's notion about intellectuals as a 'flawed universal class', however, universalism does not arise from the internalization of a normative 'culture of critical discourse', but from a set of sociological constraints that force critical cross-control of intellectual products; while the 'flaws' of corporatism are not set in contradiction to, but are seen as pertinently *instrumental* to the attainment of true universality. This indeed is true *Realpolitik*, since local interests are no longer masked by global claims, but are openly celebrated as central cogwheels in the grand engine of sociological providence.

On balance, therefore, and quite similar to Gouldner, Bourdieu has difficulty in accommodating the Janus face of intellectual autonomy, according to which general interests and group interests, public benefits and private vices come together much more symbiotically than is accountable in terms of the optimistic mechanics of his latter-day invisible hand. The professionalization of intellectual labour inescapably includes a dark side of academism, symbolic violence, waste production, clientelism, and monopolistic expropriation. It is therefore debatable whether the autarky of the intellectual field should be developed to such an extent that the assignment of reputations in the intellectual marketplace remains restricted to an audience of collegial competitors. One may likewise question the capacity and wisdom of the social sciences to carry through an inaugural revolution that would definitely put politics and other social practices at large, and would reach for the homogeneity of investments and the methodological consensus that Bourdieu still conceives as a prerequisite for a truly scientific dialectic.[13] His own emphasis is rather exclusively placed upon the threat to intellectual autonomy posed by 'heteronomous' interests such as those of the new mandarins of state and economy and the media technocrats. But this once again underplays the threat to society that is constituted by the epistemocracy of the intellectuals themselves. The anarchist question posed by Feyerabend and Foucault, i.e. how society may defend itself against the privileges and the interest politics of professional spokespersons, should therefore be weighed more seriously. In this respect (though not in others), Habermas's drive to force open the esoteric cultures of expertise in science, morality, and art, and to restore an open channel of communication between them and the practice of everyday life, appears to offer a more balanced point of departure (Habermas 1987: 119ff.; 1990: 17ff.).

Autonomy and Relative Speed

There is another dimension to Bourdieu's theory of the specificity of the scientific and intellectual field that links up more immediately with my preoccupation with the temporal preconditions of academic autonomy. Indeed, a focus upon the specific timescape of science to some extent resolves some of the theoretical difficulties signalled above, since it retains the full impact of a

Unhastening Science

dual theory of 'self-interested science', while discarding the residues of the truth/power distinction which tend to draw Bourdieu back into the quicksands of 'field objectivism', perspectival totality, and epistemological universalism. In addition, an 'unhastening' perspective would compensate for some tautological elements in Bourdieu's conceptualization, which suffers from a certain under-specification of the specificity (and hence the autonomy) of scientific interaction as compared to that in other social domains. While this temporal perspective is largely implicit in Bourdieu's earlier writings, it is more prominently displayed in his more recent works. My purpose in this section is to draw it out in more explicit fashion, and to set in play a more radical pragmatism of time and space against the residues of objectivism and transcendentalism that still burden Bourdieu's approach.

As I have noted, the distinctive nature of the struggle for scientific authority or scientific capital, as contrasted with that for religious, political, or legal capital, resides in the fact that scientists are not interested in 'ordinary' profit or 'naked' power but in the extraordinary profit-and-power of *recognition*, or in the building of a distinctive *name* in the scientific field. This reputational capital can only be accumulated in a field where producers tend to have no other clients than their colleagues and competitors, who compose the effective recognition-granting group. Cultural fields, more generally, are sites of struggle between autonomous and heteronomous principles of legitimacy and hierarchization where the degree of specific autonomy depends upon the recognition accorded by those 'who recognize no other criterion of legitimacy than recognition by those whom they recognize' (Bourdieu 1993b: 38). Cultural fields are thus polarized between a *long-term* logic of *restricted* production, where producers produce for other producers, and that of *large-scale* and *short-term* production which aims at public success, and which submits to the law of competition for the conquest of the largest possible market. The tension is between producing *quickly* for a vast audience of non-producers, or *slowly* for a small audience that consists of one's critics and accomplices: the peer group of competitors that is enclosed in a circular relation of reciprocal recognition, and has the power to define its own criteria of product evaluation (Bourdieu 1993b: 37–43, 115–20).

The autonomous sector of each field systematically inverts the principles of ordinary economics, which delegitimizes the all-out pursuit of profit and power, and hence establishes an 'anti-economic economics' (and by extension, an 'anti-political politics') that cultivates an interest in disinterestedness (in this particularly relative sense). One effective illustration of this is drawn from Bourdieu's comparative analysis of publishers such as Laffont and Minuit, who are positioned at opposite extremes on a scale running from safe and

Knowledge Politics and Anti-Politics

short-term investments (in bestsellers) towards high-risk and long-term investments (in classics), and hence display a typical variation in temporal structure linked with a quick and large vs. a slow and small turnover of capital (Bourdieu 1993b: 98–101). Closer to our subject, Bourdieu similarly stretches the academic field (the 'space of the faculties') between a pole that is 'temporally dominant' and that derives its authority from a kind of social delegation (e.g. law or medicine) and one that is oriented more towards pure scientific research (e.g. natural science, with the arts and social sciences sitting somewhere in between). This continuum establishes a set of oppositions that once again differ crucially in time-economic terms: time spent in order to accumulate short-term academic power is bound to detract from time spent in order to accumulate long-term scientific authority (Bourdieu 1988a: 62–64, 73ff., 95–99).

Bourdieu's political-economic (but also specifically anti-political and anti-economic) view of science is matched by an intriguingly *intellectual* or *symbolic* view of politics as a 'field of ideological production', or as a struggle for the legitimate representation of the social world. Politics produces politically effective forms of perception and expression of social reality, and hence triggers a struggle over the principles of vision and division of the social world. Like intellectual capital, political capital is a form of *reputational* capital that is linked to notoriety, to the fact of being known and recognized (Bourdieu 1991: 172–73, 176–77; 2000b: 63–64). The symbolic struggle for conserving or changing the social world takes the form of 'a struggle over the specifically symbolic power of making people see and believe, of predicting and prescribing, of making known and recognized, which is at the same time a struggle for power over the "public powers"' (1991: 181). Although, in the former aspect, the logic of scientific mobilization does not differ from that of political mobilization, scientific and political capital crucially diverge in the sense that scientists ideally incur their reputational capital from the small circle of their colleagues, while politicians need to persuade the largest possible number of voters. Whereas the stakes of both the scientific and the political games are marked by structural duplicity, because both simultaneously file cognitive or ethical claims and calculate for power and profit, the *political* production of ideas about the social world

> is always subordinated to the logic of the conquest for power, which is the logic of the mobilization of the greatest number ... The spokesperson appropriates not only the words of the group of non-professionals, that is, most of the time, its silence, but also the very power of that group, which he helps to produce by lending it a voice recognized as legitimate in the political field. The power of the ideas that he proposes is measured not, as

Unhastening Science

> in the domain of science, by their truth-value [...] but by the power of mobilization that they contain, in other words, by the power of the group that recognizes them, even if only by its silence or the absence of any refutation. (Bourdieu 1991: 181, 190)

However, formulations such as these also confirm that Bourdieu remains wedded to a residual truth/power dualism that has increasingly made him emphasize the distinctions rather than the continuities between the dynamic of science and that of politics. In recent statements, the science–politics continuum is clearly pulled towards its ideal-typical extremes, where the scientific pole represents traditional ideals of truth, logical coherence, and compatibility with the facts, and where the political role of science is classically depicted as 'speaking the truth to power' (Bourdieu 1995; 1996a: xvi, 336; 1997; 1999: 338–40).

It therefore seems that the accumulation of reputational capital cannot in itself sufficiently differentiate between science and politics, since it precisely defines an identity of interests across the knowledge-political continuum. Nor can we rely upon the logic of group mobilization or the performativity of representation as marking the decisive differences between the two forms of social spokespersonship, since these criteria move in close conjunction with the traditional opposition between truth and power. Indeed, some of Bourdieu's characterizations of the logic of the political as a symbolic struggle 'to make or unmake groups' (e.g. 1991: 127, 221, 249–50) seem eminently transferable to scientific representations.[14] Instead, and recalling some of the parameters of the triangular model of social time-space sketched in Chapter 2, this differentiation can be more fruitfully based on the more 'ordinary' variable of the relative number of 'voters' or 'clients' who are expected to respond to these performative visions and divisions. Rather than a residual version of the dualism of reason vs. force, it is the differential volume and composition of the recognition-granting, authorizing group that specifies both the closeness and distance between science and politics and determines their weak reciprocal autonomy.

But this variable of audience size – which recalls my previous considerations about the salience of different mixtures of privacy and publicity for a felicitous conduct of science and politics – still undertheorizes the factor of differential social speed, even though it naturally implies a particular set of spatio-temporal arrangements. This time factor was already addressed in Bourdieu's thematization of the polarity between restricted or long-term cultural production and large-scale or short-term cultural production – which to some degree overlaps with the science/politics distinction. The factor of speed is more overtly introduced in a recent critique of television mediation (Bourdieu 1998b). Here, admittedly, it is not so much the speed of *politics* that is contrasted with the slow motion of intellectual life, but the swift action flow of the mass media, which are pressurized in the last instance by economic demands. The hurried pace of television, where time is an extremely rare commodity, and where production decisions are primarily ruled by audience ratings, manifestly threatens the specific nature of intellectual communication. Criteria of market success (to get a scoop, to be there first) impose a competition for time and the pressure to get things out in a hurry. But there is a negative correlation between time constraints and the cultivation of thinking – of which Plato was already fully aware when he constrasted the leisurely situation of the philosopher with the hurried existence of people in the *agora*. 'Fast-thinking' only produces clichés, received ideas, conventions, commonplaces. It is completely opposite to 'subversive' thought, which takes apart such received ideas – and which requires time to do so effectively. Television rewards a certain number of 'fast-thinkers' who share its sense of urgency, and who offer cultural fast food (pre-digested and pre-thought culture) to a large and inattentive audience. Because journalists control the access to the space of mass circulation, not merely of ordinary citizens but also of other cultural producers (scholars, artists, and writers), they are effective gatekeepers of public existence, i.e. the ability to be recognized as a public figure. All fields of cultural production are subjected to structural pressures from the journalistic field, which is itself increasingly dominated by the market model. In this indirect way, the economy increasingly threatens to absorb the autonomy of all fields of cultural production, which ideally maintain that the producers' sole consumers are their own competitors, and which are therefore fatally undercut by the law of the greatest number (Bourdieu 1998b: 28–30, 46, 56).

This account agrees with Bourdieu's more general critical analysis of *scholē* or leisure as the condition of existence of all scholarly fields – a condition of felicity that is normally euphemized and misrecognized by glossing over its economic and social conditions of possibility. The scholastic disposition is an aristocratic one that normally inclines its possessors 'to suspend the demands of the situation, the constraints of economic and social necessity, and the urgencies it imposes or the ends it proposes' (Bourdieu 2000a: 12). In the dominant professional ideology of its practitioners, philosophy and science advance towards freely chosen, gratuitous ends, which are liberated from external constraints such as time and money. The scholastic situation implies a particularly free relationship to time, 'since, as a suspending of urgency, the pressure of "things to do", of business and busyness, it inclines us to consider "time" as a thing with which we have a relation of externality, that of a subject facing an object' (2000a: 206). Time is something we 'have', as a natural property. This bracketing of time is one of the most consistent effects of the scholarly illusion, which perhaps begins with the Platonic opposition between *ascholia* or hurry and philosophical *scholē*, and which is also recognizable in the loosely structured temporality of Bohemian life (cf. Bourdieu 1998a: 128–29).

The critical task is then to unravel this allegedly free disposition and show that it is rooted in particular privileges, impurities, and interests. As is for example evidenced by the (irritating) fact that important people (for whom 'time is money', and who cannot 'waste their time') are always in a hurry, often show up late, and hence force lesser mortals to wait for them, there is an intimate relation between time and power (Bourdieu 2000a: 224, 226, 228; Schwartz 1974; Adam 1990: 121–25). What Bourdieu in other words appears to suggest is that to 'have' time, for example time to think, read and write, which is a prerequisite for all serious intellectual work, simultaneously expresses a power-permeated social hierarchy and a 'propertied' inequality. In this important respect, at least, it resembles rather than opposes the opposite condition of always being in a hurry and having no time to spare. If this is a fair interpretation, the scholar's unhastened disposition, which typifies one extreme on the knowledge-political continuum, cannot be interpreted as purity and neutrality but must always be framed in terms of *duality*, i.e. as an interested practice that does not differ from faster economies except for their different 'speed limit' and their longer-term, more extended cycle of investment. Not the sobering *fact* of the scientific struggle, but the *slow turnover* of this particular form of regulated competition emerges as the decisive identifying variable.[15]

The Anti-Politics of Unhastening

The core theme of my inquiries throughout this book is that we need to work towards a more radical 'confusion' or de-differentiation between science and politics, without giving up the search for effective criteria of intellectual autonomy; but that these criteria can no longer be consecrated by the dichotomy of truth vs. power as residually upheld by Bourdieu, but may be better served by a more pragmatic and graded distinction between unhurried and hurried time dispositions and between spaces of relative privacy and relative publicity. This accentuation of the differential politics of time and space removes some of the tautological and objectivist features of Bourdieu's approach and substantively enriches the phenomenological description of what he has called 'the specificity of the scientific field'. In the concluding section of this chapter, I shall use this chronopolitical view as a baseline from which to develop an alternative 'politics of reason' and a vision about the political role of intellectuals that does not re-embrace a form of universalism but undertakes to sharpen Bourdieu's defence of corporatist interest politics in the direction of a 'dual' conception of the interests of *slowness*.

Bourdieu's sociology of knowledge, as we have amply seen, evens out the profanation of philosophical conceptions of rationality and truth by means of a social demarcation of science as an autonomous line of work. The *epistemological* fusion between knowledge and power appears to go together well with an *institutional* separation between science and society, even though the former principle tends to blunt the sharp edges of the latter by replacing the former singular boundary by a cascade of weaker ones. It is this precarious balance between the principles of 'knowledge politics' and 'anti-politics' that also suggests a fresh definition of the role of intellectuals, since it finally dispenses with the traditional alternatives of 'ivory tower' detachment and the involvement of a fully politicized culture. In traditional conceptions of the relationship between reason and power, the polarized options share an important characteristic, since they embody two forms of 'knowing better': two forms of elitist self-understanding on the part of the academically educated (cf. Habermas 1989: 80, 86–88). Both the idea of an 'ivory tower' and the drive for full politicization support a notion about the supremacy of the rational mind that invites the intellectual to look down in condescension upon the low life of politics and politicians. In the first case, intellectuals derive their superiority from their privileged husbandry of universal values and truths, and castigate any colleague who descends into the political marketplace as a traitor to the Spirit. In the reverse case of politicization, intellectuals eagerly invade the *agora* in order to *take the place* of their dilettante rivals – dreaming the historical dream that connects Plato to Hegel and Marx, and Marx to Sartre and Heidegger. The project of legislative reason, which in the former case leads towards renunciation of the world, turns political, and the accusation of intellectual treason is now reversed upon the uncommitted and the inactive (cf. Pels 2000: 131ff.).

The idea of an 'anti-political politics' effectively dodges this traditional dilemma. In Konrád's characterization, it is the political engagement or obtrusiveness of those who *refuse* to become politicians and, at the end of the day, prefer to leave the routine exercise of power to professionals other than themselves.[16] Crucial to this view, of course, is a performative awareness that doing politics is *different from* the display of cultural creativity; that the work of 'organizing ideas' does not and should not coincide with the work of 'organizing people' and 'organizing things', but answers to a distinct logic of its own. Konrád likens the relationship between professional politics and cultural anti-politics to that between two mountains: neither of them is interested in taking the place of, or is able to eliminate or substitute for, the other (Konrád

1984: 230–31). It is this very separation of tasks and interests, which 'neutralizes' the knowledge-political marriage between culture and politics by means of their anti-political divorce, that defines the promise of a society of democratic difference (cf. Pels 1993).[17]

Bourdieu, in deploying partial interests as providential mainsprings of the universal, still tends to contribute to an overly reverential and sublimated (self-) image of the professionals of culture, by underwriting something like their agonistically constituted but collectively ensured guardianship of truth and rationality. In a more thoroughly 'dual' approach to the problem of autonomy, however, there is no longer any single profession or corporation that may claim privileged access to a universalistic cultural grounding. All professions are equally suspect in their claims to speak for and represent the general interest, and must balance the modest recognition of their own capacities by a recognition of the distinct rationality and autonomy of others. In this perspective, intellectuals and politicians are embroiled in something like a generalized, society-wide version of the agonistic logic that Bourdieu considers a characteristic and beneficial feature of the development of science and other restricted fields. While inside the field a critical game of competition ranges scientific rivals against one another, so likewise in the broader societal field checks and balances are installed between professional powers with a vested interest in keeping a sharp eye upon each others' every movement.

In this constellation of institutionalized rivalry, intellectuals are not so much 'paradoxical' or 'bi-dimensional' beings who jump across a sharp divide between culture and politics, but mediating messengers who move at variable speeds back and forth along the knowledge-political continuum in order to accelerate and publicize the circulation of scientific knowledge, but also to slow down the pace of communication in politics and journalism. They distinguish themselves from media professionals, commentators, policy pundits, and political visionaries because they continue to find their home base in the university or the research laboratory, and only incidentally march out into the open field of political publicity in order to add their voice to the larger chorus of public debate. By intervening in politics without turning into politicians or media beasts, they continue to represent *the interests of slowness* that are concomitant with cultural production more generally. In this sense, they are ambassadors of a 'slow corporation' who defend the corporatist interests of a secluded and unhastened culture of research that delivers a specifically detached or estranged perspective on the world. This particularly 'withdrawn' or marginal perspective has always been conducive to cultural creativity and social innovation (Pels 2000). However, as the principle of duality would predict, it simultaneously comes with a dark side of esotericism, irresponsibility, over-production (publication waste) and, indeed, the negative slowness of under-production (literal 'idleness' or sinecure). Once again it seems that the bad marches in step with the good, and that such a slow-paced economy of free intellectual experimentation cannot function without incurring the risks and flaws that are the flipside of its advantages.

The temporal framework that I am developing here meanwhile suggests an interesting spin on the notion of an 'anti-political politics'. In order fully to appreciate this alternative view, we need to recall the strong, totalizing definition of the political canvassed by thinkers of the 1930s 'Conservative Revolution' such as Schmitt, Freyer, and Heidegger (Pels 2000: 104ff.), who anticipated the radical proximity of rationality and decision or of reason and force that is also incorporated in my own view of the knowledge-political continuum. In expanding the scope of the political and turning it into the grounding dimension of sociality, these thinkers simultaneously characterized politics as 'the intensive life', or 'life in the danger zone', dictated by the logic of the exception and the 'extreme case'. Politics was viewed as necessitating ultimate decisions in a condition of sharp confrontation with an existential enemy, approaching more widely shared views that war was 'the mother of all things' and was intrinsically related to speed of movement and totality of mobilization (cf. Marinetti 1972: 41–42; Virilio 1986).[18]

Against this backdrop, 'anti-politics' can now be interestingly redefined as the considered delay or the systematic stalling of decisions and actions – a 'holding pattern' that is equally a form of acting, and hence partakes of the decisionist continuum rather than falling outside of it. This halting of decisions is still a species of politics and is not generated by a principle of agnosticism or liberal neutrality. As a form of politics 'continued by other means', it typically employs the means of deceleration (such as writing and other recording technologies) that increase procedural insecurity and doubt, and accordingly render it more difficult to uphold any construction of an enemy (cf. Beck 1997: 169). Intellectual anti-politics *refuses* to think under the dictate of the *Ernstfall* or the logic of existential struggle, and resists the pressure and pace compelled by strong notions of intensity, emergency, and mobilization.[19] If, as Latour has suggested, it is the particular strength and obligation of politicians to round off arguments, to bring about closure, to take quick decisions under high pressure of time, and thereby to divide the collective into enemies and friends (1999b: 11–16, 201), it is perhaps the institutional privilege of intellectuals to be able to suspend such decisions, to delay the making of enemies, to refuse – at least for the moment – the certainty of distinction between true and false and right and wrong.

CHAPTER 6

The Politics of Symmetry

Symmetrical Delay

This chapter returns to a more in-depth investigation of some of the core methodological principles in the social studies of science and technology – a family of approaches that we have previously encountered as forming part of the Wittgensteinian impulse in the modern social theory of knowledge. As I have argued before, SSK and STS were drawn towards a form of value-free or 'a-critical' relativism that subdued the dilemma of reason vs. power but failed to rise conclusively above the residual dualism of facts and values. Their passion for detached ethnographic redescription and for 'following the natives' incurred a systematic normative deficit that diminished their critical flair, and undercut their capacity to tackle successfully the issue of the proximity or distance between observers and observed. It might then be interesting, at this point, to look more carefully into the various ways in which SSK and STS have sought to 'translate' the conventional demands of scientific detachment and neutrality in terms of their signature principle of *symmetry*.

In its initial incarnation in the Edinburgh Strong Programme, the principle of symmetry required social students of science to adopt an even-handed, agnostic approach to winners and losers in the scientific game, or to what scientists themselves generally considered as established truth and what they rejected as scientifically obsolete or erroneous. The sociology of knowledge should be 'impartial with respect to truth and falsity, rationality and irrationality, success or failure', which required it to be 'symmetrical in its style of explanation. The same types of cause would explain, say, true and false beliefs' (Bloor 1991 [1976]: 7; 1983: 2, 5; Barnes 1976; 1982: 4ff., 58–63). For Collins, these tenets defined the main thrust of the Strong Programme, focusing his own radical programme of empirical relativism in SSK (Collins 1981; 1983;

The Politics of Symmetry

1991a). Others, such as Lynch and Roth, have likewise encircled symmetry and impartiality as constituting the methodological heart of radical science studies (Lynch 1992a; 1993: 75–77; Roth 1994). Operating in close alliance, they mesh into a type of 'value-free relativism' not too dissimilar from that of mainstream twentieth-century ethnography as it has evolved from Boas and Malinowski to Mead and Lévi-Strauss (Lemaire 1976; Stocking 1982 [1968]; Fabian 1983; Clifford 1988).

My purpose in this chapter is to take issue both with the residual neutralism that is implied in this levelling of the epistemological field and with the residual 'methodism' that has made the symmetry principle into a freely extendable a priori of sociological research far beyond the expectations (or indeed the approbation) of its initial supporters. My advocacy of a more straightforward *politics* of knowledge, however, is not meant to converge with a Feyerabendian 'anything goes', since it retains an interest in the pragmatic reworking of the demarcation problem and in a reflexive rewriting of the problem of critique. It is not sufficient to stop using the boundaries between science and society and between truth and error as self-evident resources and turn them into topics of research; beyond this, we also need to construct new, perhaps weaker boundaries (or *a*symmetries) that define the specific contribution of scientific rationality to the world and, beyond the evident tasks of description and explanation, also authorize its capacity for critique.

Once again a time-space perspective may come to our aid. Indeed, a temporal or chronopolitical translation of the principle of symmetry – or rather, of the grounding intuition buried inside it – would preserve the ethos of professional 'ignorance', perplexity or estrangement that defines the core scientific attitude, while getting rid of the agnosticism, neutralism, and methodological voluntarism in which it is presently encased. The 'uncertainty of distinction between the true and the false' is indeed the gist of 'symmetry' in its initial formulation. If we translate it as a refusal to know or to decide in the face of entrenched dualisms and divisions, we are again close to the notion of intellectual delay or unhastening that is so central to my present argument. It does not imply the reiterability of method or the neutrality of impartial spectatorship, but encourages a temporary suspension of belief that prepares for *another* decision and *another* division. As a politics of waiting, temporizing, or stalling in front of an imperative either/or, the principle of symmetry, pragmatically conceived, rationalizes the inability to take either one of two existing sides, which is a necessary apprenticeship on the way to articulate and side with a 'critical third'. In this fashion, the dialectical figure of the 'third space' or 'third way' also sets an important temporal condition.

In order to pursue this line of thinking, I shall first draw on a comparative discussion of three recent knowledge-political debates in the science studies field. The first became known in the late 1980s as the 'Epistemological Chicken' debate, and exposed the rifts and battle-lines between two 'generations' in the social studies of science and technology: the 'Bath' school of SSK as represented at the time by Collins and Yearley, and the 'Paris' school of STS headed by Latour and Callon (Pickering [ed.] 1992: 301–89). The second debate is a more vicarious or indirect one, consisting of various superimposed layers of 'translation' of the historical disagreement between Hobbes and Boyle (Shapin and Schaffer 1985; Latour 1993a). The third and final debate more immediately feeds into the broader framework of the present study, in thematizing re-emerging tensions between political neutrality and partisanship within the new constellation of a constructivist politics of knowledge. The issue of STS's normative-critical role and its capacity to 'take sides' in the face of symmetrical analysis (or its inevitable enrolment on the side of the weak) was explicitly taken up in what was referred to during the early and mid-1990s as the 'Capturing' debate (Scott, Richards, and Martin 1990; Richards and Ashmore 1996).

Playing Epistemological Chicken

For many if not most of its practitioners, the history of the field of science and technology studies can be plausibly narrated in terms of successive extensions and radicalizations of the symmetry principle (Woolgar 1988a: 23–24; 1992; Collins and Yearley 1992; Bijker 1993; Pinch 1993). While earlier paradigms in the sociology of knowledge, such as those of Mannheim or Merton, might already be construed as arguing for a symmetrical analysis of science and other social institutions, the initiators of the Wittgensteinian turn in science studies condemned such programmes as 'weak' precisely for their refusal analytically to draw in the contents of natural science, and their concurrent failure to strike a radical equivalence between the sociological explanation of true and false beliefs (Bloor 1973). Pinch and Bijker's SCOT model extended the symmetry principle from science to the study of technological systems and artefacts (Pinch and Bijker 1984). Reflexivists such as Woolgar and Ashmore pushed symmetry one stage further by systematically refusing to exempt the analyst's position from the analytic repertoire that they applied to the object of analysis (Woolgar 1988a, 1992; Ashmore 1989). Callon and Latour have more audaciously pleaded the symmetrical inclusion of non-human actors or 'quasi-objects' into a general anthropology of hybrid networks or 'cultures-natures' (Callon 1986; Latour 1988b; 1991; 1992; 1993a; 1993b; Law 1991). As Woolgar notes, actor-network theory's call for breaking down the human/non-human divide 'is one instance within a larger dynamic: the successive rooting out and dismantling of fearful symmetries [...] in a glorious bonfire of the dualities' (1997: 250).[1]

Perhaps the most telling illustration of the extent to which symmetrism is adopted as a self-evident principle is found in the broad dissemination of moralizing psycho-talk about a 'failure of nerve', 'lack of determination', 'chicken-heartedness', 'gerrymandering' or 'self-betrayal', which each new radicalizer of the symmetry principle hatches against his supposedly more timid and faint-hearted predecessors. Not only the Epistemological Chicken debate itself, but the entire two-decade history of SSK and STS bristles with such debunking attributions (cf. Bloor 1973: 175; 1983: 3; 1991 [1976]: 4–5; Woolgar and Pawluch 1985; Woolgar 1992: 328; Grint and Woolgar 1997: 95ff.).[2] As a consequence, each new forward push of the symmetry frontier capitalizes upon the instantaneous appeal of unmistakable virtues such as even-handedness, fair dealing, honesty, intellectual rigour, and of course, *audacity*. Radical frontiermen of symmetry typically pose as heroic democrats and levellers, who stand firm in the face of the many complacencies and seductions offered by a world riven with asymmetries.

Without doubt, it is this democratic accessibility and transferability of the symmetry method that defines its broad attraction. The beauty of method is to provide work. Its radicalism is methodical, and can therefore be extended indefinitely; if properly socialized, everybody can play the game. This turns symmetry and impartiality into tools-for-all-users, while anything that can be construed as a form of asymmetry – any kind of opposition, polarity, demarcation, dualism, or categorical boundary-line – is in principle vulnerable to symmetrical deconstruction. With a perspicacity typically granted to losers in the epistemological chase, Collins and Yearley saw that there was no definite limit to this proliferation, since existing symmetries could always be overtopped with new hyper-symmetries (1992: 379). In a symmetry democracy, the option is always open to cut across someone else's symmetrical balance, stake out a binary that is posited tangentially to it, and go on to escape from it 'orthogonally'. Effectively, this is what is implied in the more general postmodernist celebration of 'undecidability', 'in-betweenness', 'hybridity' and 'third spaces': down with all the binaries!

Indeed, the protagonists in the Epistemological Chicken debate quarrelled not so much about the philosophical anchorage or legitimacy of these principles, but typically about how far their jurisdiction could be *extended* or *generalized*. While supporting the Bloorian injunction to treat true and false theories symmetrically, Collins and Yearley for example warned against the indefinite proliferation of the principle, which appeared imminent in the work of both

133

Woolgar and that of Callon and Latour. While the latter pleaded the necessity of 'completing' Collins and Yearley's symmetry principle with their own, their adversaries complained about a 'misconceived extension of symmetry' that took humans 'out of their pivotal role', and dismissed this overextension as a form of radical chic (Collins and Yearley 1992: 351, 322).[3] A similar competition was staked out with respect to impartiality or agnosticism, escalating the anti-normativistic impulse that initially motivated Bloor's radicalization of Kuhnian naturalism. Both contending teams obviously shared an anti-authoritarian resolve to strip from science its extravagant authority and to level it down to the status of an ordinary cultural enterprise. But whereas Collins and Yearley took care to retain a residual normative commitment, Callon and Latour objected to their adversaries' 'moral and deontological' tone, in order to do away with all remnants of a denunciatory critique in favour of a more rigorously empirical and descriptivist strategy of 'following the actors'. They also turned against SSK's moral prohibition to crisscross the venerable divide between humans and things, which might be a respectable position in a humanistic fight against technocracy, but was 'uninteresting as an empirical tool' to describe the ongoing negotiations about this very differentiation that was daily conducted by scientists and engineers themselves (Callon and Latour 1992: 360).

In this fashion, the game of Epistemological Chicken was at once a 'symmetry race', or a tug-of-war between convinced symmetricians, and an 'impartiality contest', or a race between moderate and more radical anti-normativists. But surely, if there is something amiss with the impartiality and symmetry postulates in themselves, our troubles are only magnified if we attempt to extend or radicalize them. Both principles, I will indeed argue, tend to methodologize, and hence to present as generalizable procedure, what are in fact contingent *knowledge-political* attempts to reposition various discursive fields. In this respect, radical science studies suffer from a methodism that we have otherwise learned to associate with its traditional philosophical foes.[4] This methodism is a consistent feature running from Bloor and Barnes through Collins to Latour, who all tend to *prescribe* methodological rules on the basis of new (relativist and pragmatic) *descriptions* of scientific activity.[5] On the basis of 'close description', Collins and Yearley prescribe not only symmetry but also the need to 'stand on social things' or be social realists, in order to explain natural things (1992: 302, 308–309, 382). On the basis of an even more radical descriptivism, Callon prescribes generalized agnosticism, generalized symmetry, and 'free association' of natural and social factors (1986: 200). In the appendix to *Science in Action*, Latour stipulates no less than seven Rules of Method and six Principles (1987: 258–59). Whence this universal prescriptive urge? Whence all these rulings, if they do not channel ambitions to *rule over the field*?

The Politics of Symmetry

Against this background, I want to introduce four programmatic propositions. First, I suspect that the strong Bloorian conception of symmetry unwarrantedly singularizes a multiplicity of intellectual situations, operating as an abstract container for a great variety of epistemological positions. Manifestly, the symmetrical approach works out differently with respect to controversies or disciplinary fields in which one dominant approach strongly marginalizes others, as compared to situations in which paradigms or schools more nearly balance each other out and pluralism is to a larger extent inherent in the field.[6] This also defines a continuum of disciplinary regimes ranging from relatively 'hard' and 'hierarchical' sciences such as physics towards relatively 'softer' and 'democratic' human sciences (cf. Fuchs 1992; Dehue 1995). It also works out differently with regard to disputes in relatively esoteric scientific fields as compared to more controversial and politically acute topics. In political theory, for example, my own studies of right-wing and left-to-right 'crossover' intellectuals have ventured a more symmetrical interpretation of fascist ideology, meant to escape the crippling alternative of left-wing denunciation and right-wing apology. In this case, it is far more obvious that any 'revisionist' amendment of liberal and social-democratic Whiggism remains intensely political, and can never aim at neutral equilibrium (Pels 2000). Symmetry, in other words, is rarely symmetrical. It presents a methodologized and utopian version of what remains a rather exceptional case in a larger class of what I shall generically refer to as 'third positions'.

Secondly, there is no methodological constraint that dictates a 'natural' or pre-ordained extension of the symmetry principle towards a postmodernist 'bonfire of the dualities'. Each individual extension is a contingent knowledge-political move in the face of a situated theoretical antinomy. To methodologize is to decontextualize and depoliticize such local attacks upon established symbolic hierarchies. Third positions also *reconstruct* the field of controversy itself, so that, in a sense, both the first and the second positions and their local antagonism epistemologically depend upon the performative self-positioning of the third. The third position is an interactive one: its addition induces a lateral extension of the discursive field, and by introducing a different partiality, modifies all the other parties. The risk of the symmetry principle, then, is to present as dictates of method what are in fact contingent knowledge-political manoeuvres, or context-bound reshufflings of true and false or right and wrong. Asymmetries do not 'really exist' but are retrospectively defined from the vantage point of their prospective dissolution.

This means, thirdly, that analytic third positions are not external to the field of controversy but are included and implicated in it. They are not value-free or dispassionate but situated, partial, and knowledge-politically committed. In a

135

Unhastening Science

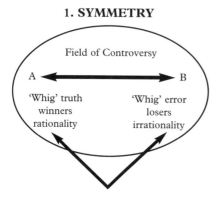
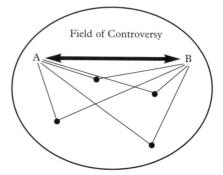

1. SYMMETRY

1. Equidistance
2. Impartial spectator
3. Disinterestedness

2. THIRD POSITIONS

1. Difference mixtures of involvement and detachment
2. No privileged point of view
3. Third positions reconfigure this field

Figure 6.1: Symmetry and Third Positions

field of unequally distributed symbolic power or symbolic capital, a symmetrical approach invariably subverts the dominant view and strengthens the side of the weak and the marginal. Symmetry is often a 'coolly' detached way of siding with the oppressed. In this fashion, it still conspires with the established authority of value-free science even while moving to attack it.[7] Fourthly, however, such commitments must not be taken to mean that the third position necessarily collapses into the first (dominant) or the second (marginal) positions in a particular controversy. It implies variable distances from both. It is partial and political, but it does not coincide with the contending parties. It is interested, without conniving with the quarrelling interests. Epistemological neutrality, then, is a misconceived methodological cloak for a very real knowledge-political phenomenon: the situated *distance* and interested *autonomy* of the third position (cf. Pels 2000: 214–22) (see Fig. 6.1). Beyond spatial metaphorics, this logic of critical 'thirding' also implies a sense of *timing* and a tactics of delay, which, by breeding an attitude of 'indifference' to the first and the second positions, opens up unfamiliar territory beyond the familiar difference that they make.

The Politics of Symmetry

Hobbes versus Boyle Revisited

Let me presently focus upon a strategic example, which may begin to illustrate some of the dilemmas set out above. It is taken from Shapin and Schaffer's (1985) widely acclaimed study of the historical controversy between Hobbes and Boyle, which has been hailed as paradigmatic by virtually every school in science studies, and has been appropriated by Latour as a prophetic anticipation of his own radical project of a semiotic anthropology of science (Latour 1993a).[8] Shapin and Schaffer introduce Boyle's experimental philosophy as still offering the great paradigm for the world in which present-day scientists live and operate. The success of the experimental programme is commonly treated as its own explanation. In the controversy with Hobbes, historians, like modern scientists, have usually sided with the winner, repeating Boyle's condescending judgment of Hobbes' natural-scientific findings, while maintaining a judicious silence about what Hobbes actually had to say. In this fashion, they have handled rejected and accepted knowledge asymmetrically. Shapin and Schaffer, by contrast, set out by adopting a stranger's perspective with respect to the experimental programme, and 'something close to a member's account' of Hobbes' anti-experimentalism. Although they argue that Hobbes' scientific reputation is seriously undervalued, their purpose is not to take Hobbes' side; it is first of all to break down the aura of self-evidence that surrounds the experimental way of producing scientific knowledge. Their stated purpose is not evaluative, but descriptive and explanatory (Shapin and Schaffer 1985: 12–13).

The balance of the chapters that follow, nevertheless, is to philosophically rehabilitate Hobbes over against Boyle. 'Matters of fact' are not mirrors of nature but social constructions, as Hobbes is supposed to have held against his adversary. Ignoring Hobbes' ubiquitous naturalism, and following a rather strained interpretation of Hobbesian contract theory advanced by Callon and Latour (1981), Shapin and Schaffer credit Hobbes with a form of actorial constructivism that they themselves mobilize against received realist legitimations of the experimental programme, both past and present:

> Neither our scientific knowledge, nor the constitution of our society, nor traditional statements about the connections between our society and our knowledge are taken for granted any longer. As we come to recognize the conventional and artificial status of our forms of knowledge, we put ourselves in a position to realize that it is ourselves and not reality that is responsible for what we know. Knowledge, as much as the state, is the product of human actions. Hobbes was right. (Schapin and Schaffer 1985: 344)

In this way, Hobbes and Boyle are recruited to fight a quintessentially contemporary fight, which yields something close to the opposite of the prevailing Whig history that routinely presented Boyle as the victor. The non-evaluative principle of symmetry results in a re-balancing, if not reversal, of values and positional strengths. Boyle has become less credible, Hobbes more so than before. But this also implies that the authors commit a presentism closely similar to that which they castigate in rival historians. On one level, they have only reversed the positions and rewritten Boyle from a Hobbesian standpoint. Their new Whig history is Hobbesian, not Boylean; it is constructivist, not realist. It is a *presentist* historiography written by *dissenters*.

But of course the intrigue is more complex. Shapin and Schaffer do not completely side with Hobbes against Boyle but actually waver, as Latour has correctly spotted, between Hobbesianism and Boyleanism (Latour 1993a: 24–27). This is because their description of Boyle's epistemological politics remains characteristically undecided about whether and to what extent Boyle himself must be rated as a constructivist *avant la lettre*. On the face of it, the authors reveal a classical discrepancy between thinking and doing, consciousness and being, or language and practice. Boyle *says* that experimental facts are an immediate and exact reflection of nature, but *in practice* he mobilizes an impressive arsenal of techniques, conventions, and social relations with and within which experimental facts are precisely *produced* as such. His interpreters show how much human agency is invested in these supposed 'mirrors of nature'. But they refuse to play up the discrepancy between (realist) actor's image and (constructivist) observer's image into a critical theory that unravels this reification of the facts, evading the basic issue of whether Boyle consciously deployed this constructivist politics of knowledge.[9]

But this is reading a duplicity into Boyle that tends to duplicate the presentist projection. *Both* Hobbes *and* Boyle are recruited to fight a late twentieth-century epistemological war, while the unfortunate Boyle is now also forced to fight this war against himself. Shapin and Schaffer are strangely absent from the battle scene, like sons of the wealthy bourgeoisie who have bought exemption from military service by paying for a *remplaçant*. The constructivist reading that makes this substitution work looks as strained in the case of Hobbes as it is in the case of his classical opponent. Hobbes' 'unmasking' of the conventional and constructed nature of laboratory-produced facts is a classically non-reflexive and denunciatory 'theory for the enemy', not brought to bear upon the architecture of the Leviathan itself, which mirrors human nature in much the same way as Boyle's experimental facts are thought to copy physical nature. The human 'Art' that brings the Leviathan to life is 'an imitation of Nature', in which all the rules and dictates of right reason are derived from incontrovertible and eternal natural necessities (Hobbes 1968 [1651]: 81ff.). Hobbes' rationalistic theory of political representation thereby follows the same reifying and foundationalist pattern as Boyle's empiricist philosophy of scientific representation. The Leviathan itself is the great social Black Box. It constitutes one enormous reification. In a single, never-to-be-repeated gesture, all capacity for action is withdrawn from the participants in the social contract and concentrated in the sovereign, an abstract and absolute power that henceforth hovers above human individuals as a quasi-natural object. Both knowledge and the State, although man-made, are ordained by Hobbes absolutely, in deduction from incontrovertible metaphysical premises. The Leviathan is made once and for all, never to be unmade. In this respect, it is as little 'artificial' as are Boyle's matters of fact. Shapin and Schaffer, however, appear to 'translate' Hobbes into an early ethnomethodologist, according to whom state and society are continuously and recursively constituted by actors' performative definitions, negotiations, and interventions. They hardly care to reconcile Hobbes' alleged epistemological and sociological constructivism with his rationalistic foundationalism, either in matters of knowledge or in matters of politics.[10]

Symmetry Strikes Again

This is why it is revealing to look in some detail at Latour's 'retranslation' of Shapin and Schaffer's 'translation' of Hobbes and Boyle (1990; 1993a: 24–27). Symmetry strikes again, as soon as Latour presents them as complicitous and duplicitous Founding Fathers of the Modern Constitution, and rounds out Shapin and Schaffer's ultimately Hobbesian reading of Boyle with a Boylean reading of Hobbes. In the process, Latour craftily exposes the authors' asymmetrical deficit, in order to generalize towards Hobbes the duplicity that Shapin and Schaffer had already discovered in Boyle. Both are now set to work as epistemological opportunists who shuttle between realism and constructivism, combining 'translation' and 'betrayal' as they see fit. Boyle is presented as 'raising and solving' the very question of the social constructivists, viz. that facts are artificially created in the laboratory. Both Hobbes and Boyle 'extend to man the constructivism of God'. Hobbes is said to have 'invented' the artificial creation of the Leviathan, while Boyle 'invents' this other artificial creation, the laboratory-made matters of fact (1990a: 149, 155). Latour's staging of the dispute thus provides another occasion to follow the risky business of radicalized symmetry, because he also radicalizes the presentist fallacies that already burden Shapin and Schaffer's rendering of the episode.[11]

Shapin and Schaffer, Latour suggests, find Hobbes' macrosocial explanation of Boyle's science 'slightly more credible' than Boyle's rebuttal of Hobbes. Influenced by the restricted Bloorian version of symmetry, they adhere more steadfastly to the political repertoire than to the scientific one, which renders them less well equipped to deconstruct the macrosocial context than Nature 'out there'. But the social does not just construct the natural, the natural also constructs the social. If Hobbes speaks the hidden truth about Boyle, Boyle likewise reveals the true doings of Hobbes. While Boyle is creating a political discourse from which politics is to be excluded, Hobbes creates a scientific politics in which experimental science has no place. 'Power', 'politics', and 'society' (which Latour does not distinguish at this point) are impotent without the natural powers of science and technology, which Hobbes consistently overlooks and denies in his installation of the Leviathan (Latour 1993a: 27–31). In this fashion, Latour's ultra-symmetrical anthropology of science adds duplicity to duplicity and projection to projection. While Hobbes, in an earlier phase of actor-network theory, was already enlisted as a paradigmatic transactionalist who duplicitously revealed the performative nature of the social order (Callon and Latour 1981), Latour now symmetrically brings in Boyle, the equally duplicitous inventor of laboratory-made matters of fact, whose cementing power Hobbes in turn refuses to acknowledge. Society and the state are no longer 'merely' stabilized by the intervening and unifying force of the spokesperson. Another and greater force must be added to the balance: the force of things. If society is made only by and for humans, the Leviathan, an artificial creature of which we are at once the form and the matter, cannot stand up.

But this means that Latour's own 'associological' proposal to deconstruct the socio-political is read into Boyle and symmetrically served up against Hobbes. If Hobbes and Boyle act in concert as complicitous founders of modernity, they are *made to act* according to a typically Latourian scenario of translation/betrayal. Its reinforced presentism not only reveals a cavalier indifference towards the protagonists' stated intentions, but also misplaces some of their more interesting similarities and differences. If Hobbes and Boyle offer rival rationalist vs. empiricist versions of the same foundationalist project, they differ, crucially, insofar as Boyle allows for a greater intellectual and political tolerance than Hobbes, whose absolutism is as evident in his obsession with socio-political unity as in his urge for rational logical demonstration. Indeed, the sharp demarcation between politics and science that defines the Modern Constitution does not emerge from both sides, but only from the side of Boyle. Hobbes, of course, is a monist for whom Knowledge and Power must immediately coincide. Boyle and his fellow scientists in the Royal Society, by contrast, demarcate between the autonomous production of knowledge and 'big' politics, in order to create a realm of freedom (a 'calm space') within which scientific debates are no longer disturbed by external political interference. Boyle thus conducts politics in order to fend off science from big politics; Hobbes embraces politics in order both to politicize science and to scientize politics totally. We may rightfully wonder about the political 'correctness' of an account which thus ends up by treating symmetrically Boyle's liberal demarcation between science and politics and Hobbes' anti-liberal politics of homogeneity, even if we now find Boyle's philosophical reasons for doing so less than persuasive.

According to Latour, the first draft of the Modern Constitution is truly symmetrical because, while Boyle invents a political discourse in which politics should not count, Hobbes devises a scientific politics in which experimental science should not count. Hobbes does not care to speak of things, even though without them his State is nothing but a colossus on feet of clay (Latour 1993a: 27–28). Both assertions, however, are up for dispute. First, Hobbes seems perfectly aware of the presence of 'things', both within the framework of his general theory of power and representation and, more specifically, in his political theory of *property*. This theory stipulates that the institutions of property are ultimately constituted by Power, and that the administration of things is hence subordinate to political sovereignty. Even though science is but a 'small Power', it stands at the cradle of more important 'Arts of Publique use' such as the art of fortification, of the making of engines and other instruments of war; it is precisely for this reason that such things must be brought under the Leviathan's political jurisdiction (Hobbes 1968 [1651]: 150–51, 217–22, 295–302).

Moreover, the symmetry-induced thesis that the Leviathan is powerless without material artefacts remains closely dependent upon a fundamental (and highly contentious) a priori assumption of Latour's conception of science-in-society. Because society is ultimately conceived in performative and transactionalist terms, Latour is incapable of discovering social stabilities and constraints other than those issuing from material things (Latour 1991: 16–18; 1993a: 21–22, 31, 37, 143–44; Collins and Yearley 1992: 317). However, it is highly debatable whether social relationships are indeed so fluid as to require the cement or ballast of experimental facts, and whether this arsenal of facts indeed constitutes the most significant form of new power in modern society. The notion of 'ballasting' society by material things partly compensates for actor-network theory's structural failure to account for the thickness and weight of reified symbolic structures and social institutions (Pels 2002). It also fatally underestimates the power of political reification. Indeed, from the seventeenth century right through to the present day, the fiction of the Leviathan has offered one of the most powerful vehicles for the mobilization of (human) masses, precisely by forcibly unifying them in the body of a single dictatorial spokesman.[12]

'Symmetry problems' such as these become more manageable as soon as it is recognized that Latour consistently conflates the polarity between science and politics with that between nature and society (e.g. 1993a: 13–15, 25–27). Boyle and Hobbes are antagonistic inventors of 'the very dichotomy between *human political* representation and *nonhuman scientific* representation' (Latour 1991: 13, 15, 17).[13] However, this superimposition of dualisms is far from self-evident, and cannot very well be generalized beyond the specific historical constellation in which Hobbes and Boyle are set up as antagonistic allies. In the early modern intellectual setting, experimental science was still largely considered coterminous with *natural* science, and confronted a social philosophy that hardly undertook to distinguish between the *political* and the *social* problems of order. Hobbes' political philosophy predated subsequent developments in the scientific representation of *people* in the course of which the autonomy of the social was increasingly demarcated in critical *distanciation* from the political, focusing upon a *sui generis* level of analysis where social facts could be studied in their own right. The new sciences of the social accomplished this by postulating social structures and developmental laws that were closely modelled after the methodical example of the experimental sciences of nature. In this sense, Hobbes' political philosophy already offered a 'natural science of society', a comprehensive science of movement that was also capable of revealing the natural mechanics of socio-political life.

In their methodological self-conceptions, Shapin, Schaffer, Callon, and Latour purport merely to 'reveal' and 'display', to readjust the balance, to even out the controversy, and to break down the self-evidence of received majoritarian ideas. But their description is loaded at every turn with a knowledge-political orientation to present-day epistemological rivalries. It answers to a presentist and critical agenda which is at the same time studiously negated. The prescriptive principle that demands an even-handed treatment of correct and false science (or more radically: of the co-production of nature and society) performs a substantial transvaluation of values and a reversal of symbolic hierarchies. But the contingency and contextuality of this value-judgment and this politics of knowledge tend to disappear behind the putative rule-character of the symmetry principle itself. It is not Hobbes and Boyle, then, who are opportunistic tinkerers, divided against themselves; it is their twentieth-century 'translators' who turn them into duplicitous actants in order to satisfy their own epistemological duplicity (see Fig. 6.2).[14]

The Politics of Symmetry

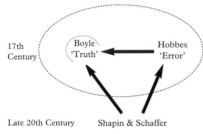

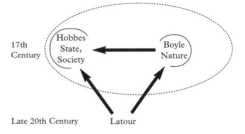

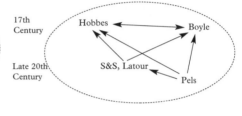

Figure 6.2: Varieties of Symmetry

Neutrality and Third Positions

This issue of the implied politics of symmetry has triggered an intriguing debate in two leading journals in the science studies field, starting off in the pages of *Science, Technology and Human Values* (Scott et al. 1990; Collins 1991b; Martin et al. 1991; Martin 1993) and continuing in a special issue of *Social Studies of Science* entitled *The Politics of SSK: Neutrality vs. Commitment* (Richards and Ashmore (eds) 1996). The intrigue of this debate was to confront directly the supposedly apolitical and quietist character of SSK/STS, and to refocus the older issue of taking sides (the need for neutrality or the inevitability of partisanship) in terms of the relativist epistemological resources more recently set loose by science studies research. Arguing on the basis of

143

personal experiences in three case studies of scientific controversy that involved matters of public concern (Scott on the importation of animal disease viruses, Martin on fluoridation, and Richards on vitamin C and cancer), the initiators of the debate concluded that the social analyst was so easily 'captured' by the controversy at hand that sustained symmetry was well-nigh impossible. If relativist symmetry, following the Bloor–Collins prescription, treated controversies as something external to the researcher, their own accumulated case evidence suggested that this separation readily broke down, and that social analysts could not avoid being implicated in the dispute. Secondly, they argued that symmetrical analysis almost always turned out to be more useful to the side with less scientific credibility or cognitive authority. The hostility towards the orthodox was matched by equally problematic embraces of the marginal. Epistemological symmetry thus often implied social asymmetry or non-neutrality. Thirdly, they suggested that the analyst's presence might subtly alter the configuration of the disputants itself and significantly change the course of the controversy. In sum, the principle of relativist symmetry, insofar as it required observers to be neutral or non-political, was little else but an illusion. The claim of neutrality was a myth that only served to distract social analysts from more overt partisanships (Scott et al. 1990).

It was a sociological commonplace, Collins argued in reply, that neutrality was considered subversive by the dominant, and that, however neutral the analyst intended to be, the work would always be drawn into the debate. But to him this did not signify that neutrality was wrong in itself: it was precisely what gave 'teeth to the idea that sociology is a critical discipline' (Collins 1991b: 249–50). There was a continuing need to distinguish between the politics and the methodology of scientific work, or between neutrality as a methodological prescription and the asymmetrical consequences it might have. To leave no room for methodological neutrality, Collins contended, was 'to make a mistake about the logic of the social sciences while accepting the politics of the dominant ideology of science' (Collins 1991b: 249–50). Martin and his colleagues countercharged that this prescription looked remarkably like the traditional positivist picture of scientific method, and that the claim of methodological neutrality might be better understood as 'a convenient myth' that served 'rhetorical and political purposes in dissociating the researcher from the socially contentious products of the research'. While disclaiming the prescription of neutrality as a misplaced bid for professional and academic legitimacy, they also reversed Collins' position on neutrality as platform for critique by suggesting that to challenge the possibility of neutrality itself might be far more subversive to scientific orthodoxy. Collins' rigid segregation of method from politics legitimized SSK's reluctance to engage with policy issues, and ran the risk of 'becoming only an intellectual exercise for armchair philosophers' (Martin et al. 1991; Martin 1993).

Reviewing Richards' (1991) work on the vitamin C and cancer debate, Pinch reacted somewhat differently from Collins. While Richards' conclusion seemed prima facie to go against symmetry, the situation was in fact more complex. When dealing with controversies where one side was clearly marginalized, Pinch agreed that the symmetrical analyst inevitably had to do most work in making plausible the rejected view, since orthodoxy could be trusted to look after itself. Thus, in the very act of 'even-handedly' redressing the balance between orthodoxy and heterodoxy, one was engaging in textual politics: symmetrical studies were always political. Following Ashmore (1993), a clean cut was made between the principles of symmetry and neutrality: 'one can reject neutrality, but still be symmetrical' (Pinch 1993: 371). Ashmore's text itself, a masterful re-analysis of a famous case of 'flunked' science (Blondlot's discovery of N-rays and his subsequent unmasking by Wood), similarly advocated a more political reassessment. In cases of fraudulent or epistemologically disreputable knowledge, where the weight of consensus appeared to lie exclusively on one side, symmetry inevitably had to take the form of 'just retribution' or 'positive discrimination' of the marginal view: even-handed analysis simply did not go far enough in disturbing the cognitive consensus. Ashmore's strategy was self-consciously partial, aiming 'to attack the credibility of the major historically accredited agent of the social destruction of N-rays in a rhetorically conscious effort to alter the grounds of consensus', without going so far as to adopt Blondlot's view (since it was 'clearly much too late for N-rays') (1993: 70). Symmetry as methodological *input* thus ceded before a more political interpretation aiming for symmetry as desired *output* of the analysis, while its achievement remained unaltered as an ultimate goal (cf. Ashmore 1996: 310; Richards 1996: 346–47).[15]

The original implication of Scott et al., on the other hand, had been that the double prescription of symmetry and neutrality constituted a double illusion and that they should be discarded as a pair.[16] Methodological *values* such as neutrality or value-freedom were not *in fact* considered feasible or realistic since, whatever their intentions, analysts could not avoid being drawn into the fray of the controversies they were studying. In the classical gesture of sociological realism, *facts* were thus adduced in order to refute *values*. Next, this empirical reality was taken as basis for the derivation of new rules of scientific conduct, suggesting that the analyst 'should be critically involved, in the role of citizen'. If all analyses were *de facto* interventions, the question was not whether to intervene but 'what sort of intervention, what audience, who benefits, and who loses'. That was why 'the choice of whom to serve' defined one of the key issues of science (Scott et al. 1990: 491; Martin 1993: 255–56).

In this manner, the Martin–Collins debate on the politics of symmetry resurrected some old intellectual phantoms. On either side, the issue tended to be cast in terms of involvement *or* detachment, partisanship *or* neutrality, determining a logic of either/or that new conceptions about the nature of science-as-politics have increasingly rendered obsolete. A Weberian dualist claim about the principled heterogeneity of methodology and politics as residing in the 'intrinsic rationale' of the social sciences still encountered its classical opposite: a Marxian case for 'external' political commitment. Politics, for Martin, was 'big' politics, 'in the familiar public sense', where large-scale political and economic interests were involved. In this framework, 'small' politics or the micropolitics of knowledge were only discernible as the politics of individual promotion of career academics who were more preoccupied by personal matters than by public interests. Hence Martin did not hesitate to debunk all 'internal' critique of fact-construction and all reflexive analysis of scientific discourse as academically antiseptic, disparaging the study of what was merely 'intellectually interesting' over against what 'the people involved' considered important, urging scientists to do action research or even to join a social action group (Martin 1993: 251–56).[17]

Weak Asymmetries

It is high time to lay such old phantoms to rest, and to escape from the stifling alternative of 'internal' detachment versus 'external' involvement. As I have argued in the preceding chapter, this requires the initially paradoxical coupling of two apparently antithetical claims. If, on the one hand, symmetry and impartiality are considered illusory, and science is viewed as a performative intervention in the world, it remains important not to be lost in a seamless web but to differentiate between different *forms* of partiality and different *fields* of politics, and acknowledge the classical intuition about *critical distance* that the protocols of symmetry and impartiality inadequately encode. Is there a way of describing symmetry as intrinsically contextual and political, while simultaneously taking seriously the symmetrists' own professions of relative impartiality and distance? I think there is, although this will require another hard look at the internal/external divide, the relationship between knowledge politics and 'big' politics, and a more substantive discussion of what I have called the 'third epistemological position' (cf. also Richards 1996: 340, 342).

In their attempts to overcome dualisms such as those between cognitive vs. social, internal vs. external or science vs. politics, constructivists tend to confuse two levels of analysis. Recognition that science is social, political, and interest-

The Politics of Symmetry

committed through and through does not necessarily dictate a radical agnosticism concerning the boundaries between science and politics as functionally differentiated subsystems in the social division of labour. As we have seen, the claim that science is social or political acquires its full significance precisely when it is *specified*: the scientific field is a field similar to others, and similarly subject to laws of capital formation and competition, but it is simultaneously a 'world apart' that obeys its own specific logic of functioning (Bourdieu 1981; 1990b). The *epistemological* separation between knowledge and power does not stand or fall with the *social-institutional* separation between science and society. While one may fruitfully refuse to discriminate cognitive from social structures in the first sense, one is still capable of discriminating them in the second sense. I have suggested before that the classical issue of demarcation can best be reinscribed by projecting a knowledge-political continuum that levels down the Great Divide, but still separates the 'logic' of science from that of politics by spreading between them a gradient of weak institutional differentiations and autonomies. I have added that this weaker demarcation can be further substantiated by drawing attention to incremental differences in cultural speed, which single out the timescape of science as one that systematically brakes the pace of action and communication that energizes faster practices in the social triangle.

With this in mind, we may return to the case of Hobbes vs. Boyle. If we agree with Shapin, Schaffer, and Latour that all solutions to the problem of knowledge have a political status, and that there cannot be a strict boundary between natural philosophy and political debate, we must nevertheless combat their tendency (most explicit in Latour) to collapse the domains of science and politics into one another and regress towards a Hobbesian monism of knowledge and power that tends to generalize the speed of the political towards other institutions.[18] There are both epistemological and political risks involved in setting aside the actual *a*symmetry that obtains between Hobbes and Boyle on this point, and in disregarding some crucial differences between their political hopes and ambitions. Indeed, the Boylean separation between science and politics and the Hobbesian politicization of science are precisely the two alternatives between which we must stubbornly *refuse to choose*. If traditional Boylean criteria of demarcation such as universal truth, reflection of natural reality, and value-freedom are no longer philosophically credible, it is nevertheless crucial to sustain the autonomy of the republic of science over against the realm of professional ('big') politics. This neither/nor position differs from symmetry in two major respects. First, although distanced from both, it remains ultimately more partial to Boyle's liberal society of differences than to Hobbes' absolutist utopia of closure and homogeneity. Secondly, it is not

147

agnostic or neutral, but operates from the overtly partisan standpoint of a 'critical presentism'. Which society would you rather live in: one with free spaces and liberal differences, even though facts go undeconstructed and institutional boundaries are strictly maintained, or one in which knowledge and power coincide absolutely, and in which the Leviathan subsumes all differences and liberties under itself?

A similar asymmetrical distance must be taken where epistemological differences are concerned. Once again, my third position sticks closer to Boyle than to Hobbes, even though it doubly distances itself from their parallel quest for objectivistic certainty, and critically opposes a constructivist epistemology to both. The liberal corollary of Boyle's methodological separation between matters of fact and theories, however, is that there exists only a partial compulsion to assent, and therefore room for opinions to differ, whereas Hobbes' philosophical authoritarianism does not settle for less than absolute consensus. If one may legitimately object to Hobbes' and Boyle's shared foundationalism, it should also be admitted that their styles of persuasion differ in kind as well as degree, and that Boyle's fact-finding naturalism offers a larger measure of epistemological tolerance than Hobbes' imperative method of definition and deduction. Once again, our commitment cannot be symmetrical and cannot be agnostic. It does not just follow the actors but also criticizes them. In so doing, it simultaneously criticizes the duplicity of those symmetricians who disavow their own politics of knowledge, while simultaneously acting it out by proxy, sitting on the backs of defenceless historical actants.

Analysts, if I may anticipate the central theme of the following chapter, are not 'apart' from the controversy they study but are reflexively implicated in it. But this does not imply a passive 'being captured' by external forces that remain untouched. Their arrival on the scene alters the discursive configuration and subtly modifies the object of investigation itself. The effect, intended or unintended, is to overturn a specific hierarchy of plausibilities. It is an active partiality. 'Politicals' such as Martin et al. are therefore correct in suspecting that 'neutralists' such as Collins are bound to enter into partisanships, insofar as they lend much more credence to marginal knowledge claims than is usually the case. They also have a case against more radical neutralists such as Shapin, Schaffer, Callon, and Latour, who systematically defer their knowledge-political interests and commitments to the parties in the dispute. However, our 'politicals' appear incorrect on two important counts. First, they tend to subsume all positions under one of the existing polarities in the controversy itself and overlook the feasibility or legitimacy of a third constitutive interest. In doing so, they are in fact seconded by Shapin, Schaffer, and Latour, who similarly obscure their analytic third position by symmetrically redistributing their critical ambitions over *both* opposites at once. Secondly, our 'politicals' fail to differentiate adequately between internal and external politics, tending to collapse partisanships in scientific conflicts with those in larger political controversies. Here they find themselves once again in the company of Latour, spokesman for the hybrids and politician of things, who tends to erase the boundary from the vantage point of a science that is closely modelled on the political.

Such an impasse appears to call for a clearer distinction between *distanciation* or *autonomy* on the one hand, which are necessary features of all third positions, and *neutrality* or symmetrical *equidistance* on the other. Detachment as a social-institutional determinant of intellectual and scientific practice no longer requires the protective girders of a philosophy of value-freedom or a reiterable methodology. What should be retained, however, is the unmasking of epistemological neutrality as a time-honoured vehicle for the promotion of professional interests, including those of a successfully established academic speciality such as science studies have meanwhile become. This is what Martin refers to as the 'taming' of science studies by its academic context. But whereas he views the professionalization of science solely in terms of its subordination to the societal *status quo* and regrets its loss of a critical calling, professional autonomy and the institutional distance it measures out remain crucially important preconditions for any kind of serious critical work. This applies both to the third position of the social student of rivalries and controversies within science itself, and to the larger autonomy that must be secured for science within the wider society. This autonomy of the observer (and of the intellectual profession at large) is not disinterested but constitutes a knowledge-political interest tied to a specific tactic of unhastening.

This political perspective on intellectual autonomy and academic freedom does not presage an end to all normative epistemology. It disestablishes not only the neutralist self-legitimations of traditional SSK and STS but also their relativism, if we take the latter to mean the agnostic injunction to provide causal explanations of true and false beliefs 'regardless of how the investigator evaluates them' (Bloor 1991 [1976]: 3–5). While symmetrical relativism promotes a sustained indifference to the distinction between truth and error, my own proposal is consistent with a more relaxed epistemology that *diminishes* or *weakens* the normative priority of the true over the false, without eroding this hierarchy or erasing this boundary altogether. It endorses a weaker conception of truth as scientific skill or competence, which, as theorists such as Foucault and Bourdieu have suggested, is no longer divorced from the quest for symbolic power or reputational interest. In such an anti-foundationalist and anti-realist epistemology, there is to be no equal treatment for truth and error, but rather a contextually specific redistribution of the credible and the less credible across a much weaker epistemological boundary. The levelling down of traditional concepts of truth (which emphatically divorce truth from interest) does not issue in a complete levelling out of the epistemological terrain. This is not symmetry, but a weaker *a*symmetry.[19]

The need to view the boundary between true and false as negotiated and constructed is not identical with abolishing it altogether. Seeing it as the outcome of 'boundary work' denaturalizes and destabilizes it, but does not rob it of all cultural footing. The boundary is redrawn across a much vaguer no-man's-land. As I have argued before, this notion of weak boundaries also constitutes our model for the intricate relationship between science and politics, or *Geist* and *Macht*. While the epistemological boundaries are lifted, the institutional boundaries remain in place, precisely because the intermediate zone is expanded and differentiated, levelling the Great Wall in order to make room for many lesser fences.[20] I think we should adopt a similar attitude with regard to the radical *démarche* advocated by Callon and Latour concerning the dualism of society versus nature. In this dimension, we may likewise settle with a weaker *a*symmetry, or a weaker notion about the permeable boundary running between humans and non-humans.[21]

The Uses of Inscription

In this final section, my aim is to return to the idea of the 'time differential' of science and the 'chronopolitical' interpretation of symmetry by paying closer attention to Latour's views about the political nature of science. While pleading the erasure of any firm demarcation between science and politics, Latour's work simultaneously includes a number of implied references to a politics of time-space and to the specific spatio-temporal difference that science makes. Some of his formulations about the 'seamless web' that is woven between science and society suggest a broad affinity with what I have called the knowledge-political continuum (cf. 1993a: 107–108; 1999a: 157–56, 305); while some of his pragmatic distinctions between science and politics (see below) have actually inspired the phenomenological description of their weak differences that was ventured in Chapter 2. Nevertheless, the main body of his work takes an angle on scientific practice that emphasizes the ubiquity of political trafficking and calculation, the 'ordinariness' of building networks of friends and allies, the sovereignty of the scientific spokesperson, and the universal imperative to stabilize scientific claims by means of reificatory inscription. In doing so, it places far less analytic weight upon the *length of time* invested in the labour of reification than upon the *solidity* or *certainty* of the facts that are its intended outcome. His political and military metaphorics consistently push towards a focus upon acceleration, mobilization, and hardness, leading away from a more 'scientific' appreciation of the sheer *sluggishness* of the work of representation, let alone of the *weakness* that envelops the claims resulting from it.[22] However, some of Latour's most recent writings on political ecology and the relationship between technology and morality are far more sanguine, not merely about the time factor itself, but also about the need to *decelerate* analytic decisions and to cultivate uncertainty, perplexity, and weakness. In my view, this new emphasis represents a welcome reversal that goes intriguingly against the 'managerial' or 'Machiavellian' grain of earlier constructions of actor-network politics.

Summarizing the results of preceding microstudies of laboratory practice in a landmark article from 1983, Latour reaffirms the symmetrical and levelling view that 'nothing extraordinary' is happening inside the sacred walls: science is not singled out by the presence of special minds, special methods, special norms, or special social relations (Latour 1983: 141, 160–61). This puts the social firmly back into science and science back into society. As the primary site where social forces are modified and changes in scale are performed, laboratories destabilize the very difference between inside and outside; they are precisely the places where inside/outside relations undergo a reversal. There is no outside of science; but there are long narrow networks that make possible the circulation of scientific facts. The micro-politics of the laboratory generate sources of fresh politics, which are effective precisely because they are not officially defined as political powers but are seen as non-political, technically neutral, value-free (Latour 1983: 155, 165, 167).

One might be led to think, on this evidence, that Latour favours an ontology of confusion that renders the flow between science and politics in principle seamless: 'science is politics pursued by other means'. But he also stipulates that laboratory science simultaneously *is and is not* a political process: as a technology of power, it generates fresh forms of power that arise from the mobilization of allies recruited not from the old resource of the human mass, but from new resources of non-human actors such as microbes and technological artefacts: 'What counts in laboratory sciences are the other means, the fresh, unpredictable sources of displacements that are all the more powerful because they are ambiguous and unpredictable' (Latour 1983: 168). What, then, are these 'other means' that define the 'ordinary specialness' of scientific work? What are the (inevitably weak) distinguishing characteristics between science-as-politics and big or 'politicking' politics? In Latour's view, scientists and politicians are not to be contrasted on cognitive, social or psychological grounds, but simply because the scientist has a laboratory and the politician

Unhastening Science

must do without one. The specificity of science resides in the special power of laboratories to reverse the scale of phenomena so as to make things readable, and thereby to accelerate the frequency of trials and mistakes. Scientists can afford to multiply mistakes, in order to record and compare them, sum them up and learn from them. The politician, by contrast, 'works on a full scale, with only one shot at a time, and is constantly in the limelight. He gets by, and wins and loses, "out there". The scientist works on scale models, multiplying the mistakes inside his laboratory, hidden from public scrutiny' (Latour 1983: 165).

Despite Latour's ambiguities about the 'seamless web' that is woven between science and society, these phenomenological descriptions still structure a minimal agenda for differentiation, which can be elucidated by once again tracking the two core variables of time delay and spatial withdrawal. Obviously, the laboratory, even while blurring the boundaries between science and society, remains a 'time and place apart' that is incapable of functioning if it is not in some sense removed from the urgency and publicity of everyday life, politics and business. The proliferation and systematic registering of mistakes is only feasible and affordable in a situation of diminished public surveillance, strict selectivity of issues, and a dramatic deceleration of the tempo of trial and experimentation. All of these variables define a relative contrast with the political cycle of communication, where professionals cannot escape (even actively embrace) the spotlight, and are expected to address a welter of current issues immediately with one-liners without 'trial and error' time, being swept along in the warp of a fast culture that cannot stop to reconsider or even to own up to its mistakes. Against this background, we therefore need to disagree with Latour that nothing special is happening in terms of the cognitive quality, normative structure, or social relations of science as compared with those obtaining in other social fields. Special conditions of temporality and spatiality apply, which do not issue from the essentialist specifications of rationalist epistemology, but simply emerge from the 'normative facts' of unhastening and estrangement. If science is indeed politics continued by different means, the invention of the calm space and easy pace of the laboratory precisely grants the 'different means' by which science is singled out as special (but not too special). Laboratories can clearly only exist in relatively estranged social settings and specially delayed time zones; only then are they able to generate the fresh cognitive and technical forces that turn them into such powerful levers to raise the world.

Within the context of my argument, there is a special significance in the unique capacity of the laboratory to 'make thing readable'. As Latour and Woolgar already argued in their classic 1979 study, the specificity and strength of laboratories (and scientific practices more generally) are precisely located in

The Politics of Symmetry

their material technologies of 'fact-writing'. Laboratory activity crucially implies 'the organization of persuasion through literary inscription' (Latour and Woolgar 1986 [1979]: 88); scientific facts are 'the product of average, ordinary people and settings, linked to one another by no special norms or communication forms, but who work with inscription devices' (Latour 1983: 162). Inscription follows a path in which weak subjective claims are gradually divested of their temporal and spatial modalities and divorced from their spokespersons in order finally to emerge as 'hard' and 'naked' facts, which are taken to be objective and transparent reflections of 'pure' nature 'out there'. Inscription devices are technologies of objectification, which solidify ephemeral ideas and images into material forms such as charts, graphs, maps, pictures, records, or scientific texts. As we have noticed, Latour immediately connects such materialities of writing with an epistemology of reification, which explains the strength of scientific statements in terms of a process of splitting and inversion that changes their modality from a state of uncertainty to one of self-evident natural fact. But what if one chooses to see this political work of identification/ certification as a liability rather than an asset of science? As argued before, the theory of translation is not initially presented as a *critical* theory that unmasks our own and others' reifications, but as a thick ethnographic *description* of 'what we all do'. Following what is basically an ethnomethodological impulse, it is anxious not to impose its own theoretical categories upon the sophisticated practical consciousness of the actors (in this case, the scientists) themselves (Latour 1992: 131; 1999c: 19–21). But if we want to read this account against the grain in order to rehabilitate the distrusted critical motive, we need to draw a sharper normative distinction between technology and epistemology, or between the *materialization* and the *fetishization* of scientific facts. In my own weaker interpretation of the strength of science, it is not the anticipated epistemological transformation towards 'identitarian' translation and singular spokespersonship, but the naked technology of writing itself that goes far towards defining the specific nature of scientific practice.[23] 'Paperwork' and other materialities of inscription have the inherent purpose and effect of unhastening the process of research and communication. It is this deceleration of discourse and the clearance of a protective zone for its unfolding that enable scientists to learn from their mistakes.

In Chapter 2, I alluded to traditions in the historical anthropology of orality and literacy that have consolidated the view that the invention of writing introduces a creative delay in everyday communicative action (Goody and Watt 1968; Goody 1977; 1987; Ong 1982; Finnegan 1988). Graphic forms acquire permanence by being sedimented in material things; hence they are preservable through time and space, offering the possibility of recall across different contexts (cf. Latour's 'immutable mobiles'). It is crucial to insist upon the *materiality* of the writing process, which is performed in real-time bodily interactions with the reed and the clay tablet, the chisel and the marble slab, the goosequill and the parchment, the keyboard and the computer screen. Because it allows a time interval and effects a divorce between author and audience, this materialization is a crucial evolutionary prerequisite for the 'stretching' or 'disembedding' of social relations, and hence for the increased complexity that permits society to differentiate into functionally specific domains and life spheres. Literacy also enables the accumulation of information over time, which facilitates access to large amounts of written records, expanding the knowledge and memory of individuals, generations and entire cultures. In this way, by affording the conduct of 'slow conversations' across time and space and providing a storehouse of documentation and evidence, literacy opens at least the possibility of a more critical and 'objective' view of the world, ideally facilitating a 'cumulative tradition of critical discussion' (Goody 1977: 47).

As I have also indicated, one should be careful not to over-associate the greater abstraction and detachment induced by reading and writing with any strong epistemology of objectivity or rationality (cf. Street 1984: 44ff.; Finnegan 1988: 149–52). The special nature of inscription as a material form is to afford a minimal objectivity in the sense of placing an object (a text, a permanent record) in front of an interactive reader who can visually inspect it and reflexively manipulate it (reread, annotate, or interpret it; scramble or sample it; tear it up or throw it away). This applies equally to the writer herself, who can only 'think' by means of such material delay (handwriting for example proceeds at about one-tenth of the speed of oral speech), to the addressee (e.g. of a letter) and to authors interacting with a larger reading public. In writing for a larger 'market', an author does not communicate with co-present and individualized others but with a faceless and invisible 'third', the 'generalized other' of an anonymous crowd, who may likewise experience the text as an object that visually endures and hence is available for many interpretations. In this restricted but emphatic sense, and without conjuring up the heavy connotations of the Enlightenment view of truthful knowledge, objectification is capable of inducing a greater amount of reflexivity and rationalization. Nor should one overstate the spatial discipline of the art of reading as 'an essentially withdrawn and private process', since it is more properly conceived as the consequence of new conventions of silent reading (rather than reading aloud to oneself and others) that only developed long after the invention of printing. But in cultures with a long tradition of literacy such as our own, the silent interaction with pen and paper or keyboard and screen has introduced forms of

The Politics of Symmetry

slow motion and relative stillness that distance the note-taker, however minimally and relatively, from others who are talking (shh! I'm studying/writing!).

By arresting the flow of oral conversation, and allowing a closer scrutiny of isolated segments of text, writing also enables both the proliferation and the timely correction of mistakes, and thus facilitates reflexive experimentation with different versions of the world. What Goody has called 'backward scanning' favours the elimination of inconsistencies and a more selectively focused and discriminatory choice of words (1977: 128, 49–50), whereas oral culture fosters redundancy and repetition and tends to gloss over mistakes (there is no way to erase a spoken word). The obtrusiveness of the writing medium facilitates a more careful and precise moulding of messages, while the process of producing the message typically remains concealed. Readers remain normally unaware of corrections made to the finished text, which may exude a tone of confidence and formality lacking from the first fumbling drafts (Meyrowitz 1985: 100; Ong 1982: 40, 104). In this admittedly broad and minimal sense, the gap between orality and literacy to some extent maps on to the earlier distinction between politics and science, if the latter is taken as the unhastened capacity to experiment with reality, to multiply and rectify mistakes vs. the need to gloss over them in 'fast-forward' mode.

In its 'smoothing over' faculty, the technology of writing also in a way resembles what Latour and Woolgar have described as the deliberate solidification of scientific claims into hard natural facts; even though the formalization afforded by the medium's inherent slowness must not be identified with any definite epistemology of factuality or reification. There may be an ideological or reificatory effect integral to the sheer materiality of the process of recording or encoding; but it is a weak form of reification that does not necessarily evoke the traditional ideology-critical wariness about the illegitimate naturalization of claims into objective facts. Although facts are of course solidified through inscription, the mode or style in which they are written down (e.g. as mirrors of nature, or as reflexive instances of circular reasoning) admits of a critical asymmetry that can function to discriminate between 'good' and 'bad' texts. Even though they are objectified speech, texts may continue to point reflexively to their contingent modalities of origin. As I shall explain more fully in the next chapter, reflexivity precisely means that objectified inscriptions remain chronically traceable to their spokespersons and 'rewindable' to their process of production.

In his most recent work, Latour has recovered this temporal dimension of science-in-the-making in a much more upfront manner, apparently relaxing some of his previous enthusiasm for 'letting the actors build their own space' by means of hard-nosed strategies of reification. Without immediate reference to the technology of inscription, he now pleads a critical deceleration of analytic practices, which should proceed 'at the pace of the tortoise' rather than rush for the quick fix: 'In the political philosophy of science, you must take your time in order not to waste it'. Political ecology should resist the urgency of action and its resultant oversimplification, in order to reveal the *uncertainty* of its founding categories, which express a deep crisis not of Nature but of objectivity (Latour 1999b: 11–13). Surprisingly, he now also hints at the *normative* implication of the profound contrast between such 'short-circuiting' and what he calls 'the slow pace of the work of representation', setting it in contrast with the obligation politicians are under to close off arguments, to take quick decisions, and thereby to divide the collective into enemies and friends (1999b: 186, 200–201) – a view squarely at odds with his former 'military' or 'Schmittian' conception of science. Notions about technical mediation as a 'detour' or 'deviation' that is triggered by perplexities in the face of natural and social risks are likewise reformulated in this temporal frame (1999a: 189, 191; 2002).

Morality, which never sat comfortably with the militant agnosticism of an approach that meticulously refused to censor the actors for using any means that could contribute to the end (of reification), is now refocused as a mechanism for *slowing down* the process of conjugating means to ends, for promoting *uncertainty* about their proper relationship, and for instilling concern about the *quality* of the means (Latour 1999b: 211–12): 'Wherever we want to go fast by establishing tracks so that a goal can race along them whistling like a high-speed train, morality dislocates the tracks and recalls to existence all the lost sidings. The goal-oriented train soon comes to a stop, burdened, powerless. As is often said, morality is less preoccupied with values, as is often repeated, than with preventing too ready an access to ends' (Latour 2002: 257). Disabling some of the basic reflexes of the actor-network approach, this 'unhastening' perspective highlights not so much the *force* of facts as the sheer *length* of a fact-making trajectory that ultimately produces no more than a texture of admittedly weak propositions. The style or modality in which facts are written up (objectivist, or reflexive) appears to make a critical difference after all. To further substantiate this view will be my main objective in the following chapter.

CHAPTER 7

Reflexivity: One Step Up

> And therefore in reasoning, a man must take heed of words; which besides the signification of what we imagine of their nature, have a signification also of the nature, disposition, and interest of the speaker.
>
> Thomas Hobbes, *Leviathan*

The Romance of Reflexivity

Ever since Socrates turned the logical thumbscrews on his opponents and fancifully called it 'dialogue', the injunction 'Know thyself' has exercised an enduring moral fascination on generation after generation of thinkers. Educated in such Socratic dialogues (and by Delphic oracles), many have also come to savour the insight, famously argued but also involuntarily exemplified by Auguste Comte, that the science that comes closest to our own existential concerns is the most intricate and difficult of all.[1] For Kant, who first championed the idea of reflexive philosophical judgment, the quest for self-knowledge was nevertheless equal to a descent into hell – an insight that found abundant corroboration in the heroic but also painful posturing of Nietzsche, who descended to psychological depths where no Kantian would ever dare enter. In Nietzsche's informed judgment, every great philosophy, including the Kantian system, was actually 'a confession on the part of its author and a kind of involuntary and unconscious memoir' (1990 [1886]: 37). Perhaps it was for this reason that Foucault chose to resist this form of truthfulness: 'Don't ask who I am' (2002: 19).

As few others, Foucault remained wary of the many pitfalls and seductions that escort all calls for reflexive self-disclosure. Reflexivity often parades itself in a show of confessional virtues, such as the courage of 'opening up', the candour of 'telling where you come from', the correctness of 'taking responsibility for your roots', and the consistency of 'not making an exception of yourself'. Reflexive sociology, whether in the autobiographical and personalistic style practised by Gouldner in the 1970s or in the more rigorously impersonal and objectifying version advocated by Bourdieu during the 1980s, typically flourishes such a rhetoric of moral rectitude and epistemological panache. Ethnomethodology has been censored for its 'skew to the right' in its failure to sustain a radical referential reflexivity (Pollner 1991). As the previous chapter has described, rival schools in the constructivist studies of science, each trying to be more self-consciously reflexive than the others, have routinely accused one another of committing the 'self-excepting fallacy' and of a 'failure of nerve' in facing the consequences of this radical methodical principle. The romance of reflexivity typically celebrates those knowledge subjects who are not afraid to look into the mirror of nature and discover in it their own image and likeness; who are not satisfied passively to reflect the world but dare actively to reflect back upon it. But, as I shall elucidate more fully below, such efforts at self-clarification may also degenerate into subtle blackmail (of which Plato's Socrates was such a past master), insofar as the self-exposure of the observer comes with a pontifical prerogative to expose the self-contradictions of the observed.

In its most elementary form, reflexivity presupposes that, while saying something about the 'real' world, one is simultaneously disclosing something about oneself. In refusing to separate knowledge of things 'out there' from knowledge of the self 'in here', the reflexive knower, while reading the Book of Nature, simultaneously writes a piece of her autobiography. According to Fabian, for example, anthropological writing is inherently autobiographic: 'The object's present is founded in the writer's past. In that sense, facticity itself, that cornerstone of scientific thought, is autobiographic.' That is why in anthropology objectivity can never be defined in opposition to subjectivity, 'especially if one does not want to abandon the notion of facts' (Fabian 1983: 89). Reflexivity therefore arises as a radical concern, and as a radical threat to the traditional canon of impersonal, value-free, and dispassionate inquiry, as soon as the epistemological rift between subject and object is mended and the subject is conceived to be in some substantial or passionate way 'present in', 'part of', or 'at stake in' the object itself (Gouldner 1971; 1973; 1976; Scholte 1974; Dwyer 1979; Clifford and Marcus [eds] 1986; Bourdieu 1990a; 1993a; 1993c; Bourdieu and Wacquant 1992). This constitutive inseparability of knower and known redefines knowledge as a matter of interdefinition or mutual genesis of what the world is and what you are; as an incessant circulation between representer and represented that can find a secure foundation neither in the external world nor in the knowing subject (Maturana and Varela 1984;

Luhmann 1982; Steier 1991a). A reflexive methodology, in other words, systematically takes stock of and inserts the positions and perspectives of *spokespersons* in social-scientific reports about the world. Reflexive texts tend to reiterate the question: Who says so? Who speaks, if natural facts and social groups are unable to speak for themselves?

In thus bending back upon themselves, in self-pointing exercises, reflexive texts are critically deployed against the perennial assumption inextricably knitted into mainstream Western philosophy: that knowledge may be called 'true' solely if it can be considered a more or less accurate reflection of a world that exists in itself, prior to and independent of the knower's experience of it (von Glasersfeld 1991: 13). Reflexive knowing is usually predicated upon a constructivist or performative view of conceptualization that emphasizes the mutually constitutive nature of accounts and reality; it is critical of an 'iconic' or 'mimetic' account geared towards the faithful observational recording of allegedly pre-existing facts. It therefore also supports a broadened, more robust notion of experience and experimentation that emphatically includes rather than methodically disqualifies the situated particularity of the experiencing and experimenting observer. In contrasting a constructivist with an objectivist epistemology, reflexivity also provides the basis for critical or therapeutic judgments (even, as I shall argue, for a form of ideology critique).[2] Notwithstanding the risk of reflexive posturing, I still presume that it is both feasible and important to talk about something and simultaneously talk (at least a little) about the talking itself; and that it is better for your epistemological health to be reflexive rather than non-reflexive.

In this chapter, I shall defend a view of reflexivity that I call 'one step up'.[3] This exercise aims at freeing some initial conceptual space between two unattractive epistemological alternatives: the 'flat' or rectilinear discourse of straighforward naturalism, which finds strength and certification in the object, in the world as it is, and the infinite spiralling of meta-discourse, which adds layer after layer of reflexivity in order to recover a final grounding in the subject. In double opposition to these divergent efforts at gaining epistemological certitude, 'one step up' reflexivity proposes to add only one level or dimension of self-reference, not more, in order to display the narrative's hermeneutic point of departure and point of return. It ties just one loop, adds one level of self-exemplification, in order to bend the rectilinear story into a curvilinear or elliptical one. In doing so, it attempts to hold both representer and represented fully into view, continually monitoring their similarity and distance, their connectedness and tensionful difference. It accordingly resists all tendencies towards identification or reification that allow the spokesperson to disappear into the object (materialism) or the object into the spokesperson (idealism).[4]

Unhastening Science

It is an obvious move, within the framework of this book, to deepen this picture of a spatially layered or 'topped up' discourse by means of a *temporal* interpretation of the principle of reflexivity. Standing back from and critically observing one's own position, which initially suggests a topological distance, simultaneously requires a specific 'downshifting' of the gears of ordinary reflection: a 'calming down' that disrupts the frantic pace of position-taking and decision-making in the everyday world. In this sense, reflexivity is only a more radical incarnation of the methodical doubt that has been identified, at least since Descartes, as the hallmark of science. The 'benefit of the doubt' requires not merely an autonomous space but also autonomous (free) time in order to be cashed in. The long history of philosophical scepticism might perhaps be reincarnated as a story about intellectual unhastening (cf. Popkin 1979). Relativism and reflexivity 'simply' take this deceleration a few gears down, taking their time to 'wait and see' in the face of lingering Cartesian certainties and other transcendental residues. In extending the scope and aggravating the posture of doubt, they further stall the pace of decision that drives more feverish cultures, both inside and outside of science.

Between Meta and Infra

Let me begin to explicate this reflexive 'one step up' (or 'one gear down') by critically drawing on Latour's discussion of meta- and infra-reflexivity (Latour 1988c), since my proposal also intends to steer clear of the *meta* and the *infra* as he chooses to understand them. It braces itself equally against the endless loops of hyper-reflexivity and against the 'flatness' of naturalistic storytelling – the single-storey story that is 'just written' without any need for reflexive precaution or therapy. My argument here is that, notwithstanding Latour's conviction that reflexivity is unavoidable in some form, his notion of infra-narration skirts dangerously close to 'flat' naturalism, and flirts with the postulate of representative identity that should be anathema to a truly reflexive epistemology. My addition of one meta-layer (of one storey to the story) usefully preserves the belief that reflexive writing offers *something more* than 'just another story'; that it is in some way nicer, truer, more interesting, and more exciting to have than stories that are epistemologically straight. A single loop or curve is required precisely in order to clear a platform for critical validation, for (why not?) 'ideology critique'. Let's take only one step up (and that's enough). No more, since we need to halt the rise of reflexive skyscrapers reaching into the clouds; no less, to prevent sinking into the positivistic flatland. Let's shift one gear down, in order to stall the pace of narratives that do not stop to think about their own logistics of representation.

Reflexivity: One Step Up

We may notice right away that Latour's juxtaposition of meta- and infra-reflexivity is characteristically rhetorical. Texts, in his conception, have the problem of being believed either too much or too little. Meta-reflexivity addresses the first quandary, and expresses an attempt to *avoid* a text being too much believed by its readers, while infra-reflexivity concerns the attempt to avoid a text *not* being believed (1988c: 166). Evidently, for Latour, the major problem for texts is being believed too little, not too much. His effort in ditching the *meta* and pleading for the *infra* is primarily directed at how to muster maximum credibility, how to make our texts stronger, not weaker. Meta-reflexivity, or the attempt to render texts 'unreadable' by adding self-referential loops, is judged both counter-productive and suicidal; ordinary readers are anyway much too devious and cunning to be seduced into thinking that texts actually refer to a stable and objective something 'out there'. Since readers are never naïvely duped by claims about referential objectivity, but routinely translate them to suit their own tastes and purposes, we had better follow their example and 'muster all possible allies at hand' in order to render our texts believable, interesting, and persuasive. Instead of methodologically self-referring, infra-reflexive texts should be *self-exemplifying* by virtue of their distinct rhetorical style. They do not claim superior critical status, but exemplify and display what everybody does anyway, namely extending networks, tying together translations, stabilizing associations – following a quasi-political logic in which 'settings strive to become centers by mobilizing everything at hand and tying their claims to as many resources as possible' (Latour 1988c: 161, 169).

There is a problem about levels of analysis here, and an ensuing circularity that I consider vicious. As we have noticed before, actors are partly defined as cunning opportunists or devious translators who mobilize all conceivable resources in order to extend and stabilize their networks of people and things. This view of actorial politics asks no privilege for itself, since we scientists are no different from them. There is no need to criticize anyone, since we are all engaged in the same trafficking and networking. Some of us, however (unlike most others) display it, are able to show how it works. This generalized ontology of networking is empirically dubious, however, insofar as it projects a constructivist enlightenment into the doings of ordinary actors and readers that effectively transforms them, if not into accredited actor-network theorists, then at least into effective actor-network practitioners. In my reading, however, Latour's constructivist ontology must be a *critical* ontology that can only be deployed from the 'estranged' and 'unhastened' meta-position of an outside observer. But this is precisely the position that Latour wishes at all costs to avoid, since it comes with the terrible suggestion that his observer's view might in some way *improve upon* that of ordinary actors, might offer a critical platform for assessing alternative knowledge claims. However, if one measures the dominant hold exercised by objectivist beliefs upon both science and common sense, there remains something to be said in favour of rigging our texts in such a way as to disrupt such beliefs. Infra-reflexivity is not enough; we need an 'outside' leverage point that makes available a place for critical distanciation.

In addition, Latour's ontological redescription of 'what we all do' is too little diversified, too abstractly singular, to be fully convincing. It tends to assume that all network-building is equal to the building of large political *empires*: an essentially quantitative process of 'total mobilization', which ultimately aims at empowering texts or translations in such a way as to transform them into black boxes. This impoverished one-dimensional logic of spokespersonship assumes something like a universal drive towards identification and reification, which assembles all allies under a single spokesperson and closes the lid on them. It brings Latour paradoxically close to the traditional ideal of transcendental objectivity, in which the spokesperson similarly strives to be 'everywhere and nowhere', to become an 'absent presence' by fully coinciding with what he speaks for (Pels 2000: 1ff.). His conceptualization eschews some crucial normative and epistemological distinctions between various forms or modalities of translation, especially between stronger and weaker ones (or reificatory and non-reificatory ones) and, in doing so, impairs our ability to evaluate them, viz. of choosing the side of the weak. Not all texts are equal in this respect; some are better than others.

Of course, we want our texts to be believable, appealing, and persuasive; but we do not want them to be so at all costs, and deploying all conceivable means; nor do we need to persuade *anyone out there*, with the purpose of making our networks as long and strong as we possibly can. For example, we may plausibly forgo or even actively resist the impulse to strengthen our alliances by objectivistically stabilizing our claims in the form of black boxes; we may seek to establish smaller networks and weaker alliances that do not rely on such reificatory power/knowledge play. If I write *weak* texts that attempt to 'point to their own margins', they stand against epistemologically closed *strong* texts that are presented as black-boxed mirrors of nature. Such texts are not written for just anyone, but for and against significant others. They are written against those who perhaps will never be persuaded anyway (or may I seduce one or two?), and for the benefit of those who are perhaps already convinced (some exceptions allowed here as well). They tend to preach to the converted, and against the unconvertible, in a showbattle for the floating vote. But they do not aim at global conquest.[5]

Interestingly, Latour's text actually includes a suggestive but unacknowledged critical standard that differentiates between stronger and weaker forms of association, and hence precisely focuses the 'one step up' that I argue is needed in order to make our texts reflexively believable. It is offered in a critical characterization of the conventional idea of causal explanation as 'holding many elements by one', as 'acting from a distance' or 'working at empire-building'. Explanations, according to Latour, typically resolve the problem of presence and absence by tying as many distant settings and resources as possible into explanatory centres of calculation. He suggests two ways of displaying the inner workings of such a politics of explanation. In the first display, power is reinforced because the distance is abolished and the represented elements disappear in their representatives. In the second display, power is weakened because the initial elements are maintained in full view. While the first view starts with *equivalences*, the second starts from *translations*, and also accounts for the *work* of rendering elements equivalent; it adds the work of reduction or reification to the reduction or reification itself (1988c: 162–63). Characteristically, Latour resists any presentation of these two displays in terms of a critical, polemical, or judgmental opposition. His aim is not to object to (to 'accuse', 'denounce', or 'scapegoat') reificatory identifications, but 'merely' to show the hard work that is required in order to render them successful. As I have repeatedly noted, actor-network theory is far from explicitly criticizing this reifying work, which is ontologically assumed to be something 'we are all engaged in anyway' (Latour 1988c: 156–57; 1993a: 43–46).

In my approach, reflexivity remains significant as a critical tool against all epistemological identitarianism, both of the naïve naturalistic variety and the much less naïve knowledge-political variety that is adumbrated by actor-network theory. Reflexive writing is precisely concerned to hold in view the critical difference between representers and represented over against attempts to collapse the one into the other and to black-box their relationship. The issue is not whether to explain or not explain, to network or not to network, to translate or not to translate; it is to demarcate critically between stronger and weaker translations and alliances. Such translations are weaker precisely because we eschew objectivistic modes of persuasion, and 'do not wish to fight with the weapons' of the conventional sciences (cf. Latour 1988c: 165, 174). Infra-reflexivity does not sufficiently capture such a critical differentiation between constructivist and naturalistic accounts; it describes ordinary practices as already deeply constructivist, and hence cannot see itself as in any way improving upon them. Hence it is satisfied to remain at epistemological ground level.[6]

Unhastening Science

Circular Reasoning

Reflexive statements, to repeat a rather vague and preliminary definition, are propositions or texts that in some way take into account their own manufacturing conditions. This formulation is clearly unhelpful, as soon as one realizes that there exist as many reflexivities as there are substantive analyses of how texts are made or how observers figure within their objects and their forms of inquiry. Woolgar classifies them along a broad continuum that ranges from 'benign introspection' towards 'radical constitutive reflexivity' (1988b: 21; cf. Lynch 2000). Roughly, his distinction is between reflexivities which are designed to improve observational accuracy and strengthen the objectivity of research, and more radical relativist views that are no longer concerned with representational adequacy or epistemological security in the realist mode. In another formulation, the continuum of reflexivity can be said to run from strong objectivist variants towards weaker constructivist or performative ones.[7] My own (experimental) sympathies lie – as do Woolgar's – with a radically anti-foundationalist and performative view located towards the far end of this spectrum (see also Ashmore 1989; Pollner 1991). I think it can be interestingly rephrased with the help of a sharpened notion of *narrative circularity*, or of the *constitutive circularity of accounts* (cf. Ashmore 1989: 32, 95–96; Herrnstein Smith 1997; Kim 1999). While objectivist reflexivities are insufficiently prepared to countenance their own circularity, constructivist or performative reflexivities consciously activate it and and actively celebrate it.[8]

If reflexivity is the curvilinear movement of a text that attempts to bend back upon itself, the metaphor of the circle or spiral is never far away. We are talking about a circular process 'in which reflexivity is the guiding relationship allowing for the circularity' (Steier 1991a: 2; 1991b: 163). Once again, this predominantly spatial image can be extended to include a temporal dimension. The admission of circularity in reasoning, indeed, has proven to be a particularly effective way of bringing down the speed of action and decision. According to most traditional epistemological views, it quite simply halts all credible persuasion, if not all intelligible conversation, since it is identified as the site and symbol of intractable and possibly suicidal paradoxes (such as that of the famous Cretan philosopher who proclaimed that all Cretans were liars). In the face of such epistemological puzzlement, I want to assert that all inquiry, both everyday and scientific, inevitably runs in a circle, even if it is not reflexively acknowledged as such; and that this circularity is not debilitating or self-detonating, as anti-relativists presume when deploying their logical puzzles, but is on the other hand liberating, precisely because it installs a constitutive *weakness* in the heart of all our accounts of the empirical world.[9] It is another way of disrupting the apprehension of texts as objective accounts, and of injecting some instability in their textual organization (Woolgar 1981: 260; 1988b: 30). Reflexive circularity is therefore not by definition vicious, but may become virtuous (in a literal normative sense) as soon as it acknowledges its own circular movement. Instead of delegitimizing all normative criteria of true (or better) knowledge, reflexive circularity thereby introduces and (self-)exemplifies a new weaker criterion of truth, which is offered as an alternative to the strong objectivist criterion of mirror-like representational adequacy. While traditional objectivism attempts to break (or square) the circle in search of an independent grounding in external reality or a transcendental subjectivity, the new performative conception welcomes it, and consequently views as less adequate, interesting, or aesthetically pleasing all views of knowledge that refuse it or attempt to escape from it.[10]

Of course, and necessarily, this proposition cannot be offered as a straight ontological generalization that describes a 'universal feature of accounting procedures' (as ethnomethodologists would say), but likewise stands as a circular claim. Its critical impact cannot be divorced from, and does not claim any protective foundation beyond, the second-order circle in which it turns (Pels 1995b: 1031–34; 2000: 26, 221). Since it is argued from within its own performative framework, and remains bounded by its own epistemological form of life, it claims no privilege or special status, and undercuts all suspicions of domination or persuasive compulsion which injunctions to (meta)reflexivity easily induce. In doing so, it fully accepts the contingency of being 'just another story', which is also an important motif in Latour's defence of infra-reflexivity. Once the circle begins to move, however, everything can be sorted into it in order to accelerate its spin: facts, trends, patterns, things, objects, machines, values, interests, politics, contexts, and institutions. Passing into the circle of representation, we can be as 'realist' as we want to be, and freely offer descriptions and explanations, dualisms and dichotomies, rules of method and logical proofs, criticisms, grounds, and foundations: we may populate the circle with as many 'actants' or 'delegates' as we like to mobilize.[11]

Circular reasoning hence offers a perfect example of persuasive 'take it or leave it'. Its closure paradoxically defines its openness, non-compulsiveness and essential fragility. References to the empirical world never escape from the integrated circuit, while critique (what Latourians like to denounce as 'denunciation') can be as argumentative and evaluative as required. Critique, indeed, happens when my circle enters yours, or your circle enters mine, in order to bisect it, displace it, rearrange its elements, push it off course. My own earlier objections to Latour offer a good example of such circular criticism. In suggesting that his *infra*-position actually conceals a *meta*-position, which goes unacknowledged precisely because acknowledging it would create a critical distance between observer and observed, I reveal its circularity; but this criticism clearly presupposes the external standpoint of one step up/one gear down reflexivity, which Latour would consider both inappropriate and suicidal.[12]

Vicious Reflexivities

Recognition of the constitutive circularity of accounts is also excellent therapy against one particularly nasty and treacherous aspect of reflexivity in the objectivistic mode. An insistent problem of the rhetoric of reflexive duplication, which wants to bend back the tools of scientific method upon the scientist himself, is that the substantive theory to be duplicated in this manner is readily taken for granted. This backloop produces a peculiar *petitio principii* or sleight of hand not too dissimilar from that which is operated by more traditional forms of ideology critique. The show of 'honesty', 'bravery', 'self-evidence' and 'non-contradiction' that often frames such reflexive duplications easily conceals the delicate blackmail involved in applying one's own analytic framework at once to oneself and to one's adversaries. The method of 'confessional purification' subtly ensnares and downgrades those who are incapable of 'opening up' and 'revealing' themselves after the imperative example of the confessor. In such cases, the call for reflexivity indeed becomes vicious – in the emphatic sense of insinuating 'real' or 'objective' motives, interests, or impulses that allegedly act under the surface of those whom we do not wish to take at their word (which is somewhat like injecting a lethal virus into a strange body and watching it self-destruct). The imperative to become self-aware thus becomes a policing demand, issued by a theoretical exhibitionist who has previously set all the cognitive and moral conditions for its emergence or its suppression: self-exposure turns into a devious way of exposing the weaknesses of others. In this situation, reflexivity acquires a nasty moralistic bite, since the adversary is effectively but invisibly accused of a failure *to adopt one's own theory*. I, the reflexive sociologist, knowing myself, also know who you are, where you come from, what your deepest interests are, why you remain unconscious of what you actually do, and why you entangle yourself in performative contradictions. If you are unprepared to 'know thyself' on my theoretical conditions, you are an unreflexive bastard, and I must tutor you in my explanatory theory, which will liberate us both.

In this scenario, the game of self-knowledge does not graduate much from the constitutive perfidies and treacheries of traditional ideology critique. Typically, ideology critique transforms the direct, upfront normative confrontation between 'correct' and 'false' views into an indirect naturalistic confrontation between objective social forces – which enhances the suggestion of disinterested rigour and ontological necessity (cf. the Marxian analysis of 'necessary false consciousness'). The naturalistic turn tends to erase the judgmental or dialogical character of the critique, and turns differences of opinion into supposedly objective contradictions (e.g. between what you *say* and what you *actually do*). 'I, the observer, contradict you, the observed' is subtly transfigured into 'you objectively contradict yourself' or 'you are incapable of accounting for yourself'. The spokesperson naturalistically masks his intrusion into the world of the other and hence cashes in all the profits of figuring as an 'absent presence'. But in all cases, the content of reflexive critique remains logically and circularly dependent upon the substantive theory that is projected in advance.

A first inevitable example is supplied by the (in)famous anti-relativist self-refutation argument, which obviously presumes and takes for granted a specific rationalistic theory of true and grounded knowledge. Another familiar instance is offered by the reflexive paradox that appears in the logical positivist argument that only empirically verifiable statements are meaningful – a statement that is itself not empirically verifiable (Lawson 1985: 19). An example of a different sort is found in the insistent sociological critique of the Marxist theory of 'proletarian' science, which is presumably caught in a reflexive paradox because it is incapable of locating its own point of origin in the relatively privileged stratum of the bourgeois intelligentsia (cf. Gouldner 1985: 3ff.). Mannheim already made quite a show of reflexive consistency when he undertook to generalize the Marxian critique of ideology towards the sociology of knowledge and intellectuals (Mannheim 1968 [1936]). However, this contradiction is not so much an internal one as it is one between Marxism (which is perfectly capable of reflexively accounting for itself[13]) and a theoretical alternative that runs at right angles to it in its different but equally naturalistic assessment of the sociological gravity of the stratum of intellectuals. As we shall see below, this analysis also applies to the scientific reflexivity of Bourdieu, who has a habit of suspecting of a reflexive deficit those fellow intellectuals who have not yet adopted his own critical sociology of the intellectual field. But it equally applies to Gouldner, who was in the habit of pontifically accusing of a failure of reflexive nerve those who were unable to support his own vision of the 'personal presence' of intellectuals and of the sentiment-relevance of social theory (cf. Gouldner 1973: 87–89, 123–24, 148–50).[14] In this fashion, the accolade of reflexivity easily transforms into a furtive means of celebration or impeachment: Hooray for myself! Down with the others!

Let me give a final illustration of this conundrum by citing a recent Schmitt-inspired account of the inescapable necessity of 'political' decisions, which nicely dovetails with my repeated allusions to the knowledge-political continuum, and which is arguably representative of many other postmodernist assertions about the ubiquity of the political (Rasch 2000b). Following Beardsworth on Derrida, Rasch distinguishes between decisions 'that recognize their legislative and executive force' and decisions that 'hide it under some claim to naturality'. The violence of a decision can be acknowledged and reflected, or it can remain unthought. Decisions, in other words,

> can reflexively affirm their status as decisions, or they can silently deny their contingency and assume the gesture of logical subsumption. For Derrida and Lyotard, as for Schmitt, it is the unthought violence of this latter possibility that poses the greater danger, the violence that camouflages itself as innocent neutrality or universal reconciliation. It is a violence that masks itself as peace and hence criminalizes opposition [...] What agitates Schmitt is not the force, but the deception. (Rasch 2000b: 23, 10)

The point I wish to make here is that this critical decisionism and the 'unmasking' gesture that it validates – which have been largely shared and elaborated in this book – can themselves only derive from circular arguments that tautologically loop back to their own a priori assumptions. In my own case, I must therefore step up the reflexivity of this critical constellation by saying that there are arguments and intellectual decisions that reflexively affirm their circular self-enactment, and others that remain 'unthought' as such and as a result administer forms of symbolic violence. Once again, what agitates *me* is not the circularity but the deception – knowing full well that this is likewise a circular observation.

In this fashion, the idea of constitutive circularity at least has the virtue of restoring the direct, up-front polemical relationship between the critic and who or what she criticizes. What renders the game of objectivistic reflexivity treacherous is precisely the fact that it offers a circular argument that is not identified in so many words. An alternative way of phrasing this would be to say that the spokesperson who critically imputes objective impulses or interests to other actors, and accuses them of a reflexive deficit, conveniently disappears from the circle, which is disrupted as a result of it. As a result of this disappearing act, a circular or performative account turns into a linear or objectivistic one. Methodological reflexivity, however, cannot but presuppose a substantive theory (of rationality, of science, of intellectuals, of self and society) that frames the conditions for its own reflexive duplication. The performative circle is still there, even though its trajectory is partly erased.

In the following, I wish to adduce a few more detailed examples of circular reasoning in self-proclaimed reflexive science that do not sufficiently 'own up to' or calculate their own circularity – which becomes vicious as a result. These arguments for reflexivity are invariably oriented towards the strengthening of objectivity and the enhancement of scientificity. From my point of view, however, they instead supply a set of involuntary counter-arguments for a more radical epistemology of circular reasoning. They all strengthen the case for a *weaker* notion of objectivity in which the spokesperson is more definitely included and positioned within the performative circle of representation itself. I shall not dwell here on the Strong Programme's 'fourth tenet' of reflexivity, which stipulates that the social study of science must simply copy what 'science itself' does and adopt 'its' tried and tested patterns of explanation – which clearly takes for granted a strong naturalistic and causalistic view of science and hence invites a vicious circularity (Bloor 1991 [1976]: 5; cf. Woolgar 1988b: 21, 24; Ashmore 1989: 36–40). Instead, let me briefly discuss the pairing of 'strong reflexivity' to 'strong objectivity' in the feminist standpoint epistemology elaborated by Harding, as well as the reflexive sociology offered by Bourdieu, who rigorously upholds a 'scientific' reflexivity against the despicable forms of 'narcissistic' and 'nihilistic' reflexivity deployed by (for example) Gouldner, reflexive ethnography and radical science studies. The circular reasoning is in both cases the same, since a substantive sociological theory (a standpoint theory of feminine experience, a quasi-economics of intellectual fields) is largely taken for granted as the medium for its own reflexive duplication.

Squaring the Circle: Harding and Bourdieu

Harding's work may strategically illustrate the circularity that afflicts all standpoint epistemologies, from classical Marxism through the Mannheimian sociology of knowledge up to more recent feminist and postcolonial versions, which arises from the fact that situations for 'situated knowledges' are inevitably framed by and hence logically depend upon prior *definitions of the situation* (Pels 2000: 156ff.).[15] The initial justification for a more reflexive feminist politics of identity was that feminists should no longer speak on behalf of abstract entities such as humanity, reason, or justice, or speak for *other* classes or categories than their own (such as the proletariat or the nation), but should 'merely' speak *for themselves* and further the cause of self- rather than other-emancipation (cf. de Lauretis 1990: 138; Harding 1993: 62). Despite such professions of reflexive modesty, however, feminist standpoint theory has displaced rather than resolved the problem of spokespersonship, as a result of its continual semantic slippage between the broader category of 'women' and the narrower one of 'feminists' (i.e. feminist intellectuals). It is critical theory that has furnished the vehicle for this processing of lived experience into consciously articulated standpoints. Women's lives or experiences may offer a privileged standpoint for theorizing and research, but they must not be taken as given; they must be mediated and informed by feminist struggle and feminist science (Hartsock 1983: 284–88). Black feminist thought, in Hill Collins' revealing formulation, is called upon to 'clarify a Black women's standpoint *of and for* Black women' (1986: 16, 24; 1991: 22). A standpoint, Harding concurs, is not a perspective; it takes science and politics to achieve one (1991: 276n, 150n). Female experiences, or the things women say, are in themselves not reliable grounds for knowledge claims; they may be good places to begin research, but it is 'the objective perspective from women's lives' that gives legitimacy to feminist knowledge (1991: 123, 167, 282, 286–87).

This silent sovereignty of politicized theory (and of politically committed feminist theorists) also emerges in a more generalized version of standpoint theory, which focuses upon the epistemologically privileged situation of women as 'strangers' or 'outsiders within' (Hill Collins 1986; Harding 1991: 124ff.). While such positional exclusion may offer an initial epistemological advantage, it is feminism that must actually *teach* women (and men) how to see the social order from the perspective of an outsider (Harding 1991: 125).[16] Feminist political activism and theorizing enables us 'to see beneath the appearances created by an unjust social order to the *reality* of how this social order is *in fact* constructed and maintained' (Harding 1991: 127, my emphasis; cf. Hartsock 1983: 304). Hence it is militant method that both intellectualizes women and appropriately estranges them, shaping the 'traitorous identities' able to generate 'traitorous analyses' of the existing social order. Not only is the standpoint ultimately theory-dependent; theory, apparently, also potentially *liberates* from standpoints, since it can 'make us strange' and help us to 'reinvent ourselves as Other' (Harding 1991: 268ff.).[17] A 'strong' objectivity is thus reclaimed over against the 'weak' objectivity of value-free, disinterested, and impartial science, which effectively cloaks the self-interest of social elites and that of status- and power-seeking scientists. In using their own historical location as a resource for obtaining greater objectivity, standpoint theories also lay down stronger standards of reflexivity against the weaker reflexivities of (for example) the ethnographic studies of science, which incorrectly bracket larger social, economic, and political variables (Harding 1991: 163).

Shows of strength such as these only reinforce my craving for epistemological *weakness*. This urge is further nourished by a second instance of strong reflexivity, which is likewise geared towards enhancing objectivity and scientificity, and is equally dismissive of gestures towards relativism (let alone circularity). Pierre Bourdieu offers a classical version of the self-duplication argument, insofar as reflexive sociology must systematically redouble its enterprise by complementing and deepening its view of outside reality (the socio-analysis) with an inside turn (the auto-analysis). Sociology does not only objectify the 'others' but also objectifies the objectifying glance itself; while investigating the object, it also investigates the relationship between subject and object, in order to avoid the danger that this relationship (e.g. unconscious aspirations or resentments) is projected onto the object in uncontrolled fashion (Bourdieu and Wacquant 1992: 68–69; Bourdieu 1993a: 3, 10; 1997; 2000a; 2001). In refusing the idea of an 'immaculate conception', and in discarding the ideal of scientific neutrality, Bourdieu takes the same polemical point of departure as that adopted in Harding's standpoint epistemology (Bourdieu 1981 [1975]: 278; 1993a: 10–11; 20–21; Harding 1992). A crucial difference arises, however, with respect to Bourdieu's new calculation of the mediating role of critical intellectuals, and his concomitant emphasis upon the 'refracting' mechanism of autonomously operating intellectual fields. In this manner, Bourdieu offers a more sophisticated and stratified analysis of the performative logic of spokespersonship, which radicalizes positional epistemology by inserting the second-order positions and positional rivalries that structure the scientific and intellectual field. Intellectuals typically act as 'dominated dominants' in the social field as a whole, and occupy a position that in this respect is homologous with that of the dominated *tout court*. But this partial similitude between spokespersons and the groups that they speak for must not be mistaken for a condition of identity – especially since intellectuals are generically drawn towards and indulge in such mistaken identity games (1990a: 140–46; 1991: 243–45).

Bourdieu's urge for a strong scientific reflexivity most obviously surfaces in his harsh and indiscriminating crusade against what he calls 'narcissistic' reflexivity; the title of one of his programmatic statements on the issue offers a knowledge-political programme all by itself (Bourdieu 1993c). The almost universal decline of the positivistic *certitudo sui*, he argues, has unfortunately incited a dangerous epidemic of rampant reflexivity which has nothing whatever in common with true scientific reflexivity. For Bourdieu, the most fruitful species of reflexivity is paradoxical because it is deeply anti-narcissistic. Against the banal curiosity of 'intimism', the self-satisfied and complacent turn towards the private person, it mobilizes a rigorously objectifying study of the social constraints and determinations that weigh upon the individual thinker, who is first of all conceived as an emanation of the logic of the intellectual field. Against the charismatic self-conception of the intellectuals as free-floating agents, who think they are exempt from all social determination, Bourdieu emphasizes that there is nothing unique or extraordinary about their structural position; they simply belong to the social category of researchers, who together make up a definite social field. The point of a reflexive sociology is precisely to lay bare the impersonal behind the personal, to discover the social in the heart of the individual person (1993c: 369). If he never ceased taking himself as an object, Bourdieu insists, this was 'not in a narcissistic sense, but as one representative of a category' (Bourdieu and Wacquant 1992: 203).[18]

Bourdieu further distinguishes a number of levels or circuits where reflexivity operates, and which may therefore give rise to possible distortions. First, one needs to objectify the social conditions of production of the producers of culture, i.e. the attitudes and interests that result from social, sexual, ethnic and other determinations. In this domain of inquiry, one easily courts the danger of 'short-circuit' explanations, which suggest an immediate linkage between cognitive contents or intellectual patterns and larger social variables of class, gender, or race, and misrecognizes how such outside determinations are 'refracted' by the determinations of the relatively independent field of cultural production itself. Hence the simultaneous need to objectify this cultural microcosm, the autonomous social world in which intellectuals struggle for specific stakes and interests – which appear disinterested only if they are compared with more narrowly conceived economic or political ones. In addition, Bourdieu thinks it imperative to address another, most essential source of reflexive distortion, which results from the ensemble of presuppositions built into the 'distanced' positions of scientists as scientists. In forgetting to calculate that social theory is the product of a contemplative gaze, and is predicated upon the withdrawal from practical activity and from the pressures of action and decision, professional observers of the social world easily submit to a 'theoreticist' or 'scholastic' bias, to an 'ethnocentrism of the learned gaze', which facilitates an intellectualist projection of their own way of thinking and behaving into the heads and hearts of the subjects they study. Hence the necessity of complicating all theories of practice by means of a reflexive theory of the relationship between theory and practice (Bourdieu 1993c: 369–71; 1997; Bourdieu and Wacquant 1992: 69–70).

According to this programme, the reflexive sociology of sociology does not lead towards scepticism or nihilism but towards a stricter application of scientific method. But this urge for scientificity once again conjures up the threat of vicious circularity. This circularity is started up, so to speak, by Bourdieu's evident prioritizing of the objectivist moment of sociological construction over the subjectivist or representational one (cf. 1990a: 123ff.; 1991: 229ff.; Bourdieu and Wacquant 1992: 106–107). Points of view can only be fully understood by first construing the field of objective structural positions that determine and constrain them, which requires us to break with the subjective representations of ordinary actors. In a second step, these subjective representations must be reincorporated in the analysis, because the stable reproduction of structures significantly depends upon the recurrent performative definition by the actors involved of the structural situations in which they act (Bourdieu 1988b: 782).

The complication that arises here is that the epistemological break or the 'objectivist moment', in which the sociologist constructs the (intellectual) field, is not itself reflexively recognized as a performative operation, as a definition of the situation that co-produces the situation that it describes, but still claims a privileged grounding in the objective reality of structured social relations (see also Chapter 5). Nothwithstanding recurrent expressions that suggest the radically constructed nature of the social space and the essential performativity of all accounts of it (e.g. Bourdieu 1988b: 782; 1990a: 125–27, 179–81; 1991: 220ff., 229ff.; 1996a: 18), the narrative voice of the sociological spokesperson (Who says so? Who defines these fields? Who construes these determinations that determine us all?) is not sufficiently focused and problematized. The performative circle of representation, while being subtly and critically traced in the domains of political delegation and of juridical, religious, and everyday classification (e.g. Bourdieu 1990a: 182; 1991: 203ff.; 1993a: 22–23; 1996b: 20–21, 25), is not recursively adopted for Bourdieu's own project of the scientific representation of the social world, which regularly relapses into a conventional reality discourse and a holistic posture of transcendence (e.g. 1981 [1975]: 283; 1990a: 181, 184; 1996a: 34; Bourdieu and Wacquant 1992: 259–60).[19]

As is evident in *Homo Academicus*, his exemplary study in the reflexive sociology of intellectuals, Bourdieu indeed takes care to include himself in the sociological picture; but it still is Bourdieu himself who sets the explanatory conditions and defines the 'objective' parameters and constraints within which he 'objectively' positions both himself and the others (1988a: xv–xxvi). If reflexive sociology may claim epistemological privilege for being able to reinvest its own scientific gains in scientific practice (1988a: xiii), this clearly presumes a prior privileging of Bourdieu's own field theory of science. If, indeed, the sociology of science is called upon to realize that it itself functions in accordance with the laws that govern the operation of any scientific field (1981 [1975]: 283), it is once again Bourdieu's explanatory sociology of science that defines the terms for this reflexive duplication. Sociologists, in Bourdieu's view, can only escape the determinations of the social if they train their own scientific weapons against themselves, and partly neutralize these determinations by objectifying them. But does Bourdieu not mask the circularity and fragility of his own constructions by first defining them as objectively real, in order 'freely' to submit to such self-imposed constraints?

Unhastening Science

Does he, the anti-Sartre, not repeat the logic of Sartrean *mauvaise foi* by inverting it?[20]

Sociology may therefore jump out of the vicious circle of historicism or sociologism by objectifying the objectifying subject (Bourdieu 1988a: xiii). The unattended-to circularity and performativity of everyday world-making, which appears to describe social reality but actually constructs it and thus creates the conditions of its own verification (1996b: 25), is approached from a critical outsider's point of view, as something that must be remedied or surpassed. Bourdieu specifically wishes to combat the impression that sociological discourse is itself performative, even though it is easily interpreted as such. In contrast to ordinary discourse, which routinely mixes facts and values, descriptions and prescriptions, sociological analysis should attempt first of all to understand and explain the social world, to 'take it for what it is' (1993a: 22–23). Its task is to engage in 'realist construction' (1996a: 27). My contrasting proposition is to retain this critique of reification – insofar as common-sense rationality indeed unwittingly creates social facts by defining and enacting them as things – but to cheerfully adopt the circular logic of performativity, fictionality, and fact/value mixture for sociological inquiry itself; and hence to duplicate the critique of reification for Bourdieu's own version of 'relational realism' (cf. Pels 1999).[21]

By carrying on the reality discourse of a rigorously scientific sociology, Bourdieu still appears caught in the same *vicious* circularity as Harding, who is likewise seduced into defining the situation for 'situated knowledges' in terms of a pre-existing objective matrix of social identities. The apparent confessional honesty and ascetic logic of the reflexive duplication, of the imperative of turning 'the' instruments of science against 'the' scientist himself, surreptitiously presumes both a unified set of rules of sociological method and the prior validity of a particular sociology of science. This circular movement remains a vicious one as long as its essential circularity goes unheeded.[22] Bourdieu's recurrent denunciation of the self-deception of the intellectuals, his reiterative claim about the superiority of his self-explanatory sociology, and his incessant and often quite narcissistic display of reflexive iconoclasm and hard-working heroism, all come together to suggest a sociologically purified gaze that somehow breaks the magic circle. While his reflexive sociology is merely one (although an admittedly 'strong' one!) among many competing intellectual performances, Bourdieu still purports to reveal the truth of the game 'as a whole', as it 'really is' (1990a: 180–84). But such totalizing constructions never depart from their own performative circle, even if (especially if) they are designed to circle high above it.

Reflexivity: One Step Up

Reflexive Critique and Intellectual Rivalry

Both Harding and Bourdieu, we have noticed, attempt to square the circle by claiming strong reflexivity for a form of positional thinking that still supports a realist ontology and a comprehensive view of the whole. In my own weaker conception, positional thinking inevitably issues in a *virtuous* circularity, which defies all transcendent reality judgments ('circular realism') and does not permit or require any conclusive synthesis of partial views. For Harding, who wishes to protect standpoint epistemology from charges of relativism, such perspectivism is not sufficiently strong to cut a figure in political debate: 'No critics of racism, imperialism, male supremacy, or the class system think that the evidence and arguments they present leave their claims valid only "from their perspective"; they argue for the validity of these claims on objective grounds, not on "perspectivalist" ones'. Standpoint epistemologies must start research from marginal positions, to generate not 'ethnosciences' but *sciences*, which provide less partial and distorted accounts of the entire social order (1992: 576, 582–83).

Bourdieu similarly emphasizes that reflexive self-analysis does not issue in a nihilistic or irrationalist attack on science. Sociology *can* indeed escape from the vicious circle of historicism or sociologism through its distinctive method of 'participant objectivation' (1988a: xiii; 1988b: 784; Bourdieu and Wacquant 1992: 253ff.). It is admittedly difficult for intellectuals to escape the logic of struggle in which everyone willingly becomes the sociologist of his enemies, while at the same time turning into his own ideologue, 'in accordance with the law of reciprocal blindness and insight which governs all social struggles for truth'. Only when she apprehends the game as a whole is the intellectual able to monitor her interested implication in it and to extricate herself from it, and thus to escape the play of partial and mutually reductive objectifications. The sociological observer may escape from the inevitable positionality and partiality of his classifications by 'raising himself to the second degree' and 'conceptualiz[ing] the space of the struggle for classification as it really is' (Bourdieu 1990a: 180–84).[23]

In my alternative approach to reflexivity, I have resisted such scenarios of duplication, according to which sociology must routinely reverse its analytic resources back upon itself in order to avoid becoming a 'standing refutation of its own claims'. As we have seen, this 'duplication mode' tends to take for granted the objective validity of the theory offered for reflexive duplication, and readily connives with forms of 'strong objectivity' that promise a last-minute escape from the logic of perspectivism itself. Against this, I have attempted to define reflexivity as a self-conscious exercise in *circular reasoning*, which breaks with the unending quest for a transcendent objectivity, and rests satisfied with merely partial and partisan perspectives, even if these are informed and systematized by a scientific sociology of knowledge.[24] Why, indeed, is it so imperative for sociology to transcend the *logique du procès*, the adversarial logic of mutual objectification and reduction that allegedly disfigures everyday communication? What except a residual scientistic prejudice (an urge to be epistemologically strong) makes a *problem* out of the risk that the sociology of knowledge might be 'no more than the most irreproachable form of the strategies used to disqualify rivals'? (Bourdieu 1981 [1975]: 283; cf. 1999: 334) Why not acknowledge and epistemologically capitalize upon the fact that every sociologist is a fairly good sociologist of the opposition, and a much less insightful observer of his own position?

In effect, this amounts to a more consistently reflexive affirmation of Bourdieu's own insight that it is the *scientific field* that is the actual subject of scientific knowledge (and thus of reflexivity) – or to a radicalized version of Popper's long-standing conviction that scientific objectivity is not a product of the individual scientist's studied impartiality, but of the 'friendly-hostile co-operation of many scientists' (1962: 217, 220). The ultimate aim of a radical sociology of knowledge must indeed be to institutionalize reflexivity in the field through the regulated mutual surveillance and control of products that is induced by interest-committed rivalry.[25] However, while Bourdieuan reflexivity claims a miraculous last-minute suspension of the 'law of reciprocal blindness and insight', my own idea is to retain the essentially polemical nature of objectification (cf. Sloterdijk 1983: 652ff.), in order to rehabilitate the game of critical 'unmasking' or 'disqualification' that both Bourdieu and the radical constructivists wish to bring to an end – although for diametrically opposite reasons.

In my conception, the epistemological break between science and (scientific) common sense, or between observer and observed, does not turn upon the conventional distinction between 'merely partial' views and views that claim the ability to totalize, but upon the somewhat unconventional distinction between views that are self-consciously circular and views (both everyday and sociological-scientific) that deny this circularity in order to claim a 'straight' transcendent foundation. The demarcation from other forms of belief (and the attendant claim to 'truth') is not annulled or completely levelled down but displaced to another dimension of difference. This also rejects an abstracted agnosticism that seeks to remove all distance between observer and observed, who, as constructivists like to repeat, are involved in 'essentially similar' epistemic practices, so that second-order accounts cannot be granted greater authority than first-order ones (cf. Woolgar 1991: 39; Latour 1993a: 44;

1999c: 20–21). Circular reflexivity, in recursively offering its own criterion of demarcation, is a standing critique of linear narratives, both everyday and scientific, which bank upon an objectivist ontology and a conventional logic of representation. We may indeed be no better than our subjects (and collegial competitors) in playing games of mutual objectification, behind-the-scenes explanation, and polemical attribution of responsibility and interest. But we (who actually is this 'we'?) may still claim epistemological advantage if we acknowledge – unlike most of 'them', most of the time – that all facts, values, distinctions, interests, and identities are embedded in circular reasoning; which installs a radical uncertainty in all our accounts of the world and in all our critical manoeuvres in the agonistic space of science.

Reflexive circularity, as such a self-grounding 'autological' operation, has the effect of critically defrosting or liquefying some of the most revered and taken-for-granted propositions and oppositions in present-day social theory. As we have seen, it reveals for example that the famous Weberian postulate of value-freedom, in proclaiming the 'absolute heterogeneity' of factual and evaluative statements, breaks its own constitutive ground rule in its very act of enunciation; but it also acknowledges, in symmetrical fashion, that the opposite and more epistemologically interesting claim about the 'natural proximity' of facts and values suffers from the same 'essential' circularity (see Chapter 4). It is even more disconcerting, perhaps, that ubiquitously promoted constructivist axioms (which I fully share) such as that an a priori account of knowledge is impossible, that a 'view from nowhere' is forever out of human reach, and that, accordingly, timeless and standpointless knowledge is an illusion, 'squarely' rest upon circular reasoning, because they invariably presuppose the prior validity of a perspectivist epistemology (while the same is true for their conventional universalistic opposites). A similar bracketing is required for the equally popular Foucaldian presumption (which has of course framed my entire discussion of scientific demarcation and autonomy) that power and knowledge are intrinsically linked; and that (mark the typical self-validating closure!) *denials* of this inevitable condition (for example by normativist political philosophers) themselves 'perform the most insidious ruse of power' (Butler 1992: 6). Finally, one may point to the 'fact' that all sociological critiques of reification 'necessarily' move in a circle: to define as thinglike what is 'really' relational, as static what is 'really' processual, as objective what is not an object but is 'really' socially or textually or symbolically constructed, tends to presuppose knowledges about ontological states which themselves cannot be verified outside the circle they describe. Which, once again, is a circular statement. Which, I repeat, is not necessarily a bad thing. Because all of these beliefs and propositions, many of which I would like to sustain, are able to move about in the world and make a difference to it without the support of transcendental, circle-breaking arguments.

I have argued for reflexivity as 'one step up' or 'one gear down', in order to tie the story back to the narrator and display the performative, projective relationship between the spokesperson and that which is spoken for. Because this reflexivity is circular, and instils a constitutive weakness in the heart of all practices of representation, it is also 'essentially' insufficient, inexhaustive, or flawed. It could well be argued that accounts that continually foreground their constructed and circular character are, if not physically impossible, at least communicatively dysfunctional: stories (such as the current one) need to be told in somewhat linear fashion, and cannot be continually sidetracked to attend to their own conditions of possibility. Due to the spectral law of objectification, all observations have their blind spot, and remain to some extent naïve with respect to their own point of departure. From this it follows that reflexivity can never be comprehensively executed by the individual knower, but remains a distributed process that must partly be left to the agonistic play of forces in the scientific and the general public marketplace. Others may be better positioned, and more strongly motivated, to discover a critical gap between my sayings and doings, or between my self-understanding – which they may refuse to take at face value – and my 'objective' position, as they construe it from their polemically distanced points of view. I can still further the reflexive complexity of my own accounts by tracing their circularity and taking responsibility for it; but at some point, critical others must step in to take over some of the reflexive burden and cross their different explanations with mine. The old question 'Quis custodiet ipsos custodes?' (or: 'Who is reflexive for the reflexivists?') can be now answered: *other* guardians, *other* reflexivists. Reflexivity is also something we must delegate to our friends, or rather: to our best enemies. If politics is about making strong enemies, science might be about making *weak* enemies who are interested to follow us around and critically measure us up in the slower conflicts that rage in the intellectual marketplace.

CHAPTER 8

Intellectual Autonomy and the Politics of Slow Motion

Philosophical Stammering and Monastic Silence

The first aim of this concluding chapter is to provide a broader historical background to the minimalist and pragmatic identification of science that I have attempted so far, in order to deepen the description/justification of its weak autonomy over against other professional pursuits in what I have called the 'social triangle'. My narrative has repeatedly distinguished between the philosophical (dis)guises of traditional demarcative exercises and the more pragmatic operational conditions that determine the sources of variation and the dynamics of interaction between the professional roles of scientists and politicians. Rather than emphasizing the more conventional spatial dimension of autonomy, I have drawn attention to the peculiar timescape or 'time flow' of intellectual work as providing such a minimal touchstone of distinctiveness. In the following, I want to enrich this critical (because anti-essentialist) phenomenology of scientific work by tracing a few lines of genealogical descent of the notion of unhastening (and of the estrangement it both requires and produces), retracing the brittle knowledge-political compromises between the academy and 'big' politics that were instrumental in gradually institutionalizing a margin of intellectual autonomy. As an informal history of the *temporality* of doubt and distanciation, this narrative leads on to a summary of how the various more interesting epistemological principles that we have encountered in this book (such as anti-politics, symmetry, and reflexivity) can be more fruitfully translated as chronopolitical principles of deceleration. After having reclaimed the differential 'time geography' of the social triangle, I shall finally outline a new interpretation of academic freedom and of the 'idea of the university' that turns the 'unhastening' argument into an explicitly normative and political project.

Let me embark upon this episodic genealogy of the times and spaces (the silences and seclusions) of science by recapturing the canonical moment in Western thought that is represented by the Socratic dialogues, which are traditionally identified as the theatre of invention of the philosophical life and the contemplative ethos more generally. In its dialogical form, philosophy is initially invented as a distinctively 'slow' manner of speaking. At this historical turn, writing still appears effectively subordinate to both public speaking and private conversation; indeed, the stated purpose of the dialogues is precisely to *separate* semi-private philosophical talk from the political talk of citizens in the *agora*. However, even though Plato's thought is recorded in dialogue form and even features an explicit *rejection* of writing as an unnatural, reificatory, and unresponsive technology, its exquisite philosophical precision is entirely unthinkable without the effects of visually available inscription (Goody and Watt 1968: 49–52; Ong 1982: 79–80, 94, 105).[1] Recording these dialogues (which were of course Plato's own *inner* dialogues and slow conversations with absent others), and editing them in order to make them readable, is still a long way from the centrality of literacy and reading that is only inaugurated by subsequent revolutions in intellectual technology. But in shifting the emphasis from public to semi-private speech, and by following an implied logic of inscription, the dialogues already perform a pragmatic time-space operation that is predicated upon a double act of temporal unhastening and 'audience reduction'. They section off a becalmed and protected arena where the speed of everyday and public life in the city is effectively reduced. Such social-institutional 'conditions of felicity' define a pragmatic autonomy for thinking, research and dialogical testing which is left standing even if it is stripped of the transcendental and foundationalist scaffolding that Platonic philosophers erected in order to hold it in place.

We do not require the Platonic purities of truth, justice, and beauty in order to ascertain the specific difference of intellectual work. As repeatedly noted, it is not the ascetics of objectivity, neutrality, or factuality we must seek to cultivate, but the politics and economics of intellectual autonomy. What we do require is a specific situatedness, a novel arrangement of time and place that promises a reflexive deceleration of action and the clearing of a 'calm space' where time may flow more relaxedly, topics may be studied one at a time, conversations may be conducted in a more leisured and egalitarian fashion, and pressing decisions may be postponed. This chronotopics of unhastening and estrangement is clearly retrievable from the 'surface' of the dialogues, if we are prepared to remain at this ephemeral level and ignore the solemn 'depth' of their formal foundationalist intent. While for the Platonic philosopher ascetic truth-seeking transcendentally dictates the need for philosophical autonomy, the pragmatist reverses the argument in order to contend that the philosophical

Intellectual Autonomy and the Politics of Slow Motion

techniques of slow motion and *agoraphobia* are in themselves sufficient to introduce a perspectival truth interestingly different from the truths that emerge from more hurried cultures (cf. Pels 2000: 194–201).

Indeed, the dominating distinction between the contemplative life and the public business of the *agora*, and the corresponding one between philosophers and politicians (lawyers, orators), effectively turns upon a pragmatic contrast between *scholē* or leisure and *ascholia* or hurry (cf. Bourdieu 1998a: 128–29; 2000a: 226).[2] In the *Theaetetus*, philosophers and political orators are contrasted as free men and slaves, primarily because the former always have time on their hands: 'They carry on their conversation unhurriedly, in peace and quiet [...] The duration of the discussion doesn't bother them, it could be long or short. Their only concern is to reach the truth.' A speaker in court is different because 'his allotted time is slipping away and forcing him to hurry his speech' (*Theaetetus* 172c-173a).[3] The adversarial and public situation in which orators find themselves, which is pervaded by personal interests and anxieties, lends their speech a 'tense and neurotic' character that is absent from the conversational style of the philosopher, 'whose upbringing has been genuinely free and unhurried', and who has 'never known the way to the *agora*'. He talks at leisure and peacefully, practising a detachment and an aloofness that outsiders may easily interpret as unworldly impracticality (*Theaetetus* 173d-e; 174c; 175e).[4]

This awkwardness is turned into a matter of boastful pride in the *Gorgias*, where Socrates happily concurs with Callicles' dismissive description of the philosopher as someone who makes a complete fool of himself when turning to a practical activity such as addressing and trying to win over a public meeting. The philosopher cannot deliver a proper speech to the councils or make a persuasive appeal because, with his head spinning and his mouth gaping open, he wouldn't know what to say (*Gorgias* 484d-e; 486b).[5] It is most unfortunate that the philosopher 'is avoiding the heart of his community and the thick of the *agora*, which are the places where, as Homer tells us, a man "earns distinction". Instead he spends the rest of his life sunk out of sight, whispering in a corner with three or four young men, rather than giving open expression to important and significant ideas' (*Gorgias* 485d–e). On Callicles' recommendation, philosophers must stop their childish cross-examinations and instead practise 'the culture of worldly affairs', taking up the kind of occupation 'which will make your wisdom famous'. But Socrates is not interested in worldly fame, or in ideas whose significance is defined by their popular appeal or their rhetorical alacrity. He admits he is not a politician, because the Truth is not something that can be put to the vote. Producing a large number of witnesses is worthless in the context of the Truth: 'My experience is restricted to producing just a single witness in support of my ideas – the person with whom I am carrying on the discussion – and I pay no attention to large numbers of people; I only know how to ask for a single person's vote, and I can't even begin to address people in large groups' (*Gorgias* 486c; 474a).

The starkness of the Platonic contrast between political rhetoric and philosophical conversation initially opens up deep epistemological distinctions such as those between 'flattery' and 'true speaking', between power-mongering and the selfless midwifery of the truth, between persuasion and rational understanding, between moral pleading and dispassionate description, and between performative and reflective speech – all of which emerge out of the absolute dichotomy that separates *epistēme* or true knowledge from *doxa* or mere belief. Pragmatic forms of deceleration and displacement are euphemized and psychologically legitimated by philosophical self-denials that idealize a disembodied (indeed, Platonic) love of wisdom that boasts a profound indifference to lust, luxury, fortune, and fame. The epistemological dichotomies force a strong essentialist boundary upon an institutional context that rather operates a knowledge-political *continuum* of ways of speaking (ways of persuading, ways of performing, and even ways of putting ideas to the vote) where relative and graded distinctions indicate differential audience size (the crowd, or just three or four young men), site of delivery (the *agora* and the lawcourts, or a small corner, a doorstep, a fenced-off garden) and tempo of speech (rhetorical soundbite or conversational turn-taking punctured by reflective pauses). If Callicles emphasizes the childish immaturity and inappropriate playfulness indulged in by philosophers, and likens their talk to 'stammering' (*Gorgias* 485b–c),[6] what else is he describing but a new breed of men who reinvent themselves as being in 'permanent education', who deliberately stretch the time interval that protects them from the world of gravity and busy-ness, whose awkward impracticality bespeaks a studied indifference to the world of practice, and whose pitiful stammering might just be a novel and hence unrecognizable way of *trying to speak slowly*?[7]

The intimate collusion between epistemological absolutism and the pragmatic 'renunciation of the world' that defines intellectual autonomy is also sharply highlighted by the famous argument about the philosopher's life-unto-death in the *Phaedo*. It is not right for the philosopher, Plato imprecates, to pursue the so-called pleasures of food, drink, sex, clothing, or other ornaments and material possessions. The true philosopher despises these bodily urges; he must free his soul from its association with the body, because the body is a hindrance to the acquisition of knowledge and only the soul is capable of attaining the truth. It can best reflect on reality when it is freed from all distractions such as hearing or sight or pain or pleasure, in avoidance of all physical contact and social association. Absolute truth, goodness, beauty, and the real nature of things are not to be apprehended through the body; it takes pure and unadulterated thought to see the pure and unadulterated object (*Phaedo* 64a–66a).[8] Since pure knowledge is 'impossible in the company of the body', the philosopher must train himself throughout his life to live in a state as close as possible to death: 'True philosophers make dying their profession' (*Phaedo* 66e, 67e). The moral ideal of *theoria* is once again framed in terms that do not merely separate individual body and soul, but simultaneously separate the philosopher's body from other bodies that pose a threat to his quest for purity and tranquillity. With typical Platonic exaggeration, the time-space distanciation that this physical separation requires is black-boxed by means of absolutist dualisms between body and soul, interest and disinterest, truth and belief, and privacy and publicity. The relative isolation of a few bodies assembling in a secluded space, adopting an unhurried disposition towards learning and education, is absolutized in terms of an ascetic mortification of the body. With equal radicalness, the tranquillity and relative silence ('whispering in a corner', in Callicles' scornful gloss) associated with this leisurely way of being are similarly absolutized as a stillness towards death. A relative social withdrawal is philosophically dramatized as the final disembodiment that marks the boundary of death.

There are interesting parallels to be drawn between the 'art of isolation' practised in these liminal locations (corners and thresholds, the semi-private garden that housed the Platonic academy) and the experience of heremetic and monastic asceticism that has often been seen as its Christianized adaptation (cf. Mannheim 1943: 158–59; Kaelber 1998). From its origins in the hermit colonies of the Egyptian and Palestinian deserts, the early ascetic movement combined *anachorēsis* (retirement, or physical retreat from the world) with various forms of *askēsis*, the systematic renunciation of worldly goods and pleasures in order to attain purity of belief and methodical self-control. Physical austerity, chastity, poverty and charity were seen as contributing to a continuous process of self-mortification which aimed at contemplative unity with God through work and prayer.[9] The great hermits, such as Euthymius the Ascetic, Anthony the Great, Simeon the Stylite, or Benedict of Nursia, were lovers of solitude and silence, retiring to deserted villages, mountain caves, or the top of ruined columns in order to escape the density, pace and noise of city life. If they attracted communities of followers, they often repeated Anthony's radical example of retreating ever further into the desert in order to escape their growing fame and regain the solitude that was lost within the charismatic community. Well-educated ascetics such as Eusebius of Caesarea or Benedict of Nursia viewed the retreat to the desert as the highest demonstration of the gospel and a logical condition of the Christian elite's search for solitude and stillness in a new 'philosophical way of life' that placed them beyond society and above nature (Goehring 1999: 17–18).

In this framework, Anthony the Great's famous injunction to 'pray without ceasing', i.e. to redefine prayer as life and life as prayer, might be interpreted as a way of deliberately redirecting the flow of conversation upwards (to a Single Interlocutor) and inwards (away from the busy chatter of fellow humans), and simultaneously as a way of slowing it down by means of ritual repetition and silent meditation. This new temporal economy was further institutionalized in the early medieval monasteries, where the Rule of St Benedict imposed a highly structured schedule that divided the day into collective prayer, reading, and physical work. The emphasis on (silent) *lectio* or reading, which included both the copying and study of texts and private meditation, turned the monasteries into treasurehouses of collective memory and repositories of learning; until the first foundation of universities in the eleventh century, they were known as the foremost centres of scholarship and literary production in the Western world. Physical work was not simply the 'active' form of prayer but was also required in order to secure the cloister's economic self-sufficiency, which demanded a concentration of all material necessities (such as water supply, a mill, vegetable gardens and orchards, animals) and all basic crafts within its enclosure. More generally, the monasteries, while inverting the order of interest found in the outside secular world, also reflected the cyclical rhythm of the surrounding, predominantly agrarian economy, which alternated between intense workloads during harvest time and more leisurely or irregular work schedules at other times. Despite their methodical and ordered way of life, however, the monasteries operated a time schedule that was still far removed from the 'capitalist haste' that would subsequently impose itself in the modern factory regime (Mumford 1967: 270–72; Zerubavel 1981: 31ff.; Kaelber 1998: 64–65; Adam 1995: 54–55).[10]

Cartesian Exile and Baconian Settlement

New inner-worldly forms of the art of isolation were invented in a more secularized age, concurrent with the early stages of the scientific revolution as its unfolded in its two major branches of 'continental' rationalism and 'English' empiricism. On the rationalist side, the most famous experiment in distanciation was perhaps contained in Descartes' almost monastic quest for 'assured leisure in peaceful solitude' (1968 [1637]: 95). The paradigmatic experience, which offered itself as a template for generalization and imitation, was of course Descartes' famous day of undisturbed meditation in a stove-heated winter room, which, by removing all distracting company or disturbing cares or passions, gave him 'complete leisure' to meditate on his thoughts.[11] Generalized in more practical terms by means of his 'retirement' to Holland, Descartes clearly echoed the anchorite's longing for the desert in a more secular, city-based frame, when reporting that he was able to live there among a great crowd of busy people 'as solitary and withdrawn as I would in the most remote of deserts'. Repeating both the philosophical and the monastic rejection of fortune and fame, Descartes piously stipulated that such things were contrary to the repose and the 'perfect peace of mind' that he valued above everything. Having no desire 'to be of consequence to the world', he held himself more obliged to those 'by whose favour I enjoy my leisure unhindered, than to those who might offer me the highest dignities on earth' (1968: 35, 52, 88, 91). As in Platonist philosophy and Christian theology, a rigorous dualism of mind and body privileged the mind and reason over more lowly impulses and interests (cf. Descartes' implacably intellectualist slogan: 'I think, therefore I am'). This disembodied and de-materialized view of rationality was focused in the individual reasoning mind, which was divorced both from the thinker's own body and from the bodyminds of other human beings. God was seen as in strategic retreat from the natural world, although He still guaranteed the unity and stability of reality and the absolute distinction between truth and illusion. In this more secularized account, it was not an abstract godlike essence or a single deity but a deified Nature that was simultaneously scrutinized for its recondite purposes and laws and worshipped as foundational and determining. It could only be approached by those who brought an 'unadulterated mind' to an 'unadulterated object', which means that it only yielded its secrets to those who adopted the methodical asceticism of solitude and silence. The great irony of this idea of 'Cartesian doubt' was that it precisely turned an ascetic experiment in social estrangement and 'leisure' into a universally prescriptive method for science which, while being made democratically accessible to all, nevertheless insisted upon a strong Platonic demarcation between science and society (cf. Pels 2000: 210–14).[12] Stripped of the philosophical guarantees that steered Descartes to his indubitable first principles, however, what else was this methodical doubt but a resolve to *reduce the speed* of intellectual decision?

There may be a world of difference between the art of hand-copying manuscripts in the stillness of the *scriptorium* and the more secular, surprise-oriented 'conduct of discovery' of the first experimental philosophers in the laboratory. But this view neglects important continuities between the two social sites, which effectively turned the monasteries into the first scientific laboratories in the history of the West, creating in advance the 'calm space' so famously required by the founders of the Royal Society of London and those of other scientific societies and academies. In the interpretations of Mumford and Collins, for example, the monasteries' highly structured work schedules and their predilection for technological experimentation clearly promoted societal rationalization, even anticipating some of the developed characteristics of capitalism (Mumford 1967: 263ff.; Collins 1986: 47, 54, 58; cf. Adam 1995: 64–65). Both authors presume a connection between social and technological innovation and the relative isolation of the cloister as a human and technical laboratory, although they also emphasize its connection to secular spheres such as the economy and technology and its ambition to actively *master* them.[13]

Relative seclusion and bounded autonomy also defined the compromise that was struck between the new power of experimental philosophy and the established powers of religion and the state. Indeed, if one reviews some of the historical literature on the rise of scientific societies and academies in Italy (where the Florentine Academia del Cimento was founded in 1657), England (where the Royal Society was chartered in 1662) and France (where the Académie Royale des Sciences was established in 1666), one is struck by the recurrence of two potentially conflicting themes: first, the *compromise character* of scientific autonomy, which cleared a space of relative freedom in return for self-censorship in matters moral and political; and secondly, the unabated sense of *intellectual superiority* and high cultural calling on the part of the scientist, who felt himself on a mission to lead his fellow men out of the darkness of superstition by the light of reason (Ornstein 1938; Hahn 1971; Proctor 1991; Delanty 2001). The 'Baconian settlement' that created the Royal Society was indeed a tension-ridden truce, which provided scientists with funds and freedom from censorship, but which simultaneously asked them to trade away all rights to moral disputation and political engagement. Charles II's ordinance to the Royal Society stipulated that its fellows were free to share knowledge and intelligence with 'strangers and foreigners [...] without any molestation, interruption, or disturbance [...] provided [this be] in matters of things philosophical, mathematical, and mechanical'; while the famous 1663 Statute proclaimed that 'The business and design of the Royal Society is – To improve the knowledge of naturall things, and of all useful Arts, Manufactures, Mechanick practices, Engynes and Inventions by Experiments – (not meddling with Divinity, Metaphysics, Moralls, Politicks, Grammar, Rhetorick, or Logick)' (Ornstein 1938: 108–109n). The French Academy similarly acknowledged its dual allegiance to both science and the crown (Hahn 1971: 3ff.). Once again, a marriage of convenience brought together the aspiring professionals and the state, which, under Colbert's direction, followed a deliberate policy of scientific advancement, technical education, and mercantilist expansion. Accepting state pensions and research funding, the scientists proved willing to serve the king and the national good, even though they were forced to abdicate a considerable freedom of action in doing so. As a close affiliate of the French Academy, Huygens reported that its meetings would never allow a discussion 'of the mysteries of religion or the affairs of state' (cit. Hahn 1971: 12).

This self-interested political settlement between budding science and established politics was characteristically sealed by essentialist philosophical distinctions that consecrated a rigid separation between the spheres of power and knowledge, of ideas and things, and of values and facts. Following both Platonic and Cartesian rationalism, the Baconian empiricist charter reaffirmed the ideal of the disinterested quest for truth; and in step with Cartesianism, it also began to sharpen the emerging distinction between moral-political and natural-scientific knowledge that had been constitutively absent from the Platonic tradition, which united the True and the Good (see Chapter 4). Anticipating Hume and Kant, Bacon advised scientists not to mix 'the proud knowledge of good and evil' with 'the pure knowledge of nature and universality'. A political compromise, which provisionally settled a local, interest-ridden, and unstable form of neutrality, was hence derived from and legitimized by a set of universalist claims about a putatively 'essential' separability of the realms of factual explanation and moral evaluation. This logic also informed the foundational categories that provided philosophical coverage for the Royal Society's effort to prise open a 'calm space' in which its members could differ civilly, without the hectic animosity that pervaded religious, moral, and political disputations. The solution, as we saw earlier, was to draw a firm boundary around the domain of 'matters of fact', which were seen as mirroring nature and hence as dictating absolute agreement, separating them from items that might be otherwise and from which absolute and permanent certainty should not be expected. Natural facts were seen as the products of discovery rather than invention, whereas scientific theories and hypotheses (and, by logical extension, ethical and political beliefs) were man-made and therefore remained forever contestable. In this manner, the idea of factuality served a double epistemological and social-political purpose, delineating a secluded form of life where dissent could be managed and kept within safe bounds (cf. Shapin 1984)[14] – or, in the terms of my temporal argument, where experimentation, deliberation, and conflict could be slowed down to a 'civil' pace, in a special time zone that remained free of the pressures, emotions, and uncertainties of moral and political dispute.

As noted, the Baconian view of the neutrality of 'matters of fact' (like the Cartesian idea of deductive universal truth) sealed a political truce between science and society, which like all truces continued to nurture an internal tension. Temporarily bowing to the factual predominance of the state, the philosophers and academicians gradually filed a far more pervasive claim to social responsibility and cultural leadership, on the strength of the 'last instance' theory of cultural determination that was always implicit (and often explicit) in their expressions of the calling of 'true philosophy' to rationalize the world. This anticipated progression from enlightenment to power constituted a general feature of all the scientifically grounded and philosopher-led utopias that emerged from the Enlightenment, from Bacon's sketch of intellectual rule in *New Atlantis* (1627) via Condorcet's *Fragment sur l'Atlantide* (1804) up towards Marxist scientific communism. In this optimistic view, the universal diffusion of true knowledge would dissolve all ignorance and prejudice, inaugurating a vast spiritual and moral regeneration of humankind. As the embodiment of truth and rationality, the philosopher self-confidently stood out from his fellow men, placing himself above the ordinary duties and obligations of civil life; superior because unprejudiced and altruistic, he was able to control his baser emotions and to resist the corrupting influences of immediate and ephemeral gain (Hahn 1971: 37–38, 42). This purity of motive and monopoly of truth turned the intellectual into 'someone special' and his trade (philosophy, science) into a special practice, which assumed a dignity and sovereignty that lifted it high above the more haphazard and indisciplined reasoning of ordinary mortals.

The Idea of the University

Tracing further passageways in the history of scientific autonomy, we may follow the unfolding of this double theme of political compromise and intellectual arrogance in the classical conception of *Bildung* and the equally classical 'idea of the university' (McClelland 1980: 101ff.; Liedman 1993; Wittrock 1993; Rothblatt 1997; Delanty 2001). From Burke and Coleridge to Humboldt and Newman, this 'idea' precisely functioned in essentialist Platonic fashion: pure and lasting, elevated above all apparent transformations, it defined an inherent purpose and a supreme dignity for an institution that would otherwise be seen as 'just another' corporate body, advancing its guild-like interests in mundane competition with other institutions and professions, and changing its identity in reponse to utility and circumstance (Rothblatt 1997: 1–10; Habermas 1989). While the ideas of truth and *Bildung* made special people out of philosophers, scientists, and educators, the 'idea of the university' turned their institutional habitat into a special 'transcendental' place, invoking a unique 'spirit' that had the power to transmute an ordinary occupation into a sublime calling with a compelling moral authority and an irresistible civilizing force. As in the case of the scientific societies and academies, therefore, the demand for institutional autonomy was put forward in absolutist epistemological terms, which contravened the uneasy pact that was simultaneously struck with the powers of the state. In this respect, the modest promise of political quietism, which secured a provisional freedom for reason (and for philosophy, its guardian angel) from political surveillance and moral censorship, was never divorced from more grandiose political claims about ultimate intellectual and cultural leadership.

It is remarkable, for example, how philosophy continually hovered between opportunistic modesty and conquistadorial presumption in Kant's famous tract on *The Contest of the Faculties* (1798). The basic (and explicit) structure of irony of this intriguing text featured an almost biblical reversal according to which 'the last shall some day be the first': philosophy, which presently found itself in a subordinate and marginal position, should by natural right emerge at the centre of academic life and at the apex of the cultural hierarchy. Still reflecting a political context in which the state exercised a large amount of discretion over university teaching and research, Kant pleaded for a dramatic reversal of the academic (and political) status quo, through which philosophy would attain the cultural sovereignty it could rightfully claim on the strength of its missionary guardianship of truth and right reason. This foundational project, which intended to transform philosophy into the supreme arbiter of the sciences and of culture as a whole, would to a large extent be institutionalized in the post-Napoleonic university reforms initiated a decade later by von Humboldt, Schleiermacher, and Schelling.

From the outset, Kant was concerned to clarify that the customary division of the university into two ranks, separating the three upper faculties of theology, law, and medicine from the lower faculty of philosophy, was undertaken from a 'governmental' rather than a strictly scientific perspective. The upper faculties were only called such because the government took a direct interest in their teachings, while the faculty that 'merely' concerned itself with the interest of science, and which used its own judgment about its teachings, was referred to as the lower one. The subjects of the three upper faculties (eternal, civil, and physical well-being) were those through which the government secured the strongest and most durable influence on the people; it hence reserved the right to sanction their teachings, while those of the lower faculty were left to the reasonable judgment of the learned themselves. The government did not teach directly itself, but commanded those who, in accepting its offices, were contracted to teach what it wanted (whether this was the truth or not). It was therefore essential that there was also a faculty that remained independent of governmental commands; that itself issued no commands, but was called upon to evaluate them all; that enjoyed the freedom connected with the 'interest' of science, i.e. of Truth; and in which Reason was legitimated to speak publicly. Kant gleefully speculated that the reason why it was still called the 'lower' faculty had to reside in human nature, because someone who ordered others around usually thought of himself as more eminent than another, who might be free, but commanded no one (Kant 1992 [1798]: 25ff.).[15]

Clearly, in this presentation, the jurisdictions of reason and government were still zoned off by means of a principled demarcation between truth and power. The case for autonomy was grounded in a disinterested 'truth interest' sharply opposed to the utilitarianism and 'political correctness' that characterized the teaching of the three higher faculties. In Kant's judgment, the faculties of theology and law were not based on reason and truth but on faith and positive law, while medicine was more ambiguously situated (and stood closer to philosophy) in deriving its procedures not from the authorities but from 'the nature of things themselves'. Since truth was the main thing, the utility that the three upper faculties promised to government was of secondary importance; hence philosophy was justified in turning all forms of knowledge into objects of investigation and criticism (Kant 1992 [1798]: 45). This opposition between truth and utility was extended into an equally sharp differentiation between scientific rationality and everyday belief. The people were not naturally inclined to use their reason but lived according to their inclinations; they did not primarily see their welfare in terms of freedom but in terms of salvation, security of property, and health (the more immediate utilities furnished by the 'higher' faculties). Hence the people had no patience for philosophy; they wished to be led (even to be duped, as the demagogues said), not by scholars but by the more practical men (clergymen, lawyers, doctors) who were educated by the higher faculties. In a progressive chain of enlightenment, the philosophy faculty was to instruct the higher faculties, which educated the professionals, who in turn would lead the people. Kant's optimistic expectation was that this regular advance of both academic ranks towards greater rationality would eventually prepare the way for a complete removal of all governmental restrictions upon public deliberation and judgment.

Despite the acknowledgment of a strong political presence within the academy (and of the necessity of public censorship; 1992 [1798]: 55), the Kantian compromise also separated university and government by emphasizing that the core conflict was an internal, intra-academic one (cf. Readings 1996: 57–59). The conflict of the faculties was essentially a conflict about the boundary between science and politics *within the university itself*. In Kant's (political) metaphor of the parliament of learning, a 'governmental' party on the right faced an opposition party on the left that 'spoke the truth to power', but strictly within the limits of the academic field. The government could therefore never be a direct object of academic criticism; in return, it should not endeavour to intervene in scholarly debates. Since the government was not interested in the truth of the teachings of the higher faculties but only in its own utilitarian advantage, it was beneath its dignity to decide about truth and play the part of the scholar (1992 [1798]: 57). Another reason for governmental reticence was that these debates were not played out on the public stage. If a conflict was brought before the judgment seat of the people (who, as Kant certified, were not competent to judge in scholarly matters), it simply ceased to be a scholarly debate. 'Neologists', who handed over scholarly questions to the decision of the people, effectively renounced the learned professions and promoted anarchy. Even if the last were to become the first, and philosophy were to rise to academic prominence, it would therefore not be interested in a direct exercise of power, but merely in the counselling of the powerful (the government) who might 'find in the freedom of the philosophical faculty, and the insights gained from it, better means of achieving its ends than its own absolute authority' (Kant 1992 [1798]: 59–61).[16]

The Kantian compromise thus reinforced a double demarcation between science and the political and science and the popular, sheltering 'truth' and its favourite spokespersons in an autonomous space distanced from all concerns of utility and urgency, and protected from the anarchy of the crowd and of the 'neologists' who pandered to it. Kant's stated goal was to create the conditions of felicity for a *concordia discors, discordia concors* (a legitimate rivalry between philosophy and the other faculties and between philosophy and the government, although perhaps not between philosophers themselves), structurally similar to the 'civil dissent' that the members of the Royal Society had sought to institutionalize within the calm space of the experimental laboratory. According to Kant, this limited competition was not equal to a war, or a heroic and final conflict 'about the Mine and Thine of learning'. It could do without divisive polemics and attacks upon traditional ways of teaching, since the conflict was settled by the verdict of truth, the decision of a judge (reason) that had the force of law (1992 [1798]: 47, 53–57). Remarkably, therefore, the settlement that officially separated the laws of government from the laws of reason did not conceal the fact that in both registers there was talk of laws, legislation, decision, and the implementation of force, and that the philosopher, who vicariously owned the only 'lawful' form of rationality, emerged as the ultimate legislator in the realm of culture. The modest plea for a limited autonomy and freedom of judgment imperceptibly shaded into a far more arrogant foundational project that consecrated philosophy as a master discipline that put the other sciences in their place and sat in judgment over all knowledge claims and all cultural expressions (Habermas 1990: 1–4).

Unhastening Science

Announcing a new episode in the institutional separation between politics and science, and a new framework of intellectual autonomy, the academic project of von Humboldt and his fellow founders of the new University of Berlin in 1810 sought to 'externalize' this Kantian conflict between philosophy and the 'politically correct' faculties to set up a more radical barrier between the university and the state (cf. Readings 1996: 62ff.). Von Humboldt fully shared the Kantian conviction that science should be pursued for its own reward and in its own tempo, not for short-term political commitments or economic-utilitarian applications, and similarly based the need for its autonomy on an epistemology of unitary and legislative truth. The notion of *Bildung* implied an apolitical theory of education that centred on the individual's efforts to achieve spiritual perfection through *Wissenschaft* (Rothblatt 1997: 22). Such *Bildung* should be allowed to develop through the 'unforced and disinterested collaboration of scholars with one another' in an atmosphere of solitude and freedom (*Einsamkeit und Freiheit*) – freedom, that is, from political and practical distractions (Proctor 1991: 71; Humboldt in Anrich 1964: 377). The requirement of the 'unity of research and teaching', when coupled with that of the 'freedom to teach and to learn' (and supported by the crucial idea that scientific knowledge was never found in a finished state but moved along an eternally receding frontier), defined an autonomy that was more principled than the Kantian one in deriving *all* academic teaching immediately from unfettered and disinterested scholarship. As a result, philosophy came to occupy the dominating position within the university that Kant had already envisaged for it. As the foundational science (*Grundwissenschaft*) that supervised and controlled the articulation of all natural and human disciplines, philosophy would guarantee both the unity and progress of scholarship and the cultural improvement of the nation as a whole (McClelland 1980: 124–25).

A state-organized autonomy of science and scholarship thus doubly shielded the institutions of higher learning from political intervention and from societal (more particularly economic) imperatives. State authorities were to avoid intervening into the life of the mind, except by providing the necessary resources for teaching and research. As Kant had already insisted, it was clearly in the interests of the state itself to grant the university this internal freedom, since a true *Kulturstaat* would only stand to benefit from the unfettered creativity and rationally unifying power of science and scholarship. The settlement hence once again included a basically affirmative, collaborative relationship between an 'apolitical' university and a funding but non-interventionist state (Habermas 1989: 108–109). Humboldt for example still insisted that the professors of the new university should be appointed by the state, not by the university itself. Himself a diplomat, civil servant, ambassador, and for a brief period Prussian minister for education and culture, he clearly sought to balance the needs of the Prussian state against those of academic liberty (Wittrock 1993: 318). From the outset, as Habermas remarks, it remained unclear how the university's mission of enlightenment and emancipation could be squared with the abstention from politics that was the price it had to pay for the state authorization of its freedom (1989: 113).

Another widely influential definition of the 'idea of the university', which similarly derived its autonomy from universalist values, was offered in Newman's description of the university as essentially a place 'for teaching universal knowledge'. Although Humboldt and his collaborators emphasized scientific research rather than instruction as the primary duty of the university, they clearly shared an emphasis upon the university's universality and the disinterested nature of true knowledge. Newman likewise limited the notion of freedom of inquiry by a demand for loyalty to the state and a strong sense of needing to avoid scandal or shocking the popular mind (Rothblatt 1997: 13). This 'duality of detachment and elite integration' (Wittrock 1993: 326) was also implied by the English 'clerisy ideal', which offered a close approximation of continental beliefs in expertocracy and the leadership of intellectuals. Spanning a full century from Coleridge, J.S. Mill, and Arnold towards T.S. Eliot and Mannheim, the notion of a 'clerisy' combined a radical claim about the intellectual superiority of the educated elite with a fiduciary agreement with the state in which it offered its unique services in exchange for a financial endowment. This arrangement shielded the clerisy from the market and from the swings of public opinion, but simultaneously turned it into an apolitical 'attached class' that agreed to remain loyal to the permanent foundations of English culture and society. Taken to an extreme in Arnold's work, the clerisy ideal claimed that art and science could offer sure guidance in a world of confusion, and that artists and intellectuals were best positioned to address the pressing moral issues of the age, enjoying a moral superiority over government not merely as citizens but also as experts in political matters (Rothblatt 1997: 389–93).

In this fashion, the 'idea' of the university, as it evolved from Kant via Humboldt and Newman towards twentieth-century philosophers such as Jaspers, remained a firmly Platonic idea, which gravitated towards the notion that institutions were the (necessarily imperfect) embodiments of essential, exemplary forms of life that centred around a unificatory 'spirit'. But this idea of essential unity progressively yielded to a recognition of fragmentation and dispersal with the gradual emergence of empirically based and specialist disciplines in the second half of the nineteenth century, which increasingly deflected philosophy from its totalizing and encyclopedic mission and undermined its traditional claim to provide a stable normative integration for the sciences and for the university as a whole. Ironically, the Humboldtian university ideal gradually hollowed out to accommodate a type of disciplinary organization of science that represented a virtual antithesis of its original unified conception of knowledge (Wittrock 1993: 315, 319–20). Beginning with the split between the natural and the cultural sciences, which was soon deepened by a further secession of the social sciences, an irreversible process of specialization and differentiation set in that gradually destroyed the holistic orientation of *Bildung* and research, prefiguring the transformation of the university as the repository of unified and universal knowledge into a 'multiversity' that loosely bundled together a great diversity of functions, interests, disciplines, schools, and methods.

Clearly reflecting these new advances of specialism and professionalism, Max Weber, in the early decades of the twentieth century, explicitly sought to counter the Humboldtian ideal of *Bildung durch Wissenschaft*, which anachronistically obliged the university to educate its clientele in a rounded conception of life and a broad set of moral ideals. For Weber, by contrast, science constituted an expert's calling and a distinctive profession (*Beruf*); and instead of cultivating the virtues of good citizenship, scientific knowledge should function as a neutral instrument in the service of value interests that were essentially external to science. While the Kantian-Humboldtian vision of scientific autonomy and neutrality, in rejecting the Baconian preference for practical knowledge and utility, had already widened the gap between analytical and normative concerns that had been opened by the rationalist and empiricist schools of philosophy, it had not separated facts from values in anything like the axiomatic sense that neo-Kantians now derived from principles such as the 'absolute heterogeneity' of empirical and value statements, the essential subjectivity of value preferences, and the undecidability of all value disputes. In this sense, Weber's new defence of the value-freedom of science sharpened up the neutrality ideal and further excavated the philosophical divisions between science, politics, and society, giving the final polish to the liberal compromise of 'unpolitical' autonomy that would govern the science–society contract for much of the remainder of the century.

The Benefit of the Doubt

The preceding historical narrative, which has taken seven-league strides linking Plato to Descartes, Kant and Weber, has once again foregrounded the way in which the differentiation between the institutional domains of science and politics was conceptually enforced by reified philosophical distinctions that separated truth from power, knowledge from interest, and matters of fact from matters of value. The truth/power and fact/value distinctions consistently operated both to certify and to reify a pragmatic and relative distinction between different cultures of speed and different routines of publicity. Absolutist philosophical categories were deployed as power tools, initially in order to wedge open an autonomous space (and ensuring a slower pace) for independent thinking and research, and subsequently to sustain and progressively widen it by means of political compromises with the political powers. By invoking the deep naturalness of ontological divisions, essentialist dichotomies simultaneously created these divisions and adopted a reificatory form to render their 'labour of division' invisible. Essentialism is an ontological politics that acts upon the world without disclosing its performative action. This critical interpretation has dramatically altered our view of the ideals of purity and neutrality that have shored up traditional conceptions of scientific autonomy and academic freedom. Such ideals, and the essentialist dichotomies that hold them in place, have never effectively performed anything but an impure and interest-ridden compromise in which political tolerance and non-intervention have been 'bought off' by an unpolitical quietism on the part of science. By silencing and disowning its own pragmatic politics of knowledge, the universalist principle of neutrality has therefore always been a deeply self-contradictory idea.

This critical approach may help to recast a core problem of political epistemology, which focalizes all disputes about neutrality and corollary principles such as agnosticism, detachment, methodological estrangement, impartiality, and symmetry: how can we institutionalize the art of doubt without turning it into an all-purpose methodology, or into a universalizable set of rules that can be adopted and applied by subjects who are in essence mutually substitutable? The point is to prise away the idea of methodical doubt from the neutrality ideal and the theme of disinterestedness, and to put it into a more contextualized and political frame. This 'particularization' can partly be accomplished by attending more closely to the *temporality* and the temporal *complexity* of doubt. It has already been suggested above that radical Cartesian doubt, if liberated from its transcendental purpose and methodological voluntarism, can be more mundanely reinterpreted as a resolve to *slow down* the process of investigation and to defer specific decisions about what the world is like. Entering a doubt is essentially a way of 'buying time' and of braking the speed of thought in order to observe the item of interest from 'close up', to 'give it another look'. This doubtful slowness escapes the conventional polarity between detachment and involvement, since it requires a passionate proximity to the object of investigation, or a form of interested absorption in it that simultaneously expresses a perplexity and an uneasiness. As a (provisional and finite) refusal to know how the world is divided up or how particular entities are categorized, it cannot be methodical in the sense of lighting an indiscriminate 'bonfire of the dualities'. Rather than a licence to doubt everything everywhere at any time, which is the secret ally of the equally extremist ambition to establish indubitable certainties, *dubitandum* is hardly ever *de omnibus* but is always contingent and site-specific: it is structured by what I have called a 'social distribution of doubt' and the precarious condition of strangerhood (Pels 2000: 43–48, 210–11). This unequal situational spread of the proclivity to doubt also has a temporal dimension. This is not merely because doubts arise, aggravate or are allayed at different times following different experiences of anxiety or reassurance, but also because there are multiple rhythms of doubt and a cascade of (weak) temporal boundaries that separate situations in which decisions cannot be postponed from situations in which doubt can creep a little further and reality can remain 'on hold'. Intellectual autonomy therefore needs a politics of time that liberates and legitimizes the art of doubt by protecting it against the threat of multiple acceleration.

Durkheim's ruling that we must consider social facts as 'things' offers a good example of a pose of doubt and 'ignorance' that is elevated into a neutral and universal prescription. To approach a social phenomenon as a thing is indeed to confront it as a *strange* and *unknown* entity, in the face of which one must adopt a radical attitude of distanciation. The closely derivative rule that 'one must systematically discard all preconceptions' is placed in immediate lineage with Descartes' methodical doubt and Bacon's critical theory of the idols (Durkheim 1982 [1895]: 72; cf. Pels 2000: 55–56). We simply do not know with certainty, Durkheim observes with more than a hint of sociological certitude, what social institutions such as the state, sovereignty, the family, property, morality, the market, political democracy etc. actually are (1958: 40; 1982 [1895]: 65). Ordinary consciousness, political ideologies (especially socialism and communism), as well as rival academic disciplines such as psychology and political eonomy have been *too hasty* in their judgments; they have not taken (or did not have) the time for proper research and reflection. But science cannot be improvised: it is intrinsically *patient* and does not precipitate itself towards its conclusions. As I critically noted in the introductory chapter, this temporal perspective on scientific work remained packaged in a neutralist conception of scientific method that not merely delegitimized all lay and competing professional knowledges but also staked out an almost limitless terrain for sociological study and for the remedial certainties it promised to deliver.

Although in many ways opposed to the Durkheimian conception of definitional truth (which derived from Cartesian rationalism) and the Durkheimian 'fetishization of facts' (which derived from Baconian empiricism), SSK and STS have relived some of these paradoxes and dilemmas in their methodological prescriptions of impartiality and symmetry. By initially targeting the demarcation between true and false beliefs or the normative boundary between science and non-science, early SSK and STS easily drifted towards the factual side of the value/fact dichotomy, withdrawing all critical epistemological judgment in favour of close empirical descriptions of actual scientific practice. Symmetry, insofar as it implied holding two normative opposites at (equal) arm's length, similarly implied and invited neutrality. When Callon and Latour decided to step up this symmetry in order to include the society/nature or human/non-human divide, their ontological shift did not divert but only hardened the principle of agnostic doubt that already stemmed from the first epistemological decision. Extended symmetry merely generalized the attitude of uncertainty from Nature to Society, and hence not only discouraged the taking of sides in natural-scientific controversies, but now also refrained from judging how actors analysed their own society: 'No point of view is privileged and no interpretation is censored' (Callon 1986: 197, 200). In the course of the Epistemological Chicken debate (see Chapter 6), this systematic refusal to know how the world was carved up produced an effective rhetoric against a sociology of science that insisted it could 'stand upon social things' in order to explain natural things (cf. Pickering [ed.] 1992: 301ff.). Latour has recently repeated this 'Durkheimian' profession of creative ignorance in a reply to French critics who reclaimed the symbolic nature of social life and the ontological privilege of humans. These 'tenants of a priori ontologies', as he calls them, refuse to do the political work of 'experimental metaphysics', believing that 'humans are born in language, live in meaning, and die in the symbolic'. While they seem to know how the world is composed, he doesn't (Latour 2001: 142–44).

The extension of constructivist uncertainty from Nature to Society hence correlates with an a-critical shift that refuses to censor the actors themselves, even when (especially when?) they violate a more traditional critical epistemology/ontology by committing natural or social reifications. In this sense, the principle of extended symmetry presupposes a strong ethnomethodological hypothesis about actorial flexibility and competence and a radical ontology of hybridity, both of which are referred to significant empirical groundings. Instead of applying a generalized protocol of doubt that detonates all the received modernist distinctions, extended symmetry more nearly introduces a *local* perplexity that simultaneously appears to premeditate and frame its own solution (cf. Barnes 2001: 343–44). Symmetries in one department are bound to engender asymmetries in another; where and when which dualisms are levelled is not a problem of how radically one is prepared to doubt them, but of which knowledge-political context offers itself *for doubting particular dualisms and not others*. As I have attempted to make clear, a more thorough dissolution of the fact/value problem must also revoke SSK's initial epistemological decision to symmetrize truth and error, because it reflexively reintroduces a critical gap between observational and actorial accounts and hence partially lifts the agnostic ban on 'outside' critical judgment. We cannot merely 'follow' the actors in order to 'display' their actions and articulations. If we keep running after them in this indiscriminate way, we merely copy their speed instead of slowing down in order to take proper analytic distance. In this minimal sense at least, we need to recover Durkheim's spirit of critical demarcation from what he wrongly identified as the 'prejudgments' of 'common sense', and *take the time* to subject them to doubt.

The symmetrism of STS therefore suffers from a residual neutralism and a methodological voluntarism, insufficiently realizing that the inability to decide what something is always relies upon a contextual decision, and is not the same as principled 'undecidability'. It is not surprising that critics sometimes interpret postmodernist 'difference talk' as an excuse to avoid judgment (Smith and Webster 1997: 106). Against this must be pitted a more overt *politics* of doubt that is also a situational politics of time. Beck makes an important gesture in this direction by advancing scepticism and insecurity as the major planks of the political programme of a radicalized modernity. In his view, we must exploit the contradiction between the system's promises of technical, social, and political security and the elemental *loss of security* that is induced by reflexive modernization. The introduction of insecurity 'may help to achieve the reduction of objectives, slowness, revisability and ability to learn, the care, consideration, tolerance and irony that are necessary for the change to a new modernity [...] Everything a couple of sizes smaller, slower and more open to the opposite, to antagonism and refutation, as is proper for self-assured doubt.' A society that is beset by productive self-doubt, and is therefore 'incapable of truth', strictly speaking 'cannot develop or uphold any construction of an enemy' (Beck 1997: 168–69). As hinted at before, this view is nicely coterminous with Latour's recent reinvention of an ethics of care, an epistemology of weakness, and a politics of deceleration with regard to political ecology (1999b; forthcoming). Interpreted in an appropriately reflexive mode, however, it still affords a *weaker* conception of truth and the naming of *weaker enemies*, who could be identified in circular terms as those who resist such a temporal politics of doubt, and who synchronize their actions with the speed and security of faster systems in the social triangle.

In the course of this book, I have made an effort to demystify and reframe the social theory of knowledge by placing it in a more pragmatic spatio-temporal perspective. This resolve has repeatedly produced critical distance from epistemological programmes that retained various residues of the truth/power and the fact/value polarities. In previous chapters I have discussed Elias's conception of the moving balance between involvement and detachment, Habermas's focus upon the normative rationality of uncoerced communication, and Bourdieu's defence of the autonomy of the scientific field, as involving such modernist and rationalist residues. I have also suggested that, insofar as these paradigms were concerned with establishing the 'conditions of felicity' of scientific doubt, or with specifying the institutional conditions for intellectual autonomy, they were translatable into the post-transcendental terms of a politics of unhastening that marked a new (albeit weaker) demarcation from the more hurried systems of politics, business, and the media. They all seemed to work from the intuition that one needs time in order to cultivate an 'untimely' attitude towards the present. But unlike Nietzsche and Foucault, who remained painfully aware of the existential 'strangeness' that must preside over such an attitude, all of these theorists appeared to want to neutralize this existential and 'fateful' context in favour of an impersonal methodology of distanciation. Detachment, for Elias, implied not so much a decision to *involve* oneself *slowly* with an object, but a resolve to eradicate all the residues of 'fervour', which Durkheim had already identified as the root cause of all intellectual bias. Habermas's ideal speech situation, I suggested, was ideal not so much because of its felicitous retrieval of the unavoidable transcendental conditions of communicative interaction, or its capacity to engender the 'uncoerced coercion' of the better argument, but because of its more mundane projection of a specific temporal delay that enabled a wider population of speakers to speak and to be properly heard and understood. Bourdieu's field theory of scientific autonomy was considered less inspiring for its residual objectivism and 'scholasticism' than for its more pragmatic emphasis upon the variable of 'audience size' and the temporal boundaries of science as they were pressurized by an accelerating market logic and by the 'fast thinking' of politicians and media professionals.

The issue of ethnographic distance or anthropological strangerhood, as it is recouped in the various ethnomethodologies of science studies, offers another illustration of this uneasy 'stand-in' relationship between an epistemology of neutrality and a pragmatics of temporal delay (cf. Fabian 1983: 71ff.). In the classic picture, of course, the ethnographer does not merely travel to a distant and unfamiliar place but is also transported 'back in time', not merely because she often chooses to observe a culture that beats to a slower rhythm than the fast-paced Western metropolis, but also because the observer herself enters into the more unhastened time-frame of participant *observation* rather than participation *tout court*. If ethnography carries the additional connotation of describing phenomena and accounts 'from the inside', as they appear to the participants themselves, it once again focuses the reflexive dilemma of whether the ethnographer must 'keep up to speed' and 'run along' with the actors, which implies running the risk of descriptively duplicating and endorsing their views (the classical pitfall of anthropological and ethnomethodological relativism), or should operate the systematic deceleration that shifts the temporal constraints of everyday life in such a manner as to enable her to 'creatively lag behind' and to follow the actors *critically*.

On the plus side, however, we may also rejoice in the fact that every interesting epistemological idea that has been recovered in the preceding pages appears translatable without too much effort into more pragmatic temporal strategies. My discussion of intellectual anti-politics in Chapter 5, which was inspired by the work of Bourdieu and Habermas, yielded a conception of knowledge politics that emphasized its downward gearshift from the relative intensity and publicity of 'big' politics. This time-sensitive view was especially effective when brought into critical contrast with radical notions about politics as the 'life of danger' or 'life on the knife-edge', which were rampant in right-wing political existentialism and other varieties of political totalitarianism. However, it also proved effective against politicizing conceptions of science such as those advocated by actor-network theory, insofar as they emphasized an ethos of total mobilization and the hardness of reified facts over the long and steady work of fact-making itself, or chose to consider the two faces of science (strong purification and slow mediation) *together* without any urge to bring them into critical contrast (or to articulate an ethics of weakness). Anti-politics as time politics also interestingly re-charges the role of public intellectuals who, according to a vision shared by Konrád, Habermas, and Bourdieu, are politically engaged but drop the ambition to become full-time politicians or to appropriate their professional place. As regular commuters on the knowledge-political continuum, they routinely mediate between different cultures of speed, simultaneously gearing up the circulation of scientific findings by importing them into a more public and popular arena, and 'calming down' the feverish intensity of political debate by taking 'time out' for a more subdued and reflective commentary.

In Chapter 6, I have similarly attempted to 'rescue' the preferred STS methodology of symmetry from both its methodism and its neutralism, introducing a chronopolitical version of symmetrical 'delay' that differs from strong forms of epistemological levelling by effectively reinstating a weaker demarcation between truth and error. The revival of this critical capacity overturned both the relativism and the residual value-freedom of what I described earlier as the 'value-free relativism' of the Wittgensteinian turn. The category of 'third positions', I argued, was better able to capture the actual multiplicity of observational placings and timings, because it offered mixed modalities of engagement and distanciation that were not easily 'methodologizable' but emerged from lived experiences of proximity and estrangement, being reflexively implicated and interested in the relevant fields of controversy rather than standing on the sidelines in the pose of an impartial spectator. In Chapter 7, this reflexivity, which itself only represents another 'step up' in the long history of the art of doubt – and thus in the art of deceleration – was finally unpacked as producing a radical but virtuous narrative circularity. The admittance of 'circular reasoning', while having an obvious 'arresting' effect on high political velocities, simultaneously offers a criterion of 'better' knowledge that remains radically immanent because it itself issues from a circular proposition. Nevertheless, it does make available a capacity for critique that, while admittedly weak, can still be mobilized against knowledges that frantically attempt to break or square the circle in search of more solid epistemological foundations.

Remixing Culture, Politics, and Economy

One of the most absorbing and challenging questions for social and political theory in the new century is how we can reconcile the modernist-liberal insistence on differentiation, boundary maintenance and institutional autonomy with the postmodernist or poststructuralist emphasis on de-differentiation, interlocking, hybridity, and boundary transgression. Is it possible to escape sideways from the conventional stand-off between the liberal-democratic 'art of separation' and the totalitarian 'art of fusion' – which may equally come from the left and take an economic form (Marx) or emerge from the right and adopt a political form (Schmitt)? How can we negotiate a 'third way' between the essentialist ontology of different value spheres that follow immanent logical imperatives (Kant, Weber, Popper) and the residual postmodern essentialism of the seamless web of fusion, fluidity, and flow, which so uncomfortably approximates the unmediated complexity of the totalitarian ideal?[18] How, in other words, do we retain a robust notion about domanial autonomies and the equilibrium of partial spheres (cultural freedom, political freedom, and market freedom) while acknowledging that their 'codes' increasingly mix and interpenetrate? To put it even more succinctly: how can we reconcile the liberal idea that culture, politics, and the economy are (and should be) *somewhere*, i.e. that they occupy a definite sociological time-space in an articulated totality, with the equally pertinent post-liberal idea that they are (and should be) *everywhere*, washing over their boundaries and spilling over into adjacent domains and time zones?

In Chapter 2, I made an initial effort to tackle this paradox – which poses an inseparably sociological-empirical and normative-political question – by drawing the contours of a social triangle that accommodates (but admittedly simplifies) the multi-directionality of transgression and boundary erosion in three converging processes of culturalization, politicization, and economization, while simultaneously measuring distances and leaving gaps that balance out the distinct institutional complexes and professional roles. The social triangle is made up of three gradients that simultaneously separate and interlock the three domains of culture, politics, and the economy across a multiplicity of weak and contested boundaries and a plurality of hybrid institutions and intermediate zones: the knowledge-political continuum, the culture-capital continuum, and the power-property continuum (Fig. 8.1). The knowledge-political continuum spans the entire institutional range from 'blue sky' academic research through strategic or policy research to politically commissioned 'intelligence work' and 'big' professional politics. The culture-capital continuum likewise stretches from academic research or independent art through editorial and publishing work, commercial information services, independent consultancy (e.g. in headhunting or career counselling), commercial laboratory science and managerial/entrepreneurial work, spanning the whole institutional range of what has been identified as the 'cultural economy'. The power-property continuum extends from politicians to commercial pollsters to shop stewards to economic consultants to NGO functionaries to private managers and entrepreneurs, and similarly includes the whole range of mixed institutions that populate the modern political economy.

This triangular cartography pictures a 'centreless' society that no longer recognizes last instances or meta-codes, and in which neither culture, politics nor the economy is able to occupy the place of ultimate constitution. Modernist social differentiation was typically both enabled and contained by an overarching and overdetermining factor such as an infrastructural economy (Marx), a totalizing politics (Hobbes, Schmitt) or shared values, representations, and communicative understandings (Durkheim, Parsons, Habermas). All three institutional corners of the social triangle have been theorized – in a definite historical sequence that has elicited sharp intellectual rivalries between the 'master sciences' that acted as their privileged stadtholders – as embodying the 'part' that represented or even constituted the whole, offering the functional anchorage that guaranteed and sealed its articulated unity. In dialectical fashion, all three 'factors' were capable of shifting between a narrower sectoral

Intellectual Autonomy and the Politics of Slow Motion

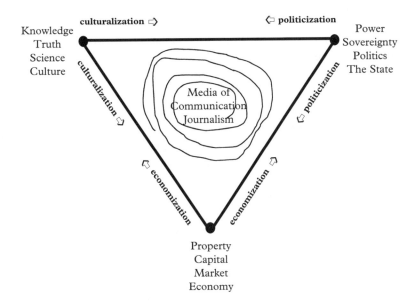

Figure 8.1: The Social Triangle

definition, in which they were one part of the distinction that generated an institutional 'other' (e.g. the state vs. civil society), and a broader totalizing definition that expressed the underlying *unity* of the distinction itself (e.g. the political as social-ontological essence). As Luhmann has stated it: while until now all self-thematizations of society have fallen into synecdoche, attributing primacy to a single component subsystem and hypostatizing it into the social whole, we must now dismiss such hierarchical theorizations and contemplate a 'horizontal' model of social differentiation (1982: 344ff.). Previous historical claims for the institutional *autonomy* or *specificity* of politics, economics and culture have invariably and typically been empowered by means of essentialist assertions of the *sovereignty* or *primacy* of either Power, Property or Truth (cf. Rasch 2000a: 165). Such priority claims have also doubled as dissimulated power bids by different 'master sciences', which conveniently interpreted the self-attributed foundational nature of their objects as proof and warrant of their own sovereign status in the hierarchy of the sciences.

This totalizing purpose constituted the shared historical ambition of both 'Aristotelian' political philosophy and 'Smithian' political economy, pushing these master sciences into a a long historic rivalry with regard to the 'last instance' status of either the polity or the economy (Pels 1998: passim; Von Beyme 1991: 93–94). Contemporary 'extension bids' by schools such as

Chicago 'economic imperialism', neo-Marxism, and rational choice theory from the economic side, or Foucaldian, feminist, and neo-Schmittian approaches from the side of the political, drag this disciplinary contest into the present. But the 'politicization thesis' and the 'economization thesis' were always compounded and offset by a third 'culturalization thesis', which identified a 'last instance' role for a variety of cultural factors, including philosophy, science, morality and art. Originating in the French philosophical Enlightenment (cf. Voltaire's assertion of the primacy of culture) and in the German Romantic tradition of Kulturgesellschaft (cf. the priority claims of von Humboldt and his fellow university reformers), this third intellectual imperialism has also been channelled through much of sociology and modern cultural theory, up to and including some of the more methodologically aggressive of postmodernist cultural studies, which intend to absorb everything into a universal aesthetic.

Nevertheless, within a 'flat' and symmetrical rather than hierarchical model of social articulation, which dismisses all last instance claims and *pars pro toto* generalizations, there must be a way of acknowledging that not only is culture everywhere, but so are the political and the economic (cf. Von Beyme 1991: 343), without jeopardizing the simultaneous ambition to protect a broad range of post-liberal autonomies. We need to multiply the Latourian slogan that science is politics 'continued by other means' by symmetrically adding that culture is politics *and* economics 'continued by other means', as politics is culture *and* economics and the economy is likewise culture *and* politics by other means. Such de-differentiations find expression in and are performatively enhanced by a whole array of currently salient 'intrusive' metaphorizations, among which that of 'knowledge politics' is merely one. The same expansive conception of the political sprawls across the recent literature on 'governance', which covers both governmental and non-governmental institutions, including those of science and market-based enterprises (e.g. Fuller 2000a). A similar and related metaphor is the Foucaldian one of 'governmentality', which similarly stretches across the entire social field to include new formations of subjectivity within micro-settings such as the family or school, or the new technologies of the self that result from the generalization of entrepreneurial practices and rationalities to all other forms of conduct (Barry et al. [eds] 1996; Dean 1999; Rose 1999). The ubiquitous talk about 'enterprise culture' itself legitimizes the use of commercial and marketing criteria and of managerial and entrepreneurial work-styles for restructuring various non-economic domains. Like the concepts listed above, it displays a typical elasticity or part/whole ambiguity that enables it to oscillate continually, in this case between a narrower conception about 'buying and selling' and the absorption of state

Intellectual Autonomy and the Politics of Slow Motion

industries and public utilities by private companies, a broader one that aims at the restyling of non-market institutions along entrepreneurial lines, and a still more expansive one that advocates personal qualities or dispositions that reflect a 'spirit of enterprise', such as initiative, energy, self-reliance, and willingness to take risks (Keat 1991; Fairclough 1991; Ray and Sawyer [eds] 1999).

Beck's views about the generalization or 'unbinding' (*Entgrenzung*) of politics into the 'sub-politics' of science and technology, health, the family, and the economy provide another example of such discursive overflow (1997: 185, 194, 231). His related model of the 'political bourgeois' may serve as an attractive template for delineating similar isomorphic fusions in other fields that still guard their relative autonomy. Environmental issues and the issue of technological risk have turned industry and business into political undertakings, in the sense that the running of the enterprise itself can no longer be executed behind closed doors. In reflexive modernity, business policy increasingly requires legitimation, justification, and negotiation with company staff, the general consumer public, and the political apparatus itself. There is a growing need to perceive and anticipate public consequences and influences, and to integrate them into internal company decisions. This situation gives rise to new types of connection between economy and politics which also synthesize the system codes of economy and *publicity*. For Beck, these are only the beginnings, 'the already documentable experiments of synthesizing business *and* politics within the economy' (1997: 127–31).[18] This formula intriguingly balances specificity and assimilation by suggesting the feasibility of a political economy that does not erase the functional independence of either politics or the economy. In the same manner, we might begin to think of a culturalization of business and a commodification of culture in a cultural economy that does not issue in the 'abolition' of either culture or economy. Or, completing the round of the social triangle, we may contemplate forms of politicization of culture and science that do not jeopardize the relative autonomy of the cultural field.

Before turning to a more detailed assessment of the place and time of science, let me briefly sketch in some of the assimilation drives along the other sides of the social triangle. The culturalization of politics has generated new forms of political culture that express a new intimacy between politics, popular culture, and the media. The result of this symbiosis is a mediatization of the public sphere and a personalization of political competition that tends to render obsolete older forms of representation by numbers and clearly demarcated political-ideological constituencies, introducing elements of an 'audience democracy' in which politicians act on the media stage before a volatile public of political consumers who are prepared to shift their trust and allegiance much more rapidly than ever before (Manin 1997; Scott and Street 2000; Corner and Pels [eds] forthcoming). Such an 'aestheticization' of politics is not confined to historical right-wing regimes, but has become part and parcel of the spectacle of the mediatized political establishment of the West. It fosters a partial integration between the cult of celebrity and the cult of power (Marshall 1997: 206ff.) that runs parallel to a commodification of political messages, party 'brands' and personality-products, creating a 'promotional' political culture (cf. Wernick 1991: 124ff.) that multiplies the practices of political advertising, commercialized opinion polling, and media-oriented canvassing and campaigning.[19]

Culturalization –in the sense of both scientization and aestheticization – also enters into the heart of the capitalist economy, which has taken a much-discussed 'cultural turn' (Lash and Urry 1994; Thrift 1999; Ray and Sawyer [eds] 1999; Slater and Tonkiss 2001; Du Gay and Pryke [eds] 2002). This shift notifies both the 'intelligence' of an innovation-driven informational capitalism and the aesthetic sensibility of a 'discursive', sign- and design-intensive mode of production that both borrows from and determines a broader 'aestheticization of everyday life' (Featherstone 1991). Schumpeter's famous description of the entrepreneur as a virtual intellectual or originator of 'new combinations' (cf. also his realistic 'entrepreneurial' modelling of the politician as a competitor for votes and political spoils) is expanded by the current enthusiasm for artistic play and creativity, and for artists as role models for business management, by the vastly increased importance of the business media industry (e.g. the magazine *Fast Company*, cf. Thrift 2002), and by the more general attraction of a 'fast' image promotion (e.g. through celebrity endorsement of products) and of the celebrity system itself for the managerial community (cf. celeb CEOs such as Gates, Murdoch, Branson or Bezos, or the continued success of management gurus). All of these mutations have rearticulated the relationship between economic base and cultural superstructure in such a way as to firmly embed them into each other and to disable all primacy claims and last instance causalities.

A prominent feature of the triangular model is its focus upon the logic of *publicity*, and thus the presence of the media industries and of professional journalism, as a conduit for all of these contagions and de-differentiations. We already noticed that, in Beck's approach, the emerging 'code synthesis' between business and politics effectively coincided with the economy 'going public' (1997: 128–29). Similar analyses of the discursive or cultural constitution of the economy and of the ubiquity of 'promotional culture' likewise point towards the increasing media-saturation of business life, and hence to the growing functional importance of new strata of designers, image-makers, spokespersons and wordsmiths within the economic field. In the political field, the most significant changes are likewise directly triggered by the rise of the media and the inclusion of media professionals as immediate partners in the political game and hence as powerful gatekeepers of the new 'emotional democracy'. The spillover of the logic of celebrity from the entertainment industries into other spheres such as politics, science, art, sport, and business is an important motive force in de-differentiating the system as a whole, and is directly related to this growing preponderance of media technologies and of those who own and operate them. Media intellectuals, business journalists, and political spin-doctors collaborate with these owners and gatekeepers in drawing their respective fields closer into the vortex of professionalized publicity, modifying and mixing the interests that the modernist 'art of separation' used to assign to clearly distinct and autonomous realms.

As will be recalled, this prominence of the media was directly linked to the all-important factor of accelerating social speed, and to the fusion of time horizons that was the contagious effect of fast media dominating the slow. I have systematically singled out the slow motion of scientific processes as their most distinctive feature as compared with the higher velocities and faster practices of business and politics, provisionally disregarding the internal time differentials which, for example, oppose the lightning speed of new globalized financial markets to the slower pulse of a 'political' or corporatist economy (cf. Santiso 2000 on the accelerated temporality of market-induced privatization drives), or those that reflect a bureaucratic or legal 'stalling' of high-pressure political decisions. In the critical phenomenology of scientific practice that was attempted in Chapter 2, I also alluded to these temporal distinctions as being distributed around the 'technological' axis of faster media of talking and imaging vs. slower media of reading and writing, contrasting the relative 'orality' of professional politics and economic management with the relative 'literacy' of scientific research. If the impact of the mass media at the hub of the triangle is to increasingly draw all other fields into the spiral of public communication, it has the important effect of rehabilitating such oral (and especially, audiovisual) cultures as compared to textual ones. This reinvention of orality and visuality in a technologically mediated form restores the sensual immediacy of face-to-face talk and the visual richness of the image, and hence tends to speed up the tempo of communication and alter the forms of public address in all domains, potentially threatening a relatively unhurried and secluded practice such as science.

Reinventing Academic Freedom

It has been a primary aim of the present book to argue that a new settlement between science and society needs to sever the time-honoured linkage between autonomy and neutrality and between academic freedom and value-freedom. In the new constitutional contract, autonomy and distanciation, or solitude and freedom, must be bargained for in terms of a self-conscious and explicit politics of knowledge.[20] As I have repeatedly suggested, this bargaining effort and boundary work, rather than rehearsing points of philosophical principle, should focus upon creating the pragmatic conditions of felicity for a politics of temporal unhastening and the spatial isolation that is protective of it. The separations between knowledge and power and between values and facts have always been a misleading (because reificatory) way of distinguishing between slower and faster technologies of communication. The specificity of the power/knowledge and fact/value structures that make up science is rather found in a special 'inertia' (an 'elongation' of time), which contrasts with the time-space mediations of more speedy practices. Such a politics (and economics) of diminishing speed and spatial withdrawal may begin to delineate a new idea and ideal of academic freedom that transcends the rigid parameters of the traditional liberal compromise. In order to complete this chapter's (and this book's) argument, we might begin to sketch the contours of this new 'idea of the university' in the following way.

The model of the knowledge-political continuum began by establishing the relative autonomies of science and politics, not by installing a single, ontologically etched boundary, but by constructing a cascade of weaker boundaries, lesser thresholds, and penetrable enclosures that simultaneously divided and connected a multiplicity of mediating institutions. In this rendering, the resistances that hold the two autonomies in place and in check are not constructed through a purification of hybrid structures but precisely by a proliferation of agencies and agents in an intermediate zone of engagement, which activate a number of protective rings that slow down and hold off direct infiltration either way. A structural politics of autonomy is not so much served by anxiously avoiding forms of fusion or de-differentiation between science and politics but precisely by *multiplying* and *diversifying* them. Filling in the middle with a strong set of intermediaries is still a good Aristotelian recipe for democratic equilibrium. This 'stretching' and 'cascading' of the middle range also diversifies our picture of science, which is now seen to include a variety of structures and roles that themselves differ in relative 'politicalness' or 'marketability', and hence run along different speed lanes (e.g. 'fast' disciplines such as engineering, biotechnology or computer science vs. the 'slower' because less marketable human and social sciences) or inhabit different time zones (e.g. vice-chancellors, heads of department, admin staff, directors of research centres and laboratories, technology transfer officers, regular teaching staff, postdoctoral researchers, lecturers on study leave etc.). The grand inside/outside boundary is broken up into a plurality of lesser ones and moved 'inward', turning the negotiation about various mixtures of science and the social (political accountabilities, stakeholder interests, market priorities) into an everyday routine within scientific establishments themselves.

Rather than identifying a single activity or purpose (the Humboldtian extreme of curiosity-driven research in 'solitude and freedom', which is unconstrained by considerations of utility or external relevance) as normatively legislative of and hierarchically superior to all others, as happens in traditional versions of the university 'idea', this alternative view enables us to acknowledge and defend the *whole range* of mixed activities and practices that make the 'multiversity' work, without distinguishing too closely between inside (e.g. laboratory benchwork) and outside (e.g. the scramble for funds), or splitting epistemological hairs over the status of pure, applied, strategic or commercial research (cf. Elzinga 1985; Gibbons et al. 1994; Latour 1987). Such 'pure' research, as I have repeatedly noted, is never so pure as to be liberated from all material interest. It remains thoroughly political in the sense of being deeply embroiled in the micropolitics of fact-making, discursive negotiation, reputational rivalry, external and institutional assessment, and jostling for funds and positions that structure the background (and foreground) of all scientific endeavour. I have called it 'self-interested' rather than 'disinterested' science. But it is also singled out by a unique relationship to time, working to an exceptionally slow schedule and a long-term cycle of investment that both fosters and requires an 'untimely' and 'peripheral' attitude towards the present and the status quo (cf. Barry et al. [eds] 1996: 5).

If we now add the complementary angle of science–economy relations along the other leg of the triadic model, while retaining this factor of differential speed, we can finally articulate the 'post-liberal' equilibrium between differentiation and de-differentiation or between code synthesis and autonomy for science in a comprehensive manner. My point is that we precisely require the 'essential tension' between these contrary processes to reach a new workable definition of what makes science special amidst other major human pursuits. For this purpose, we need to acknowledge the legitimate intrusion of both economic and political codes in the inner fabric of science, not merely as orienting metaphors but also as substantive mechanisms for structuring action. The relevant question is not *whether* political and economic elements do in fact enter into science, but *which forms* are permitted to spread *how far* into it, on the understanding that boundaries will remain porous and continually contested. In many respects, of course, the economic idiom converges with the political one in accentuating the microeconomics of scientific discourse, the investment in subjects, disciplines, methods, and ideas, the accumulation of reputational capital, and the challenges of scientific entrepreneurship. More crucially, it focuses what the political idiom would identify as a struggle for scientific authority in the central requirement of regulated *competition* in a market where intellectual assets and resources are unequally distributed. Both vocabularies also 'disenchant' the universities as special places of unified and transcendental learning, and redefine them as 'not-so-special' knowledge-producing organizations that are themselves internally divided and must now share their epistemic monopoly with a host of new contenders (cf. Bauman 1997).

More recently, a stronger view of the economization of science has taken hold, which highlights the rise of the 'entrepreuneurial university' and the entrepreneurial scientist as instances of the spread of a more general culture of enterprise (Etzkowitz 1983; Etzkowitz and Leydesdorff [eds] 1997; Slaughter and Leslie 1997; Etzkowitz, Webster and Healey [eds] 1998; Currie 1998; Robins and Webster 1999). Reflecting the general loss of distinction between 'high' academic culture and 'low' popular culture, the pressures of increased political or bureaucratic accountability in a contagious culture of audit (Power 1997), and perhaps most significantly, the force of intensified competition in intellectual markets that are increasingly globalized (and anglicized), a new honeymoon is announced between the rationalities of science and the market. This entrepreneurialism not merely champions a market-resourced financial autonomy to make up for the decline in state-derived and block-grant funding, a more businesslike structure of academic and research management, and a more consumer-friendly approach to the (fee-generating) student 'market' and other stakeholder constituencies, but also supports the cult of 'enterprising selves': energetic personalities who are self-reliant, confident, responsibility-oriented, and willing to take risks, in eager pursuit of the opportunities and rewards offered by this 'brave new world' of high competition.

If we take a longer historical view of the configurational dynamics of academy and society, we notice that this new characterization of the university's social mission performs an intriguing switch of allegiance, which substitutes an 'economic' coalition for a 'political' one, even though the vocabularies that channel these alignments are also convergent and continuous (cf. Grit 2000: 105–21). This switch illustrates how autonomy may flourish precisely by alternating strategic partnerships with political and economic forces rather than by the strict observance of an 'essential' principle of neutrality. Both the 'critical'

Intellectual Autonomy and the Politics of Slow Motion

university of the 1970s and the 'entrepreneurial' one of the 1980s and 1990s effectively sank the classical ideals of the value-freedom and cultural sovereignty of science by permitting significant intrusions of social elements into the ivory tower. But whereas the former sought to hold capitalist commodification at bay through a politicizing drive that invoked the protective assistance of a democratically reconstructed state, its entrepreneurial counterpart shifted the line of fire by adopting economic metaphors and linking up with market forces and funds in order to minimize its dependency upon a state bureaucracy seen as inert, stifling, and over-protective. In shifting its institutional allegiance, however, the university did not 'sell out' but remained focused upon the same relative autonomy that it had attempted to secure with the different means of political activism. Like the earlier Baconian compromise between science and the state, this 'new deal' with the market was a devil's pact that was meant to strike a precarious balance between social orientation and institutional autonomy.

Continuities also emerge on the 'dispositional' level. The 'critical' university celebrated a form of individualism that contrasted the self-reliant, vocal, and nonconformist attitudes of a politically conscious avant garde with the 'apolitical' collectivism of narrow-minded careerists, defining scientific expertise not as a saleable commodity but as a social asset that should be made universally accessible and democratically accountable (Grit 2000: 112ff.). The entrepreneurial university, in combating the 'culture of dependency' and the bureaucratic sluggishness that was partially attributed to previous experiments in radical democracy, extended these values of enterprising and risk-taking individualism within the new market frame, while laying more emphasis upon knowledge as a private asset that could be patented, licensed and marketed in a businesslike way by both individuals and institutions.[21] The factor of relative speed once again provides an interesting angle on these paradoxes of continuity and discontinuity. The 'critical' university evidently wished to break the inertia of a paternalist and authoritarian culture of governance by entering the political speed lane. The entrepreneurial university embraces the even faster vocabulary of managerialism and market efficiency in order to distance itself from political stagnation and adapt to the new accelerations of a knowledge-intensive and technology-driven economy.

If, as claimed, politics and the market thus legitimately and meaningfully intrude into the inner texture of academic work, what is the specific difference that delineates the latter's proper place and function? One already belaboured point is that science is politics and/or economics continued by *slower* means: those of observation, experiment, comparison, inscription, calculation, unhastened conversation, protracted conflict, and low-gear competition. In this framework of a considerably *decelerated* politics and economics, demarcation problems arise and autonomy is potentially threatened whenever the struggle for scientific authority (or the competition for scientific capital) is deflected towards the faster stakes of publicity, celebrity, profit-making, and commodification. This happens, for example, when academic politics is diverted from the representation of *idea*-forces and *intellectual* interests towards the representation of large numbers or collective interests. 'Political correctness', indeed, is basically about the (deeply worrying) application of the law of large numbers to scientific 'truth'.[22] Alternative problems arise when competition and commodification are (too much) collapsed into each other, giving rise to forms of academic capitalism or state-subsidized entrepreneurship (Slaughter and Leslie 1997) in which the search for profit increasingly dictates the distribution of reputational capital-and-power (e.g. when human capital is freely bought and sold, when publications are considered less reputable than the ownership of big grants or marketable patents, or when potential sales win out over academic merit). Similarly, if the shy and sheltered game of science is speeded up by the high visibility and velocity of media exposure (to large audiences, for big money), the fast currency of celebrity and universal fame takes over and erodes the slow honours and the local name-making that are considered typical for more 'fundamental' scientific work.

A full characterization of the specifity of science must therefore simultaneously include the transgressions that turn it into an intrinsically political and economic affair, and the specific difference of its unhastened temporal profile that balances out and 'neutralizes' these transgressions. Unhastening science without strong internal competition or a strong discursive politics, without the energetic jostling for recognition and a distinctive name, tendentially drifts towards simple idleness, leisure, or *otium* taken too literally. On the other side of the balance, the pace of circulation in the scientific republic or the scientific marketplace must be shifted down in order to escape the 'time famine' of more stressed-up polities and economies. Given the absence of 'natural' or essential restraints, this balancing act requires that the interference of these faster cultures is never direct, but is 'retarded' by the delays and translations imposed by a set of filter institutions, mediating zones, and hybrid professions. The danger of an acute flooding of the scientific agenda is averted through an intensification of the liberal 'art of separation', which produces more robust structures of autonomy by simultaneously lowering and multiplying the boundaries between science and society. In one respect at least, such a post-liberal conception of autonomy improves upon the liberal pact of neutrality and political compromise, since it *enlarges* the bounds of academic freedom by creating open zones for critical engagement and partisan research. As Root

(1993) has also argued, the liberal principle of value-neutrality, while protecting academic life from external determinations, simultaneously limits academic freedom by disciplining and repressing political speech in the disciplinary space of the sciences themselves. If this liberal ban were to be lifted, academic freedom would be enhanced by allowing the free expression and communication of critically distanced and 'untimely' views, not merely about matters of fact but also about matters of moral and political value.

This new pragmatic characterization of academic and scientific freedom returns the university to the profanity of an ordinary enterprise, while simultaneously suggesting a normative standard that may define its 'integrity' against the fast timings and public spaces of politics, business, and the media. No longer does the university float as an island of high learning in a threatening sea of commerce and banality, as it did according to the idealist Platonic vision (cf. Kumar 1997: 30); but there is something about its islandic quality and the classical demand of solitude, *Freischwebendheit* and 'leisure' that is worth preserving after the demise of such tyrannical conceptions of the academy's civilizing mission. This extreme point on the knowledge-political and culture-capital scales no longer sets up an 'essential tension' with the opposite extremes of power and profit, or marks a point of absolute normative sovereignty. But it nonetheless identifies a range of specification that holds relatively true for all scientific work, and may be used as a benchmark or warning bell that alerts scientists to counterproductive accelerations of their professional timetable. These weaker identities centre around a specific unhastening of thinking, inquiry, and debate; a predilection for long-run investments rather than short-cycle proceeds; and a refusal of the demands of urgency, which enables one to study issues selectively, one at a time, in comparative privacy, and appropriately shielded from the glare of publicity and instant accountability. As Kumar notes, universities still offer 'breathing spaces in life's course', sites of cultural exploration where there is sufficient *time* to stretch oneself, in a *place* that escapes the tempo of the 'real' world 'outside' (1997: 29, 31).

In a 'nanoculture' that glorifies speed and faces a continual shortening of its time horizons, and in a celebrity culture that adores publicity and fame,[23] the university may provide a resort of stillness and slow motion where strange thoughts may be cultivated in relative liberty and leisure. What Habermas identifies as 'the special communicative forms of scientific and scholarly argumentation', which in his view hold the various learning processes within the university together (1989: 124), may well be divested of their rationalist afterthoughts in order to make room for a less ontologically charged ethic of decelerated discourse. The 'ideal speech situation', I repeat, is ideal not so much as a result of its deep structure of communicative consensuality, but for its critical slowdown of the pace of communication. The threat of system colonization then precisely includes the 'stressing up' of scientific work by an aggressive culture of auditing, accountability, and profitability that forces the pace of research and publishing (e.g. research assessment exercises, teaching quality assessments, competitive grant-bidding, course accreditation by professional associations, short-term teaching and research contracts, campus commercialism). Political standards of user relevance and stakeholder interest, and economic criteria of profitability and efficiency, tend to shorten the investment cycle of scientific creativity.[24]

The idea of 'unhastening science' therefore simultaneously includes a new, anti-essentialist description of the special nature and niche of scientific work, and a normative plea for the 'unwinding' or 'chilling out' of science in a fast society that threatens to reduce its independence by imposing a law of universal acceleration. Obviously, in proposing this 'normative fact', and making the description continuous with the norm, the idea of 'unhastening' deliberately dodges the fact/value dichotomy. It favours a new politics of time management and speed control, in a self-interested defence of scientific autonomy against the excesses of the productivist scramble for 'output' and the political pressure for 'relevance'. It is a paradoxical proposal, because science is itself political through and through, and interestingly entrepreneurial, but still needs the protection of a special 'time capsule' in order to be able to resist unsolicited intrusions of Big Politics and Big Business.

As I intimated before, this does not mean that benchmarks such as societal relevance, accountability, outreach and transfer, marketability, or budgetary efficiency are out of place in the rarefied atmosphere of the scientific republic, since such criteria obviously monitor the success of the hybrid forms of knowledge politics that straddle the distance between pure and more strategic or commercial types of research. Instead of putting up a principled resistance against the intrusion of economic and political discourses, work-styles and performance criteria, we may cautiously welcome them while simultaneously installing speed limiters that brake the velocity of the hasty cultures and liberate areas of stillness within which research may follow a more leisurely and reflective path. The pressures of internationalization, of publish or perish, the demands of academic entrepreneurialism and independent fund-raising, the scramble for image-recognition, branding and market positioning, the growing importance of media savvy and the intellectual celebrity system, the unstoppable machinery of research and teaching assessments, the substitution of tenure-track positions for short-term contracts, the endless restructuring drives, and the resultant hypertrophy of academic management (interminable meetings, lots of nervous talk, little sustained writing) jointly produce a major acceleration that threatens the weak boundaries of 'fundamental' or long-term intellectual production, imposing the tempo and habitus of 'fast deciders' upon a practice whose product is weak and uncertain and may not be marketable for some time to come.

The 'time wars' that rage along the broader reach of the knowledge-political continuum (and within the social triangle as a whole) are replicated and enhanced by internal academic conflicts that immediately recall the Kantian contest between the 'higher' (and faster, because politically correct and user-relevant) faculties of theology, medicine, and law, and the 'lower' (and slower) one of philosophy. Bourdieu has similarly distinguished between faculties or disciplines that are 'temporally dominant' and those oriented more towards scientific research, adding that 'nothing better sums up the set of oppositions established between those situated at the two poles of the university field than the structure of their time-economy' (1988a: 62, 64, 98–99). Indeed, in the current 'entrepreneurial' situation, a new 'contest of the faculties' is emerging that pits the faster technosciences, which are more user-oriented and market-driven (biotechnology, business and management studies, new materials science, information systems, intellectual property law), against smaller brokers and slower earners (such as philosophy, the humanities, the social sciences) who 'lag behind' in a timescape that is increasingly stressed up by the culture of academic capitalism.

However, in juxtaposing these two cultures of speed, both Kant and Bourdieu assume a rift between 'scientific' reason and 'worldly' power or profit that my knowledge-political continuum has attempted to dissolve. A more pragmatic chronopolitical defence of scientific autonomy would rather accentuate the gradual variation and the weakness of the boundaries between these disciplinary timescapes, while still reclaiming the right of the slower subjects to critically analyse and resist the demands of technoscientific acceleration. The mission of disciplines such as literary studies, historiography, cultural studies and sociology is not to catch up with the technosciences, but to follow them from a critical distance at a slower analytical pace. In Eriksen's option, universities may either adapt themselves to the market and increase their speed, or 'redefine themselves as countercultural institutions that embody slowness, thoroughness and afterthoughts' (2001: 118). My argument is rather that the university's unique function is both to preserve useful variations in 'scientific speed', and simultaneously to act as an institutional placeholder for the slowest tempi of critical analysis. Time is indeed of the essence in order to be able to think 'untimely' thoughts. To foreshorten the temporal horizon of the collective imagination by bowing to a fast synchronicity critically undercuts the crucial social function of playful curiosity, 'idle' estrangement and 'lateral' thinking, and hence paradoxically risks slowing down the rate of social creativity and innovation over the longer term.

This 'temporal strangerhood' of the university still defines it as an enclave of productive alienation, where the universal synchronization and densification of social time is retarded by the cultivation of slow motion and temporal breaks. This refusal to bow to the 'tyranny of the moment' (Eriksen 2001), to the pressures and addictions of high speed, may enlist academic scientists in a broader movement of 'time pioneers' who liberate themselves from the fast programmes and the 'pace ideology' of a society for which everything urgent is *ipso facto* of the highest relevance. In choosing their own time horizons, such time pioneers are veritable heretics who seek to reverse or slow down the all-powerful accelerating trend (Hörning, Gerhard and Michailow 1995: 136–37, 164). In this way, the idea of slow motion as nonconformism invigorates an 'idea of the university' that rearticulates the critical and potentially subversive function already assigned to it by Kant. As a site of resistance against the dominant imperative of high speed, the university may indeed institutionalize an adversarial culture (Readings 1996; Fuller 2000a; Delanty 2001), which, in cultivating the art of taking one's time, liberates autonomous spaces for a further radicalization of doubt. It is in this spirit that we may slightly adapt the Marxian exhortation: Slow Workers of the World, Unite!

Epilogue: Weak Social Theory

Feel the Force!

Philosophy and science, including social science, have always been strongly attracted to strength. Reason has always sought to be a force, denying its own politics of knowledge in order to become that much harder to beat. Epistemology has been a war machine, wielding logical proof and an all-purpose methodology as its main pieces of ordnance. The whole point of Truth is, of course, that it is compelling and compulsory, that you cannot escape from it, that it has an inherent tendency to spread, and that there is something seriously amiss with the unconvinced or recalcitrant ('Surely you can't deny that ...'). It is closely allied to the classical ambition to know the world *in toto* with total certainty. Nowadays, of course, we no longer want our truths to be absolute, immediate reflections of eternal Platonic Forms; we are prepared to let our doubts creep a little further than Descartes, who doubted only for a brief moment before fastening upon great and indubitable certainties. A small weakness crept in when Popper dared to replace the metaphor of the rock-bottom of science with that of the 'swamp', but foundationalists and other truth-sayers have not been much impressed. Everybody who is anybody in epistemology still seems interested in pointing the gun at everyone else in order to forcibly extract agreement.

Classical Power Talk

Sociology has likewise been aggressively macho, loving itself for its methodological prowess, its legislative potency, and its governmental (or guerilla-fighter) power speech. It must now learn to be soft, weak and vulnerable, refusing to flex the muscle of a male-dominated epistemology. Let us look again at the

Unhastening Science

record of the founding fathers. Marx's grand ambition was to trace *iron* laws of capitalist development, while Durkheim infamously defined social facts as *things*. In both cases, these objective laws and things were taken to exert constraint on individual behaviours; but they also conveniently constrained 'common sense' and all alternative interpretations of the social world. Weber spoke approvingly of the Greek invention of conceptual thought as wrenching opponents in a logical-argumentative vice. Weber fans would probably like to exonerate their favourite intellectual celebrity on the grounds that his view of disenchanted rationality removed at least the 'value' half of all that we can say about the world from scientific certification. OK, whatever we say about the 'good life' cannot be too strongly expressed. But Weber also left the other half of the house of knowledge standing, assuming that the results of logical reasoning and fact-finding would necessarily be compelling 'even to a Chinese'. Stubborn Weber fans could now retort by citing his familiar modesty about the fate of all scientific work being overtaken in 50 years' time. But this is a silly view, not only because of the residual arrogance of claiming to last at least half a century in a single paradigm, but also because of the happy circumstance that Weber himself has already lasted (although with ups and downs) for twice that length of time.

Modern Hardness

'Strong hand' metaphors continue to proliferate copiously in modern scientific vocabularies. Evidence still needs to be hard, theory grounded, facts solid, results robust, methods rigorous, proofs decisive, arguments compelling, conclusions inescapably powerful, propositions firmly anchored in nature or reality. 'Conversation Analysis has emerged as one of the most powerful approaches to the study of human communicative interaction', says a leading voice in this eminently sceptical approach (Hutchby 2000: 55), clearly aware that he is wielding a performative that is meant to co-produce what it states. Self-styled relativists in science studies have advanced Strong Programmes, dismissing older sociologies of knowledge as 'weak', because they suffered from a 'failure of nerve' in the face of the as yet unassailed stronghold of natural science. Feminists, instead of embracing weakness as a matter of epistemological pride, have supported bitchy forms of 'strong' objectivity and 'strong' reflexivity. Habermasians crave for the 'uncoerced coercion' of the better argument, and rejoice in their ability to trap opponents in the logical vice of a performative contradiction (which is nothing else than a disguised infiltration of the critical theorist who turns a literal contra[-]diction between himself and his opponent into a hapless self-contradiction from which he is

Epilogue: Weak Social Theory

cowardly absent). Actor-network theorists routinely condone the many ruses and tricks of reification that working scientists employ in order to render their claims unassailable: total mobilization, hard facts, black boxes, singular spokespersons who speak in a big voice for all their defenceless actants. In this fashion, ANT's core political metaphorics simply reverse the conventional presumption of normative philosophers of science that truth is sufficiently strong to spread of its own accord. What these adversaries continue to share is a fascination for the pursuit of hardness and strength.

Slow Science

Speaking in comparative terms across the spectrum of social institutions, we could say that such craving for strength and certainty better fits the practices of politics and business than the production of scientific knowledge – a 'strong Britain' with a 'strong economy', shored up by a 'strong pound', as Blair and Brown would have it. In this constellation, weakness and uncertainty could be the typical contributions of (social) science to the shape of the world. It could say things that are interestingly feeble, shaky, risky, and weird. Political and entrepreneurial metaphors, work-styles, output indicators and leadership models wreak havoc when they are too diligently pursued in science. Expressed in slightly different terms: while politics and business are cultures of acceleration, inhabited by the 'fast cats' of a capitalism of speed, the economy of science reveals a much slower timescape. As I have argued in this book, science is a form of unhastening, which aims to reduce the breakneck speed of everyday and other professional communication. This argument can be developed into a new touchstone of demarcation and a new normative description of the specificity of social science: to cultivate a slow disposition in the face of the incessant gearshifts of a technology-, media-, and globalization-driven capitalism. The special calling of science is to give us the close-up, the slow-motion replay and the still frame rather than the hellish image blitz of the MTV model of entertainment.

Circular Reasoning

I have also proposed that social theory is necessarily weak and imperfect because it is deeply circular. If we accept that all our statements run in a circle (including the present one), we have hit upon a perfect formula for being in a weak spot. This is not relativism of the all-cats-are-grey variety. Let me clarify this by recalling Garfinkel's claim that people (not just scientists) use a 'documentary' method of interpretation to make sense of their everyday world.

Unhastening Science

The significance of events and actions is grasped in terms of broader structural and institutional patterns, which in turn inform people's understanding of individual events. It is a circular process in which each level is taken to account for, to derive from, or to elaborate on the other: instances are explained by patterns and patterns by instances. There is no way out of this circle: it is what all of us make do with in our everyday lives. However, as Pollner's (1987) work on 'mundane reason' has shown, we normally operate under the assumption that we live in the same underlying reality, and that any neutral observer, if placed in the same position, would see the same things as we do. Retranslated towards Garfinkel: we normally *fail* to acknowledge the deep circularity of our lived interpretations. In the routine conduct of everyday life, such documentation (of instances) and exemplification (of structures) takes on a reificatory character: we continually attempt to jump out of the circle. We are everyday essentialists. Or perhaps we should rephrase this as a relative condition and an empirical question: *to what extent* are actors in everyday life and in science aware of this documentary circularity? The issue is that this relative (in)ability to see and accept circularity supplies us with an entrance point for a normative critique (of reification and essentialism, on the level of both lay and scientific rationality). Garfinkel himself appears singularly uninterested in this, and is especially disinclined to put a normative gloss on his analysis. He appears content to 'follow the animal' and go wherever it goes, even if it begins to roar and becomes predatory.

Take It or Leave It

The new truth is weak and should be proud of its weakness. For weak social theory (WST) to say that a statement is powerful, that it carries force, or that it stands up well, is to find it distasteful or even slightly obscene. To say: 'That is a very vulnerable argument' is to pay a compliment to it. Practise the following phrase: 'I am persuaded by the delicate weakness of this account'. WST is not after a big arsenal of persuasive powers, but pleads a radical absence of compulsion and obligation; it makes no appeal to listeners that they should adopt any of these views (this includes my present proposals). You can take it or leave it. Claims to epistemological truth are like taking out an insurance policy on being able to herd all the others into conviction; you must be a very scared and deeply insecure person if you want to do that! Who do I wish to convince? Probably only those readers who want to go where I want to go, or, speaking with Latour, who want to go faster in the direction they themselves wanted to go in anyway. John Lennon charmingly compared his work to 'leaving messages', or 'sending postcards': 'Look, I'm doing this, what are you

Epilogue: Weak Social Theory

up to?' He is joined by the Dutch philosopher who recently admitted that he did not write in order to persuade others, but merely 'to find out what he thought about a subject himself'. Researching and writing as a way of persuading yourself; is that not a sublime admission of epistemological weakness?

Notes

Notes to Chapter 1

1. The term 'science' is used throughout this book in a broad sense refer to all academic subject areas including social and political studies, the arts and the humanities.

2. This is also vividly expressed in titles such as *Campus, Inc.* (White and Hauck [eds] 1998) or *College for Sale* (Shumar 1997).

3. From an early date, both Giddens and Thrift have incorporated insights from time-geography in social and cultural theory. According to the latter, it is one elementary insight of time-geography that mobility takes up both space and time, and that it makes only limited sense to privilege one above the other. Especially in the contemporary world of high mobility, notions of space as enclosure and of time as duration are reconfigured as space-time relations (Thrift 1996: 285–86). Urry similarly insists that 'mobilities are all about temporality' (2000: 105; cf. 1985; 1995: 18ff.). Hetherington and Lee (2000: 176) suggest that the 'blank' allows a switching between orders and disorders, presence and absence, but also between stasis and change. Cf. also Kern's view that speed is a natural juncture of the temporal and the spatial (1983: 3, 109ff.).

4. This is a simplified rendering of an overloaded and rather contestable typology which skips the categories of 'cyclical', 'retarded', and 'alternating' time, as well as the category of the 'explosive' time of acts of collective creation.

5. On the history of this 'great acceleration of life' (Nietzsche) see e.g. Kern 1983, Thrift 1996: 256ff. and Eriksen 2001.

6. In doing so, Smith of course also pioneered the disenchanted view of philosophy and science as 'nothing special', as ordinary utilitarian occupations in the societal division of labour. Not Feyerabend but Smith is the true godfather of the 'nothing-specialists' who populate our disenchanted age.

7. Deliberately reversing the pious connection between prayer and inaction or stillness (see Chapter 8 below), Marinetti submitted that 'if prayer means communication with the divinity, running at high speed is a prayer' (1972: 96).

8. If we want to live the 'intensive life', we cannot miss anything; we must desire to be everywhere at the same time, always ready for new opportunities, in order to turn every second into a success (cf. Jauréquiberry 2000: 258–59).

9. The Durkheimian collusion of the distinctions between fact and value, science and politics, and slow and fast paces is echoed in a typical comment by Hammersley on

Notes

Mills, who is accused of a failure to clearly separate his social scientific work from his political work, 'and one result is his conclusions are rarely supported effectively by evidence; he takes them to be valid too readily' (Hammersley 2000: 58).

10. In this context, one may also recall Weber's 'political' but simultaneously 'long' view of political economy as presuming a specific form of unhastening: 'In the final analysis, processes of economic development are *power* struggles too, and the ultimate and decisive interests at which economic policy must serve are the interests of national *power*, whenever these interests are in question. The science of political economy is a *political* science. It is a servant of politics, not the day-to-day politics of the persons and classes who happen to be ruling at any given time, but the enduring power-political interests of the nation' (Weber 1994 [1895]: 16). Expressions such as these of course simultaneously channel the professional interest of the intellectual spokesperson to replace or at least stand as indispensable advisor behind the throne of the acting politician. Cf. comparable expressions in 'conservative revolutionaries' such as Freyer (1931), who commissioned political sociology to represent the long-term interest of the German people as opposed to the short-termism of movements such as National Socialism.

11. In this sense Durkheim pursues the same trajectory of reification that Latour and Woolgar (1986) describe when they trace the stabilization of subjective claims into objective facts; even though Durkheimian epistemology is of course opposed to constructivism and assumes a discrepancy between 'actorial sense' and 'observer's sense' which is sharply dismissed by the two contemporary authors.

12. Cf. the way in which an early critic of Marxism such as Hendrik De Man anticipated what Bourdieu would later call its misrecognized 'theory effect': 'There is no human *science* of the future. There is only faith in the future; and among the forces which combine to bring this future into being, the faith in its coming is one of the most effective ...' (De Man 1928: 134; cf. Bourdieu 1990a: 181; 1991: 251).

13. As Marx wrote in 1844: 'Criticism itself has no need of any further elucidation of this object, for it has already understood it' (1972: 14).

14. Scott similarly suggests that the meshing of ethical and practical concerns in Weber's theory of action renders it paradoxical and contradictory: 'We appear to have a defence of a distinction between means and ends which itself presupposes its opposite – i.e. the entanglement of the practical and the ethical, of *Zweck* and *Wert*' (Scott 1997: 45–46, 55–56). Steding (1932) has most clearly brought out the deep contradictoriness of Weber's value-based postulate of neutrality and of the deeper *identity* of science and politics that hides beneath their formal separation. Such radical right-wing criticisms of the epistemology of neutrality offer remarkable anticipations, both in substance and form, of more recent poststructuralist and constructivist criticisms. See also the following notes.

15. Cf. already Freyer's critical discussion of Weber (1964 [1930]: 207–12). The book features a sustained polemic against Weber's *Wissenschaftslehre* and other 'logocentric' (*logoswissenschaftliche*) epistemologies, in the course of which Freyer maintains that a 'science of reality' must inescapably also be an *ethical* science, and closely approaches a constructivist or performative conception of social reality (cf. 82, 206). The Weberian theory of rationalization is interpreted as resting upon a basic *value* premise (157).

16. As Freyer expressed it in his sharp attack on the liberal university ideal, the boundaries between university and state or between *Geist* and *Macht* were not drawn 'as a result of power struggles or diplomatic negotiations but through a consideration of the essential law (*Wesensgesetz*) of both'. Not issuing from a real contest of forces, but from 'insight', it was self-evident that it was science that was obliged to expound upon these boundaries, since it recognized the freedom of research and teaching as its own

indigenous law. What could be said about its relationship to the state followed from this with logical necessity (Freyer 1932: 531). Science was hence overdeterminant in all the relationships between itself and any other field of facts or values: 'How science and politics should relate, is itself emphatically a scientific, not a political problem ...' (1932: 536).

17. This view of the politics of demarcation echoes Schmitt's fundamental idea that the decision whether a matter or a domain is unpolitical (e.g. the 'objectivity' of bureaucracy, the 'independence' of jurisprudence), is itself a political decision (1935: 17). See also his broader historical critique of the liberal 'neutralizations and depoliticizations' (Schmitt 1993). Schmitt's view is close to a Foucaldian analysis of liberal government as resulting from a *political* constitution of the distinction between civil society and the state, which does not imply a *lessening* of political concern with 'the conduct of conduct', but rather an alteration in its form (cf. Barry et al. [eds] 1996: 9).

18. If the principle of value-freedom is often summoned to support one *political* position against another (e.g. Weber's defence of *Staatsraison* and economic nationalism, Michels' defence of elitism, or Sombart's defence of racialist ideology), drives towards politicization are similarly usable to argue a preferred *intellectual* or *cultural* position over against a rival one. Such confrontations highlight ambiguities and contradictions that penetrate all forms of 'political correctness', which even in historical cases of intellectual fascism tend to preserve a (relative) autonomy of culture and a (bounded) freedom of the academy.

19. One must of course subtract the militant 'knowing better' which characterizes Marxian ideology critique ('waking society up from its dreams and illusions') and the equally militant resolve to 'make society better' which is conditioned by it. Otherwise Latour offers a constructivist version of the same dialectical agnosticism that inspires Marx, insofar as he undertakes to follow the actors of modernity in their full duplicity, displaying how the active contradictions of the modern constitution precisely work to proliferate the very opposite of what it ordains (see Pels 1995a).

20. This is why the issue of 'taking sides' inevitably re-emerges within science and technology studies (STS). See especially the 'Capturing' debate as discussed in Richards and Ashmore (eds) (1996) and in Chapter 6.

21. Competition is essentially a race against time. This might offer one reason why the sports world exerts such a deep fascination on an ever-accelerating world. How do we envisage 'slow' forms of competition, slow races, in which winners and losers may take a lifetime to emerge?

Notes to Chapter 2

1. This reminds us of the former salience of essentialist models of 'last instance' determination which fired a long-standing quarrel between spokespersons for the primacy of the economic, on the one hand, and for that of the political, on the other. Bourdieu for example robustly interweaves economic and political metaphors in his analyses of scientific, linguistic, and artistic fields, and freely proliferates the metaphor of capital, but still singles out economic capital as the effective grounding of all other types, and ultimately favours a critical *economy* of cultural and other social practices (Bourdieu 1986: 243, 252; 1991: 230; 1993b: 37–40). Foucault's writings offer an almost perfect mirror image of this unconsummated convergence, by similarly mixing the metaphors but ultimately pleading a shift from what is called an 'economics of untruth' towards a 'politics of truth', and setting to work a rather omnivorous conception of power (1980: 93, 98, 107ff.). In science studies, ground-breaking texts such as Latour and Woolgar's *Laboratory Life* (1979) or Knorr-Cetina's *The Manufacture*

of Knowledge (1981) initially adopted the gist of Bourdieu's quasi-economic analysis of scientific credibility and authority; but subsequent work in radical science studies has increasingly shifted the metaphoric emphasis in order to suggest that science is effectively a 'continuation of politics by other means'. On this tug-of-war between economic and political metaphors, see more extensively Pels 1997.

2. This differentiation is emphatically undertaken in order to neutralize the justified accusation about a 'reverse' or 'back-door' essentialism which suggests a total erasure of cultural boundaries (Gieryn 1995: 442n). The repetitive assertion of knowledge being power or science being 'nothing special' is not an analytic stop but merely provides us with a rather abstract beginning point. Mouton and Muller (1995) plot a similar continuum which describes a set of mixed attitudes of involvement/detachment, varying from pure, disinterested intellectualism through both 'weak' and 'strong' interventionism towards the opposite extreme of 'strong activism' or professional politics.

3. Foucault already stipulated that the power/knowledge coalition should not be viewed in terms of identity, as encircling one and the same thing (1994a: 132–33; cf. Barnes 1988: 169). The point of the slogan was rather to trace structural differences in the ways that various power/knowledge alignments were deployed in different fields and contexts (Rouse 1987: 111, 113; Keenan 1987: 12–13).

4. The discrepancy between 'doing' and 'saying' constitutes a foundational trope of all ideology critique and a fundamental problem of all spokespersonship. In my interpretation, the classical rift between self-consciousness and observed practice does little else but polarize *two sayings*: what actors themselves say they are doing and what I, the critical observer, say they are 'actually' doing – an imposition that is immediately reified in terms of a 'really existing' contradiction. See more extensively Pels 2000: 14–15, 21–25). We should hence be careful not to mistake the proposed critical phenomenology for a method of liberating 'objective' facts.

5. I must apologize at this point for cavalierly disregarding all differences in velocity between forms of political and economic decision-making, or, more generally, between the systems of politics and business themselves. Political or bureaucratic procedures (such as citizen participation initiatives, legal procedures, or administrative planning trajectories) are capable of seriously arresting and frustrating economic initiatives (e.g. by project developers). Juridical procedures are often resorted to both within and outside politics as effective mechanisms of deceleration. Within the economic field, it is especially the financial and speculative markets (see the high nervous tension of the stockjobber) and the high-tech and dotcom companies that whip up the pace of circulation. For a sympathetic perspective on the 'uneasy dialogue' between speed and democracy, see Chesneaux 2000. In highlighting the close association between speed and totalitarianism, Virilio's 'dromology' similarly implies an argument for the critical slowdown of the pace of politics, which has succumbed to the logic of potential catastrophe and must regain the time and space of deliberation and discussion, which are identified as its essential work (cf. Kellner 1999: 107). Santiso (2000) interestingly thematizes the temporal conflict between 'political sluggishness' and 'economic speed' in terms of contemporary drives for entrepreneurialism and privatization. Within science, one might similarly begin to distinguish between relatively 'fast' disciplines such as biotechnology, engineering, or computer science, which are increasingly stressed up by forms of 'academic capitalism' (Slaughter and Leslie 1997), and 'slower' ones such as philosophy, arts and human sciences, resulting perhaps in a new 'time war', i.e. a *temporal* version of Kant's 'contest of the faculties'.

6. If the act of writing tendentially separates the knower from the known and in this sense promotes objectivity, this feature clearly cannot be equated to epistemological objectivity as conventionally understood. The major distinctions between literacy and orality intriguingly map on to those between modernist and postmodernist philosophy

and sociology, insofar as writing fosters abstract, neutral, disembodied and context-free patterns of thought that are distanced from the lifeworld and the existential setting, categorical rather than situationally pragmatic, severed from narrative and rhetoric, and disengaged from the arena of human struggle (cf. Ong 1982: 42ff.). We may speculate that postmodernism appears to revert to some of the epistemological features of an oral culture, perhaps reflecting the rise of the 'secondary oralities' of telephone, radio and television (which adds the visual experience), which tend to reincorporate human communication into its broader existential setting.

7. Milton and other ethnographers report that members of so-called 'primitive' peoples such as Brazilian Indians, the Ituri of Zaire, and Australian Aboriginals incessantly talk to one another, which indicates the crucial importance of the oral transmission of culture as compared with the silence and privacy of reading in our own culture (Wood 1993: 44). Ong submits that 'writing is a solipsistic enterprise. I write a book which I hope will be read by hundreds of thousands of people, and that is why I must isolate myself from everyone' (1982: 101). Goody describes a typical academic workday as consisting primarily of reading, typing, and writing memos, during which virtually the only oral communication is voice-to-voice on the phone rather than face-to-face (1987: 299–300). This slow circuit of objectification is intimately tied to specific materialities: thinking is impossible without reading texts, holding a pen to paper, manipulating figures, drawing charts and matrices, scanning pictures, typing on a keyboard and looking at a screen, correcting printouts and proofs, and finally reading one's own article or book. Texts emerge from an intense visual (and silent) interaction between an object and a thinking mind, slowly taking shape through a circuit of revisions and fine-tunings; thinking is therefore inconceivable without this remarkable object-attention and this pragmatic (rather than epistemological) practice of reification.

8. See n. 5 for some necessary temporal differentiations between political institutions and political ideas.

9. The knowledge-political continuum is hence also a decision-making continuum, but it is one across which the objects, nature, and especially the pace of decisions 'decisively' differ. In this light, Habermas's 'ideal speech situation' is not a situation in which truth and power or knowledge and decision should be kept divorced in principled fashion, but a situation in which decisions can be postponed a little while longer, e.g. because one decides to 'wait' for others who so far lack a voice in the discussion.

10. The time limitations of journalistic practice typically generate simultaneous and mutually reinforcing processes of *personification* (relativity = Einstein; psychology = Freud) and *reification* (theories and concepts are caught in easily graspable metaphorical images; logical or formal relations are given ontological status) (Bucchi 1998: 6).

11. This includes the boxes-within-boxes schema proposed by Bourdieu, which still depends on a last instance economic model of determination (cf. 1993b: 38).

12. On the ontological primacy question, see more extensively Pels 1998: passim, and Chapter 8 of the present book.

13. This neo-Kantian 'value essentialism' also informs Merton's and Popper's specifications of the institutional norms of science and their ideal of scientific autonomy (cf. Gieryn 1995: 398–400, 441n).

14. Postmodernity as de-differentiation is also thematized in Lash 1990; Urry 1990: 82; Featherstone 1991; Lash and Urry 1994; Etzkowitz and Leydesdorff (eds) 1997; Etzkowitz, Webster and Healy (eds) 1998; Beck 1997: 27; Ritzer 1999: 132ff.; and Ray and Sawyer (eds) 1999.

15. Modernity paradoxically accepted the permanence of change ('all that is solid melts into air') while simultaneously attempting to tame this vision by specifying the fundamental *laws* of change, evoking a scientifically predictable and fundamentally benign process of modernization-as-differentiation. But the solidity of this vision in

Notes

turn succumbs to the melting process as it becomes unpredictable and ironically reverses its direction (Crook et al. 1992: 36–37, 47, 57, 70, 220, 228).

16. Cf. Beck's (1992; 1997) conception of 'sub-politics' or Giddens' (1991) notion of 'life politics'.

17. See Pels 2000: 193ff. for a pragmatic approach to social distances or 'orders of estrangement', which introduce a routinized stalling or halting of the ordinary flow of events and thereby stretch the reflexivity of everyday action into a methodical form of life. Such a pragmatic linkage between cultural fermentation and distanciation in time (unhastening) and space (decentring or estrangement) establishes minimal, relative, and contextually sensitive conditions for intellectual autonomy which are not fixed by the unconditional sublimity of an epistemology of truth. The much-glorified 'methodical doubt' of science is not produced by a sublimating ascetic, but results from a pragmatic resolve to provisionally postpone a specific intellectual decision. Everyone knows that there is an immediate and inverse relationship between doubt and speed of action. Science does not produce certainty but *un*certainty and hence promotes the *in*capacity to act.

18. It is interesting to note how these temporal divisions are reflected and materialized in different media of communication, differentiating daily broadsheets from weeklies, monthlies, professional quarterly journals, yearbooks, ordinary books, and encyclopedias. Different time-scales also define different topic structures, as is evident in comparisons between the daily TV news, news magazines and current affairs programmes, special reports, and well-researched documentaries. Topics range from panoramic coverage (newspaper rubrics on foreign affairs, domestic affairs, sports, business, the arts, the sciences, leisure and tourism etc.) towards a selective focus upon a few issues or just one issue. All these different media are characterized by specific 'grades of acceleration' or temporal intensities, which also distinguish between professional journalistic attitudes and modes of journalistic attention. Cf. also Meyrowitz 1985: 109–13.

19. Cf. Beck's notion of the 'political bourgeois' and the emerging synthesis between business and politics *within* the economic realm (Beck 1997: 126–31), as also reviewed in Chapter 8.

Notes to Chapter 3

1. Barnes (1982: 9, 34, 65) cites Fleck, Piaget, and the later Wittgenstein as Kuhn's primary extra-historical intellectual sources. Bloor (1983) likewise suggests strong parallels between Kuhnian naturalism and Wittgenstein's supposedly naturalistic 'social theory of knowledge'. Subsequently, Barnes (1994) has indicated that, even though Kuhn missed Mannheim, his contextualist and historicizing approach to the natural sciences represents a thought style continuous with Mannheim's critique of rationalist epistemology.

2. See e.g. Collins 1991a: 12ff., 24n, 152n; 1986: 3, 8n; 1990: 17, 20–21, 225n; Woolgar 1988a: 45–50. See also Fuller 2000b: 29. The pivotal significance of Wittgenstein is also exemplified by an exchange between Lynch and Bloor, in the course of which the former opposed an ethnomethodological and anti-epistemological reading of Wittgenstein to the latter's naturalistic interpretation (Lynch 1992a; Bloor 1992; Lynch 1992b; 1993). The lines of disagreement are anticipated in Lynch's earlier objections against Phillips' (1977) 'appropriation' of Wittgenstein's philosophy for a Bloorian sociology of science (Lynch 1985: 179ff.). Lynch (1993) provides a rich account of the development of 'strong' SSK and its many sibling rivalries.

3. The width and depth of the schism that has developed between the two traditions can also be fathomed by looking at the institutional and professional distribution of research interests and research personnel. The ISA Research Committee on the History

of Sociology and the ASA Theory Section, which include a number of 'regular' social theorists and Mannheim scholars, have hardly concerned themselves with the new social studies of science and technology, which are pursued in rival institutional forums such as the Society for the Social Study of Science (4S) and the European Association for the Study of Science and Technology (EASST). Established sociology journals, including relatively new and adventurous ones such as *Theory and Society* or *Theory, Culture, and Society*, hardly ever print contributions to science and technology studies, which tend to appear in *Social Studies of Science, Philosophy of the Social Sciences* or *Science, Technology, and Human Values*. In the Netherlands as elsewhere, science and technology studies have mainly developed outside of and in opposition to established social science departments and at new, often technically oriented universities. While the new social studies of science, in defiance of the Mannheimian proviso, have massively invaded natural and medical science and technology, the social and historical study of social science itself has somehow remained the preserve of regular sociology.

4. This Wittgensteinian methodology also closely approaches Foucault's suspicions about ideology critique and his alternative precept to investigate the empirical linkages between truth contents and power effects. Foucault likewise neutralizes the question of legitimacy/illegitimacy, which must be exchanged for the concrete study of the acceptability of knowledge claims in terms of a 'reconstruction of their "positivity"' – which is simultaneously an uncovering of their fundamental *arbitrariness*, their contingency, their 'violence' (see e.g. Foucault 1994b: 74–76; see Pels 1995b for a critical perspective). The critical issue is also (involuntarily) captured by Sharrock and Anderson's opinion that Wittgenstein's injunctions 'to describe, not explain', and 'to look, not think' were 'not a call to create a programme of empirical research, but for us to take notice of things that are staring us in the face' (1984: 386). Cf. also n. 22 below.

5. Cf. Lynch on Woolgar's and Ashmore's reflexivism as 'perhaps a more consistent application of Bloor's impartiality postulate than Bloor had in mind', and as 'an extremely strong injunction to act in accordance with the Mertonian norm of disinterestedness which in fact takes Mannheim's "nonevaluative" conception of ideology to its ultimate limit' (1993: 106–107). But Lynch himself does not abandon the search for a neutral or non-evaluative observation language, despite his critique of empiricism within current sociology of scientific knowledge (SSK) and his advocacy of a praxeological turn (115, 141ff., 303).

6. The term 'social epistemology' already occurs in Merton (1973: 107, 113, 123), whose register entry of the term retrospectively suggests that it had been present in his work since 1941. However, the exact expression does not appear to be used before 1972. The term was also employed by anthropologist Douglas (1986) and social constructionist Gergen (e.g. 1988), before being popularized by Fuller (1988). See also Schmitt 1994.

7. Stehr's 1994 book provides one of the sharpest illustrations to date of the gap between more traditional sociology of knowledge concerns and radical constructivist science studies; the latter are damagingly absent from his macrosocial narrative. Bauman's prolific writings on intellectuals and the state remain likewise uninformed by constructivist studies of science, even though he shares many of their anti-essentialist concerns. On the other side of the looking-glass, the voluminously authoritative *Handbook of Science and Technology Studies* (Jasanoff et al. [eds] 1995) is astonishingly weak on contributions from the sociological mainstream (honourably excepting Restivo 1995 and Gieryn 1995). For example, while Bourdieu's work receives only scant and inadequate discussion (cf. Callon 1995: 37–41), neither Mannheim nor Elias is considered worthy of even being indexed.

8. For example, it raises the difficult issue of the conditions of possibility of an anti-realist and anti-foundationalist ideology critique. See further Chapter 7.

Notes

9. Elias's criticism of Mannheim is most balanced in his *Norbert Elias über sich selbst* (1990: 138ff.). However, his account of the Züricher Soziologentag subtly exaggerates his own intellectual independence and plays down his own partiality to Mannheim's views in the latter's dispute with Alfred Weber. Goudsblom (1987: 69) fails to mention that the 'spiritual revolution' supported by Elias over against Alfred Weber's idealism and individualism was effectively started by Mannheim (cf. Elias in Meja and Stehr [eds] 1982: 388–89; also Mannheim 1982 [1922]: 62).

10. And including its problems: cf. the residual universalism of Bourdieu's notion of rational progress and his residual holism (cf. Mannheim 1982 [1922]: 170; see more extensively Chapter 5 below).

11. Merton's familiarity with Mannheim's work can be traced back at least to his article 'Karl Mannheim and the Sociology of Knowledge' (1941).

12. 'The ultraradicalism of a sacrilegious denunciation of the sacred character of science, which tends to cast discredit on all attempts to establish – even sociologically – the universal validity of scientific reason, naturally leads to a sort of nihilistic subjectivism. Thus the cause of radicalization which inspires Steve Woolgar and Bruno Latour drives them to push to the limit or, better, to extremes, the kinds of analysis, such as the ones that I proposed more than ten years ago, which endeavour to transcend the (false) antinomy of relativism and absolutism' (Bourdieu 1990b: 299).

13. And even further back to a 1901 work by Alois Riegl (e.g. Mannheim 1982 [1922]: 127, 233). Wittgenstein's *Philosophical Investigations* (1998 [1945–49]) advance concepts such as 'language game', 'life form', and 'family likeness', but do not feature the concept of 'thought style'.

14. Fleck appears to have quarried his knowledge about sociology primarily from Jerusalem's long contribution to Scheler's collection *Versuche zu einer Soziologie des Wissens* (1924; cf. Meja and Stehr [eds] 1982: 27ff.). Jerusalem's article refers neither to Mannheim nor to the concept of *Denkstil*.

15. Cf. Bourdieu's repeated warnings (e.g. 1990b, 1993c) against 'short-circuit' explanations, which fail to notice the inevitable 'refraction' of external social interests by the laws of the intellectual field. This accusation is also mistakenly levelled at the Strong Programme and its radical offshoots (1990b: 298).

16. In this context it is interesting that Bloor's view of a rift between Wittgenstein and Mannheim extends into a similar one between Wittgenstein and Durkheim. While Durkheim still opted to except Western scientific culture from the social explanations that he saw fit to apply to primitive systems of classification, Wittgenstein did not 'lose his nerve or betray himself in this way' (Bloor 1983: 3n).

17. In terms of the 'Epistemological Chicken' dispute about the radicalization of the symmetry principle, as conducted between Collins and Yearley and Callon and Latour (see Chapter 6), Mannheim would no doubt be considered an early example of 'chicken'. Woolgar capitalizes (on) 'Mannheim's Mistake' as general indicator of the tendency to exempt particular kinds of knowledge (maths, natural science, sociology, SSK itself) from the purview of sociological analysis (1988a: 23, 48).

18. Barnes locates a tension in Mannheim's sociology of knowledge between 'relationism' and a remnant 'epistemological' faith in the potency of reason. The idea of a *freischwebende Intelligenz* as granting access to really valid knowledge is judged 'an unsatisfactory finale to a penetrating and courageous work' (1974: 147–48). Similarly, although *Ideology and Utopia* opens with a commitment to an activist conception of knowledge, a great part of Mannheim's argument is in fact predicated upon a 'contemplative' model (Barnes 1977: 3–4).

19. Cf. Bloor (1991 [1976]: 164) who considers the social-institutional character of our best scientific achievements 'not a defect but part of their perfection'.

20. In addition, there are intimations scattered across Mannheim's work that the

dualism between natural science and historical-cultural-social science is not a final methodological state. 'Static' methodology must ultimately find a place within a more comprehensive 'dynamic' conception of science, and historical method may provide a 'point of unity' from which the dualism can be surmounted (e.g. 1952: 130–33). Cf. his parallel remarks about the 'partial' nature of traditional epistemology, which is exclusively oriented towards natural science (1968 [1936]: 261ff.). See also Hekman 1986: 78.

21. This distinction is also increasingly emergent in Collins' work (cf. 1991b; 1996), as further discussed in Chapter 6.

22. Elsewhere, Mannheim emphasizes more strongly that the sociology of knowledge has 'no moral or denunciatory intent', and pleads avoidance of the term 'ideology' because of its moral connotations. A few pages after this, however, he reaffirms that a 'particularizing' sociology of knowledge inevitably arrives at a point where it becomes more than sociological description, and turns into a critique 'by redefining the scope and the limits of the perspective implicit in given assertions' (1968 [1936]: 237, 256). Cf. also n. 4 above on Wittgenstein and Foucault.

23. As Mannheim himself concedes in a well-known 1946 letter to Wolff, his work does not conceal inconsistencies ('covering up the wounds'), but intends to 'show the sore spots in human thinking at its present stage'. Such contradictions are not mistakes but 'the thorn in the flesh from which we have to start'; they are not due to shortsightedness but result from the fact that 'I want to break through the old epistemology radically but have not succeeded yet fully' (cit. Wolff 1959: 571–72).

24. Lynch (1992a: 215n) suggests a workable distinction between foundationalist epistemology, the traditional *bête noire* of symmetrical and agnostic sociology, and a 'small *e*' epistemology that is compatible with constructivist empirical interests. If extended in a Mannheimian direction, however, a defence of small *e* epistemology would tend to undercut or at least loosen up the original Bloorian postulate about the symmetrical causal treatment of truth and error, possibly in the direction of a *weaker* postulate of *a*symmetry (see Chapter 6). Hence, and presumably different from Lynch, my small *e* epistemology would also be compatible with a minimal normativity, following the recognition of a 'natural proximity' of facts and values.

25. From a very different position, Hilary Putnam likewise proclaimed a double moratorium on ontology and epistemology (1990: 118).

26. With the exception of some recent contributions by Latour (e.g. 2002), which will be referred to in subsequent chapters.

27. But see veteran *Social Studies of Science* editor David Edge's recommendation that 'the essential inseparability of facts and values' must be confronted directly in STS research, and 'is likely to be a very fruitful focus of STS debate for the foreseeable future' (Edge 1995: 16–18).

28. A presently unelaborated point concerns Mannheim's lack of resolution, which I take to be equally productive here as in preceding cases, in the face of the micro/macro alternative. While radical constructivist science studies have tended to privilege various actor-centred approaches, mainstream sociologists such as Elias and Bourdieu have continued to emphasize the weight of institutional macrostructures. Although all parties have filed claims for the supersession of the dualism in a relational direction (e.g. Elias's figurational sociology, Bourdieu's 'constructivist structuralism', or Latour's 'associology'), ultimate preferences for actor- or structure-oriented approaches have remained divided. If Mannheim's emphasis upon a 'configurational' methodology and a 'historical structural analysis' (1952: 87, 181) cannot be hailed as a solution, it does something to counteract the more radical versions of nominalism and interactionism that science studies have introduced.

Notes

Notes to Chapter 4

1. Of course, like the assertion about 'natural proximity', this can be only a circular claim that seemingly pushes us into unending regress. However, I assume that this circularity is virtuous, not vicious, and is precisely what is needed to render such propositions appropriately 'weak'. See Chapter 7 for a more extended defence of reflexive circularity.

2. In illustration of the former, see Weber's paraphrase of Socrates' discovery of the logical concept: 'In Greece, for the first time, appeared a handy means by which one could put the logical screws upon somebody so that he could not come out without admitting either that he knew nothing or that this and nothing else was the truth, the *eternal* truth that would never vanish as the doings of the blind men [in the Cave] vanish. That was the tremendous experience which dawned upon the disciples of Socrates' (1970: 141). Note that Weber does not critically address this 'torturing' and all-or-nothing character of logical truth (screw you!). Callicles (another Platonic voice!) is well aware of such discursive violence in his famous debate with Socrates, e.g. when he states that a particular admission by his friend Polus enabled Socrates 'to tie him up in logical knots and muzzle him' (*Gorgias* 482d).

3. This scepticism is widely shared, by (for example) Austin (1962), Searle (1969), Gouldner (1973), Feyerabend (1975), Taylor (1978), Hesse (1980), Grafstein (1981), H. Putnam (1981; 1990), R.A. Putnam (1985), Doeser (1986), Herrnstein Smith (1988), Potter and Wetherell (1988), Fairclough (1995) and Potter (1996).

4. This diamond pattern is similar in its overall structure and theoretico-normative intent to the one previously applied to the theoretical antinomy between what I have called 'property theory' and 'power theory' (Pels 1997; 1998). The reflexive structure of its forward-backward movement is further elaborated in Pels 1998: 254–59. See also Pels 1990; 1991a for earlier versions of the diamond scheme as applied to the fact/value dichotomy.

5. In this fashion, the proposed matrix is deliberately 'presentist'. It is therefore not so much the Whiggishness of the standard view that I object to, but its lack of awareness that all such reconstructions are inevitably circular. Perhaps this view can be somewhat ironically described as 'Whig relativism'.

6. MacIntyre suggests that 'if the current interpretation of Hume's view on "is" and "ought" is correct, then the first breach of Hume's law was committed by Hume [...] if Hume does affirm the impossibility of deriving an "ought" from an "is" then he is the first to perform this particular impossibility' (1969: 36, 39). Cf. Searle on the thesis that no set of descriptive statements can entail an evaluative statement: 'The irony lies in the fact that the very terminology in which the thesis is expressed – the terminology of entailment, meaning, and validity – presupposes the falsity of the thesis', since obviously 'language is riddled with counter-instances to the view that no evaluations can follow from descriptions' (1969: 175, 198).

7. This claim goes beyond views such as expressed by Portis (but already clearly articulated by 'conservative-revolutonary' critics such as Freyer [1964 (1930)] and Steding [1932]; see nn. 13 and 14 to Chapter 1) that Weber's substantive works conflict with the distinction between normative and empirical theory that he so authoritatively justified on a methodological level (Portis 1983). My claim attacks the logical structure itself rather than the inferred discrepancy between principle and practice. Although Portis comes close to interpreting Weber's central sociological concepts as 'value-impregnated' (1983: 32), he does not contemplate a more visceral unity of normative and empirical theory. Proctor (1991: 150) comes closer in suggesting that Weber's view that the divorce between science and ethics was not a political choice but a division etched in the nature of things paradoxically furthers a liberal

politics of neutrality. See also Brubaker's contention that, throughout his empirical work, Weber uses richly value-laden terms in a value-neutral manner (1984: 36).

8. On performative contradictions, see Pels 2000: 232 n. 22, and Herrnstein Smith 1997: 73ff., 88ff. See also n. 19 below.

9. Austin proclaims his intention to 'play Old Harry with two fetishes which I admit to an inclination to play Old Harry with, viz. (1) the true/false fetish, (2) the value/fact fetish' (1962: 151). Following an Austinian lead, Skinner's historical pragmatics explicitly identifies a mixture of facts and values as central to a radically performative view of language. Speech act theory 'throws out as an old piece of positivist bric-à-brac the alleged logical distinction between factual and evaluative statements, concentrating as it does on a group of terms which perform an evaluative as well as a descriptive function in the language' (Skinner 1988: 111—12; cf. Tully 1988: 13–14, 17).

10. Cf. the various contributions on Confucianism, Islam, African thought and Indian philosophy in Doeser and Kraay (eds) 1986, which illustrate a variety of absolutist or essentialist forms of fact/value identity, and the exceptional status of the Western tradition in this respect.

11. It is clear that, for Weber, scientific inquiry demands a prior *value decision* to follow truth wherever it may lead. Perhaps the knowledge-political purpose of the Is/Ought distinction is spelled out most clearly by ethical socialists such as De Man, who seek to establish the primacy of value decisions over the factual determinism of (Marxian) scientific socialism (cf. De Man 1928: 441–43).

12. Alexander (1981) similarly assumes that the critique of reductionism requires a categorical separation between facts and values. Cf. also Hollis 1993, Stedman-Jones 1988 and more extensively Hammersley 2000 for recent defences of value-neutral science.

13 Recent attempts to commit the naturalistic fallacy and get away with it, such as ventured by Black (1969), Searle (1969), Kohlberg (1971) or MacIntyre (1984), who muster miscellaneous logical, linguistic, and substantive evidence in order to argue the legitimacy of deriving Ought from Is, are therefore still too much constricted by the logic of the dualism. Hesse likewise supports this logical transition: 'Hume attempted to divorce the question of truth from that of value, while certain scientific humanists have attempted to derive value from truth. A consequence of my argument on the other hand has been that, at least in the sciences of man, a sense of "truth" that is not merely pragmatic may be derivable from prior commitment to values and goals' (1980: 203). Potter and Wetherell clarify that 'factual-type description is meshed into explanation, and the whole package has an evaluative orientation. There is no sense in which one could identify statements of fact, separate statements of evaluation and then statements offering the speaker's evaluation. The discourse works as a package – a seamless texture of talk' (1988: 64–65). Potter (1996: 108ff.) offers intriguing examples of the everyday perpetration of the naturalistic fallacy as a matter of 'local identity management', i.e. when concerns about being seen as prejudiced are managed by constructing evaluations as 'mere' factual descriptions.

14. The same asymmetry is evident in Bhaskar's critical naturalism, which similarly emphasizes the 'cognate' character of statements of fact and value, assuring that we can 'pass securely from statements of fact to practice' (Bhaskar 1986: 118, 169). Cf. Archer et al. (eds) 1998: xviii, 409ff.

15. See also the exchange between Laudan (1981) and Bloor (1981).

16. Barnes (1977: 1) alludes to this rivalry logic when he remarks that the separation between normativism and naturalism gradually emerged as an 'artefact' in the debate between Lakatos and the Strong Programme.

17. In his Law of the Three Stages, Comte already 'spreads out' the criterion of truth that demarcates metaphysical from positive knowledge across the entire secular

development of human thinking (see Pels 1996: 118–19). Hegelian objectivist dialectics accomplishes much the same.

18. Interpreting Weber, Brubaker immediately matches the fact/value distinction with that between formal and substantive rationality (1984: 36). Insofar as Habermas's labour/interaction dichotomy (unhappily? see Giddens 1982: 109) marries the Weberian distinction between purposive-rational and value-rational action to the Marxian differentiation between the forces and relations of production, it likewise appears to map on to the fact/value dichotomy.

19. Habermas elaborates a rather idiosyncratic conception of performativity that is locked in the communicative side of the dualism between communicative and strategic action (cf. especially Habermas 1992). In my usage, which more closely conforms with the original direction of Austinian speech-act theory, performativity precisely undercuts or surpasses this distinction (see also Bourdieu 1991: passim). More generally, Habermas does not heed the deep circularity of all accusations or 'unmaskings' that employ the vocabulary of the performative contradiction – an epistemological condition that I happily accept for my own circular 'unmasking' of Habermas. The idea of a performative contradiction typically imports an exogenous standard of evaluation into the adversary's work, which is subsequently exposed as riven by self-inflicted tensions. The ventriloquist critic absents himself and escapes responsibility for this imposition since the contradiction is not between himself and his opponent, but between the opponent's own sayings and his actual doings. See n. 8 for further references.

20. In Pels 1995b I consider Habermas's familiar critique of the descriptivism and crypto-normativism of Foucault's genealogies of power/knowledge, which directly resonates with my earlier critical account of Wittgensteinian 'value-free relativism'. However, I argue that rather than committing another version of the naturalistic fallacy, Foucault perhaps pulls off a more intriguing trick that positions him beyond the playpen of the logical dualism, replacing both objective statements of fact and universalistic moral prescriptions by contingent fact/value mixtures so intimate that the descriptions necessarily bring along their own normativity. Rather than agnostically moving *beyond* good and evil (Gordon 1980: 234–35), Foucaldian genealogy is better viewed as permanently poised *in between* these normative extremes, and hence as affording a minimal normativity that avoids relapsing into a Habermasian universalistic framework. On this interpretation, the power/knowledge theorem escapes and surpasses the traditional reductionist oscillation between normativism and naturalism in order to position itself with aplomb at the diamond's arrival-and-departure point.

21. In this technique of defusing a normative dilemma by spreading it across a descriptive continuum, their 'micro-historical' and 'psychogenetic' model shows similarities to the macro-historical and socio-genetic model proposed by Elias.

22. Latour admits that there is always a danger in this 'perilous' distinction, which suggests that the practitioners do not really know what they are doing, while it is supposedly transparent to a detached observer. But somehow the particular contradictions of the ecological movement are taken to justify such talk of a growing contradiction between theoretical beliefs and actual practice, which merely has to be 'revealed', not remedied, in order to help it fulfil its promise (1999b: 34–38).

23. 'I have spoken of the Pasteurians as they spoke of their microbes' (Latour 1988a: 148). Double fault!

24. 'In order to grow we must enrol other wills by translating what they want and by reifying this translation, in such a way that none of them can desire anything else any longer' (Callon and Latour 1981: 296); 'Each text, laboratory, author and discipline strives to establish a world in which its own interpretation is made more likely by virtue of the increasing number of people from whom it extracts compliance' (Latour and Woolgar 1986 [1979]: 285); 'Act as you wish, so long as this cannot be easily undone'

(Latour 1988a: 160); cf. also the chapter heading 'From Weakness to Potency' in Latour 1988a (158).

25. Let me pedantically note that Latour's picture of the 'double arrow of time' offers a less systematic overview of the fact/value dialectic than the diamond scheme. The main differences are, first, that Latour's diverging time arrow and his converging one are not formally clicked together to produce a diamond; secondly, that his conceptualization tends to bury the fact/value problematic in a much larger and perhaps overburdened conceptual scheme, where various dualisms such as those separating humans from non-humans, cultures from natures, and values from facts are all registered as coincident with the apparent master dichotomy of subject vs. object; and thirdly and most significantly, that he fails to incorporate a reflexive end point (P) from which the entire story of divergence/convergence is 'retro-narrated' in circular-hermeneutic fashion.

Notes to Chapter 5

1. Cf. also Bauman 1987: 2, 8, 172–77; 1992a: 1–25. Bourdieu is highly critical of the concept of professionalism, and argues against all a priori or operational definitions of intellectuals, artists, and scientists, since there is always struggle within the field over who can be admitted to the game as a legitimate player, who deserves the title of writer, intellectual, scientist etc. (Bourdieu and Wacquant 1992: 100, 242–45).

2. Feyerabend specifies that 'anything goes' should in fact be read as 'any *method* goes': not the *only legitimate* method, but *every random method* offers reasonable chances of success (1977: 400). Without further elaboration, such a minimal demarcation is incapable of conclusively distinguishing between science and various other methodically conducted disciplines such as Christian or Islamic theology, astrology, homeopathy, Marxism, or psychoanalysis. It does not even serve to demarcate science from other professionally conducted practices or from everyday knowledge use. As I shall argue in a subsequent chapter, methodism is only 'scientific' when it performs a methodical slowing down of communication, which requires additional criteria of scientificity such as institutionalized *autonomy* and the slow turnover of *organized competition*.

3. On the Janus head of professionalism see also Johnson 1972: 17 and Robbins 1993: passim. Talcott Parsons mentions the 'rosy and less rosy aspects' of modern professionalism, but consistently emphasizes the positive side of the case (1968: 372).

4. Zygmunt Bauman speaks about a *simultaneity* of detachment and involvement, characterizing the sociological enterprise as incurably two-faced, both critical and dominatory, so that you cannot have the one without incurring the danger of the other (1992a: 90, 212). The dual nature of expertise is discussed more directly in Bauman 1992b.

5. One of the sources of this rehabilitation of interest is Nietzsche's conviction that the more affects and interests we are able to muster and let speak about a particular object, the better we are able to see it, the greater our 'objectivity' (Nietzsche 1996 [1887]: 98).

6. This 'scholastic fallacy' consists in picturing social agents in the image of scientists, and of conceiving the projected scientific contructions as actual causes of everyday practices (cf. Bourdieu 1988a: xiii–xiv; 1990a; 2000a; Bourdieu and Wacquant 1992: 69–71).

7. 'The most "dysfunctional" tendencies [...] are inherent in the very same functions which generate the most "functional" dispositions' (Bourdieu 1981: 275).

8. If actors and groups are 'defined' by their relative positions within objective social spaces in which each 'possesses' its own logic, its own principles of differentiation and its own hierarchy, who does the defining? Who attributes this particular logic? When

Notes

objective relations and positions have primacy above the individuals and their position-takings, who has accorded them this primacy? Who actually says that 'there are *general laws of fields*' (Bourdieu 1984: 113)?

9. Such a 'minimally normative' amendment of the principle of knowledge politics deflects a predictable Habermasian critique of the crypto-normative structure of Bourdieu's sociology, without reverting to an a priori separation between communicative and strategic action that once again separates the two hemispheres of the knowledge-political globe. The principle of knowledge politics is neither a form of metaphysical essentialism nor a purely empirical generalization, but a mixed normative-empirical hypothesis that advances its own criterion of truth, proclaiming a natural proximity of communicative and strategic action without falling into epistemological or ontological reductionism. For his part, Bourdieu explicitly rejects the distinction between instrumental and communicative action (Bourdieu and Wacquant 1992: 113). See also my Pels 1995b and Chapter 4 n. 20 (above) for a similar argument in the context of the Habermas–Foucault controversy.

10. Cf. the opportunistic manner in which Bourdieu draws the distinction between 'true' and 'false' autonomy in order to demarcate pseudo-science (which remains dependent under the guise of being independent) from 'real' science (1981 [1975]: 275ff.); and his tautological definition of the limits of a field as the point where 'field effects' cease to exercise their influence (Bourdieu and Wacquant 1992: 100, 232).

11. 'Renouncing the angelic belief in a pure interest in pure form is the prize we must pay for understanding the logic of those social universes which, through the alchemy of their historical laws of functioning, succeed in extracting from the often merciless clash of passions and selfish interests the sublimated essence of the universal' (Bourdieu 1996a: xviii).

12. 'I am a resolute, stubborn, absolutist advocate of scientific autonomy ... I think that *sociology ought to define its social demands and functions on its own*' (Bourdieu and Wacquant 1992: 187). Cf. Alvin Gouldner's plea that it is sociology's first task 'to establish the conditions of its own existence as a practical rational discourse', and that these conditions are of 'universalistic relevance'. Gouldner's politics of establishing autonomous theoretical collectives committed to 'true speaking' retains a normative theory of rationality and a divorce between truth and interest that edges closer to Habermas than to Bourdieu (Gouldner 1973: 97–100).

13. 'A genuine scientific field is a space where researchers agree on the grounds of disagreement and on the instruments with which to resolve these disagreements and on nothing else' (Bourdieu and Wacquant 1992: 176). In my view, scientists need only agree on a collective defence of their institutional autonomy and of the social conditions of unobstructed intellectual competition, even while recognizing that the field limits, like the structure of competition, both remain at stake in struggles within the field itself.

14. 'A truthful idea can only be countered by a refutation, whereas an *idée-force* must be countered by another *idée-force*, which is capable of mobilizing a counter-force, a counter-manifestation' (Bourdieu 1996b: 20). Politics is a struggle for ideas, but it concerns a particular type of ideas, *idées-forces*, ideas that supply force and function as forces of mobilization. The problem of intellectuals is then how to lend force to truthful ideas without entering into the field and the game of politics (Bourdieu 2000b: 68). However, as I have suggested before, ideas are themselves forces, and the specific difference between scientific and political reasoning is not that political speech is mobilizing and performative and scientific speech is not, but that scientific speech is slower, more leisurely, and more playful, and less decision-driven or constrained by considerations of 'fast' utility.

15. Bourdieu (2000a: 185–86) writes as if his analysis is not applicable to science

and sociology itself. See Chapter 7 for a further critical appraisal of the lack of reflexivity in Bourdieu's reflexive social science.

16. I distance myself here from Havel's humanistically coloured version, which identifies 'anti-politics' in terms of a practical morality that stands in service to truth (1988: 381ff.).

17. Habermas similarly approaches this notion of anti-politics when he suggests that the 'communicative power' of the intellectuals confronts administrative power 'in the manner of a siege', influencing judgment and decision-making in the political system without intending to conquer the system itself; it can therefore become effective only in an indirect manner (1996: 486–87, 489–90). Communicative power', of course, is not defined in the more radical knowledge-political terms that I adopt in this book. See also Habermas's superb account of Heine as foreshadowing the image of the 'true' intellectual as someone who commits himself to the public interest 'as a sideline' (*im Nebenberuf*), rejecting the false alternative between fetishizing the mind and turning art (or science) into a political instrument. (Habermas 1989: 87–88, 91).

18. This radical political temper is of course not an exclusive asset of right-wing thought but is also echoed in the young (and old!) Marx's conception of critique, which is not so much a 'lancet' but a 'weapon', whose object is an *enemy* that it aims not so much to refute as to *destroy*. Practical-critical activity is immediately identified with the '*ruthless* criticism of *everything* existing'. Criticism is criticism 'in a hand-to-hand fight'; it must be *ad hominem* and radical, adopting the mode of indignation and denunciation (Marx 1972 [1844]: 8, 13–14, 18).

19. Cf. also Chesneaux 2000 for a general discussion of the (adverse) relationship between democracy and speed.

Notes to Chapter 6

1. In Law's relativist semiotics, all 'essentialist divisions are thrown on the bonfire of the dualisms' (1999: 3).

2. As Roth has noted with regard to the Epistemological Chicken controversy, 'There is an aura of "political correctness" surrounding this debate, with each side suggesting that the other has broken the faith which the symmetry principle represents' (1994: 104n).

3. Cf. Latour's view that Bloor's symmetry principle is 'completely asymmetric' (1992: 278–79; 1993a: 94). Darmon argues that the empirical relativist studies of science in fact *violate* the monist and symmetrical approach that distinguishes them from rationalist philosophy of science, since they typically fall prey to the reverse bias of explaining cognitive contents as mere by-products of social contexts (Darmon 1986: 743–52).

4. Here I side with Richards and Schuster's (1989) criticism of 'methods discourse' as floating abstractly above the proper complexity of scientific fields, and as furthering myths that serve as routine resources in socio-cognitive power struggles. See also Pels 2000: 210–14 for a discussion of what I call 'methodological voluntarism'.

5. Although the four postulates of the Edinburgh Strong Programme are explicitly presented as methodological prescriptions, they immediately derive from naturalistic analysis, thus endorsing a Weberian separation between explanation and evaluation and a conception of the world as 'morally empty and neutral' (cf. Bloor 1991 [1976]: 10, 141; Barnes 1976; 1977: 25).

6. One example could be taken from the post-'1968' history of Dutch sociology, which, following an intermezzo of 'wild' pluralism, gradually settled into two dominating streams or schools: Figurational Sociology, which rather exclusively came to represent the more international paradigm of Historical Sociology, and Explanatory

Notes

Sociology, which performed as the Dutch variant and embassy of the Rational Choice paradigm (Van El 2002).

7. Cf. Gouldner's (1973: 27ff.) classical account of 'cool partisanship', as discussed in Pels 2000: 220, Lynch 1993: 78–79 and (very critically) in Hammersley 2000: 90ff.

8. For further doubts on Latour's 'translation' see e.g. Collins 1991a: 192n; extensively Pels 1995a; 2000: 10–18.

9. This lack of clarity is already evident in Shapin's brilliant preparatory study of Boyle's literary technology, the stated purpose of which was 'to display the processes by which Boyle constructed experimental matters of fact and thereby produced the conditions in which assent could be mobilized' (1984: 483). Whether Boyle himself conceived of matters of fact as at once an epistemological and social category, whether he purposefully rendered his narrative credible as a mirror of reality, whether his technologies of fact-fabrication and naturalization were consciously applied or merely 'worked' according to Shapin's external functional judgment, is consistently left in the dark. Terms such as 'strategy' effectively cover up this methodological hole in the ground. Cf. Shapin and Schaffer's view that 'each of Boyle's three technologies worked to achieve the appearance of matters of fact as *given* items. That is to say, each technology functioned as an *objectifying resource*' (1985: 77).

10. Boyle's strategy of authenticating matters of facts as mirrors of nature was explicitly geared towards the mobilization of universal assent. If Hobbes denied the foundational status of matters of fact and rejected experimental procedures, this was not to reveal himself as an early constructivist, but to claim a far greater degree of philosophical certainty than mere 'facts' could ever provide. In this respect, Boyle and Hobbes exemplify two classical objectifying 'technologies of certainty', the rationalist and the empiricist, which conspire as opposites in their similar urge to naturalize the foundations of knowledge and politics.

11. It is not a little piquant to see Schaffer accuse Latour of asymmetry in his description of the 'Pasteurization' of France, because he 'attributes life to the inanimate, silences controversy, and asymmetrically credits his hero's stories while ignoring those of his hero's powerful rivals' (Schaffer 1991: 192). Latour's interpretation of Pasteur's politics of knowledge offers the same duplicity that also traverses his analysis of Hobbes and Boyle, despite the fact that Pasteur's own texts invariably write Science with a capital S and reveal him to be a convinced positivist and realist (cf. Pasteur 1922: 326ff.).

12. On actor-network theory's 'uncovered flank' vis-à-vis the identitarian conception of (scientific and political) representation, see Pels 1995a; 2000: 10–18, 199–201; also Lee and Brown 1994.

13. The overlap between the science/politics and nature/society dualisms is also demonstrated by the fact that their constituent terms are crosswise exchangeable. Science also stands over against (social) context, while politics is regularly posited over against nature. This also holds for the seamless overlap between the dualism of citizens and facts and that of people and things. It is precisely as a result of such identifications that the dualism of subject and object can perform as the grounding dichotomy of the Modern Constitution. Shapin and Schaffer, for their part, already tend to identify science with the 'study of nature' and politics with 'the study of men and their affairs', and to collapse the dichotomy between society and nature into that between subjects and objects (1985: 337).

14. 'It is important to notice how much of what Callon and Latour describe as they "follow the actors" is preordained, as it were, rather than "found". In practice, prior external decisions fix not only which actors will be followed, but also what theoretical schemes will be used to make sense of what those actors do' (Barnes 2001: 343).

15. In a subsequent paper, Ashmore defines symmetry as 'an *epistemological*

commitment, not a tool to be selected or rejected for particular tasks. To call it a method is to ignore both its epistemological character and its status as a commitment' (1996: 310).

16. In 1996, both Richards and Martin appeared to endorse an epistemological separation between symmetry and neutrality (Richards 1996: 346–47; Martin 1996: 266–67). Martin now holds that impartiality and symmetry can also be interpreted as compatible with partisanship: 'one can both explain a belief and support it'. Even a stance of detachment, he says, involves a value choice, namely not to intervene (1996: 268).

17. In attacking the alleged 'academism' of authors such as Latour and Woolgar, Martin curiously reproduces the very separation between epistemology and power that these authors are so much concerned to extinguish.

18. Shapin and Schaffer still differentiate between the polity of the intellectual community and the polity of the wider society, but Latour tends to overlook this important nuance. Emphasizing that the solution to the political problem of order and the problem of true knowledge are closely interdependent, both in Hobbes and Boyle, Shapin and Schaffer also stipulate that Boyle's problem of order specifically regards the state of 'scandalous dissension' in natural philosophy, and that the political society that he envisages is first of all a *polity of science* (1985: 342–43). The 'civil wars' that Boyle wishes to bring to an end are primarily scientific, not political conflicts, while Hobbes envisages a definitive conclusion to the very real Civil War that wrecked seventeenth-century England. Shapin and Schaffer consistently put Boylean 'civil wars' in quotation marks, but Latour characteristically drops this typographical distinction.

19. As a conception of truth that posits its own standard of validity and justified belief, this is of course a specimen of 'circular reasoning'. However, not all circles are vicious: some are virtuous. On circularity, see also Zolo 1992: 3–9 and Knorr's suggestion about 'closed circuitries' and 'circular foundations' (1993: 557). I shall discuss the incurable circularity of theorizing and the self-referential or 'self-absorbed' nature of critique more fully in the next chapter.

20. We should note Latour's specification that the principle of symmetry 'aims not only at establishing equality – which is only the way to set the scale at zero – but at registering differences – that is, in the final analysis, asymmetries – and at understanding the practical means that allow some collectives to dominate others'. In the absence of a single Great Divide, such new differences constitute 'countless small divides' between nature/society collectives (1993a: 107–108). Although generalized symmetry thus allows us to detect new asymmetries, Latour permits them to be 'only of size' (however 'sizeable' they are), refusing all qualitative hierarchization between them. Callon's model of 'extended translation' similarly rejects broad divisions without challenging the pertinence of network differences themselves. Translation networks 'weave a socionature, an in-between that is inhabited by actants whose competence and identities may vary along with the translations transforming them. Both passive beings and genuine actors are found there, but the dividing line is not laid down' (Callon 1995: 58). This is still different from saying that the dividing line is *weaker*. My view also differs from 'just following' how agency is attributed, since it reflexively inscribes the new asymmetries in the translations made by the analysts themselves.

21. See also Eckersley 1992: 50, 54. If Man is no longer the sovereign measure of all things (natural), there remains ample reason to retain a relaxed priority of humans. Callon and Latour wish to emancipate the nonhumans by giving them a voice. It is true that people are able to speak both for citizens in the polity and for natural facts. On this level of representation, the logic of translation and spokespersonship is the same: the attribution of agency is story-relative. But it seems that the prescriptive relation and the direction of the delegation cannot be symmetrically turned around, if we wish to avoid

Notes

the trap-door of reification and slide back into realism. In terms of the Epistemological Chicken debate, I therefore tend to side with Collins and Yearley and oppose Callon and Latour. However, while refusing to be drawn into Latourian *ontological* symmetry, the former remain partisans of the *epistemological* symmetry advocated by Bloor. My advice has been to drop the notion of symmetry altogether. With this, I have manoeuvred myself into a third position which, in its defence of 'weak asymmetry', is very symmetrical (don't ask me how I did it) between the two contending STS schools.

22. In this perspective, a constructivist view that focuses the long and slow work of inscription, explanation, and fact-fabrication can also be fruitfully contrasted with a more conventional logic of discovery that highlights the sudden 'finding' of facts that are somehow 'given' by nature.

23. That is, if one adds the crucial *competitive* structure of this economy of inscription, which institutionalizes a 'slow' market for scientific reputations within a protected time zone.

Notes to Chapter 7

1. Comtean positivism tends to berate reflexive self-contemplation (and all forms of psychologizing) as an illusion, since it is 'manifestly impossible' for the human spirit to observe intellectual phenomena during the act of observation itself. The thinking individual cannot subdivide himself into a person who reasons and another who watches the reasoning (Comte 1975: 33–34).

2. Here I follow Gouldner's definition of ideology as an objectivist discourse that conceals the presence of the speaker in the speech, and of reflexivity as its critical opposite (1976: xv, 45–46, 50). I also like to preserve some of Gouldner's original inspiration that 'a Reflexive Sociology would be a moral sociology' (1971: 491, 499). This normative and epistemological approach to reflexivity contrasts with the agnosticist and levelling claims that currently predominate in science studies (e.g. Woolgar 1991: 39, 41; Lynch 2000: 26–27, and my discussion further below). See also Radder's criticism of agnostic forms of constitutive reflexivity and his defence of normative reflexivity (1996: 98–101, 183–87).

3. I shall moreover defend it in standard narrative form, that is, in a seemingly unreflexive voice that does not require intricate deconstructive loops, massive quotational bracketing, ironies running wild, carnivalesque heteroglossia, or New Literary Forms. My concern is to argue a radical proposal about the circularity of representations, and argue it 'straight', i.e. in a customer-friendly manner (cf. Potter 1996: 9, 84, 229–30). There is no particular need, in telling a story that is peculiarly attentive to its own representational practices, to adopt special literary devices designed to undercut the basis of what is told. 'Telling it like it is' does not condemn the narrator to epistemological realism. Like Fabian (1991: 194), I tend to think that New Literary Forms are an 'unnecessary escape' from our epistemological dilemmas.

4. It is not entirely clear how this relates to Woolgar's view of the 'double bind' of similarity and distance in representation practices (Woolgar 1988b: 20ff.; cf. Ashmore 1989: 49). The conventional view of representational adequacy tends to privilege verisimilitude or the mimetic reproduction of reality. But strong reflexivity, Woolgar style, likewise appears to deny any relevant distinction between representer and representer while in his view, weak forms of reflexivity maintain a postulate of distinction. I am interested in a version of reflexivity that preserves the gap and the tension between representer and represented and resists all claims of representative identity.

5. Perhaps I write this text first of all to persuade myself. Showing the result, an articulate and inspectable textual object, both to myself and to significant others (e.g. a few readers of this book) is then not so much a matter of seeking universal glory but a

way of certifying that I am not completely out of my mind, that there is perhaps something to it, that I enjoy some legitimacy as an 'interesting' voice in this particular conversation. Should I/we ask for more? Traditional truth-sayers stake out an epistemological advance claim on their proposals, and demand a right to their universal diffusion (cf. the Enlightener's happy conviction that it was 'the property of truth to spread'). I say we settle for a more modest insurance policy. Do I persuade anyone with this minimal theory of persuasion?

6. Latour and Woolgar's admission that their own account is neither superior nor inferior to those produced by the scientists themselves (1986 [1979]: 254; cf. Latour 1988a: 148–49, 171; 1999c: 20) clearly hinges upon their quasi-empirical description of their subjects as duplicitous constructivists/realists. See also Kim's (1999) criticism of Lynch and Woolgar, whose 'ground-zero' approaches arguably annihilate all cognitive distance between analyst and analysand. An interesting parallel arises with Lévi-Straussian anthropology (cf. Diamond 1974: 402–403) where 'symmetry of values' constitutes the deep meta-value of the relativistic professional: 'He may know better, but he may not admit to knowing better as an anthropologist'.

7. Fabian (1990; 1991) favours a reflexive and performative ethnography. Potter's discourse analysis likewise tends to identify reflexivity with performativity (1996: 47–48, 228–32).

8. Giddens avers that all knowledge claims in conditions of modernity are inherently circular (1990: 176–77; cf. 36–45, 49). But this applies primarily to forms of institutional reflexivity where knowledge is seen to circulate in and out of the environment it describes, and does not appear to implicate the epistemological circularity of sociological accounts. By the same token, Beck's theory of the risk society insufficiently confronts the risk of its own epistemological circularity, and as a result offers a rather unreflexive view of reflexive modernization. In a more recent contribution, however, he intriguingly suggests a notion of 'reflexive realism' that negotiates between 'naïve' realism and equally 'naïve' constructivism, and which is close to the 'circular realism' defended in this chapter (Beck 1996: 7).

9. This radical view is inspired by Heidegger's resistance against the dismissal of the hermeneutic circle as a *vitiosum*: 'Das Entscheidende ist nicht, aus dem Zirkel heraus, sondern in ihn nach der rechten Weise hineinzukommen' (1972 [1927]: 153). See of course also Gadamer 1979: 238, 270, 274–75. Luhmann notes that 'transcendental' theories tend to block the operatively closed, 'autological' circle of investigation, and consider any demonstration of circularity a lethal affair. Rather than avoiding circular argument we should precisely admit theories that are capable of such self-referentiality (Luhmann 1990: 71–72, 469; cf. Teubner 1993: 13–24). On circularity and recursivity in systems theory, see further Rasch and Wolfe 2000: 10–13 and Rasch 2000a: 16. Kuhn has likewise observed the circular nature of scientific paradigms (1970a: 80, 90, 94, 109). See also Chapter 6 n. 19.

10. Woolgar closely approaches this idiom of reflexivity as circularity (1981: 382–85), but characteristically does not debate it in critical-epistemological terms (cf. also 1983: 260). His equivocation on this point is the same as that which transpires from Garfinkel's view that ethnomethodological studies are not meant as 'correctives' or 'ironies', and that the essential reflexivity of accounts is basically 'uninteresting' to their proponents (1984: viii, 7–9). But Woolgar's question how it is 'that in most instances of practical argument no significance is attached to its defeasibility on grounds of circularity' can be answered rather plainly (and critically): because hardly anyone in everyday practice actually sees it as circular!

11. Latourian networks of representation might be better thought of as circles that are started up and reflexively bend back towards the spokesperson than as 'galleries' that extend outwards in linear fashion. In addition: circularity of connection between

Notes

representer and represented does not signify their identity or similarity, but highlights the separateness of the two positions; resisting the positivist/idealist conflation of spokesperson and object, it articulates the difference between them in order to calculate their joint contribution to the 'reality effect' of statements. See more extensively Pels 2000: ch. 1.

12. Latour's argument about Pasteur's knowledge-political trafficking issues in a reflexive quandary when he declares that no account escapes the relations of force that he has described – his own even less than that of the others: 'What I cannot attribute to the Pasteurians I do not claim to attribute to myself. My proofs are no more irrefutable than theirs, and no less disputable [...] I have spoken of the Pasteurians as they spoke of their microbes' (1988a: 148). The 'duplication mode' surreptitiously validates a prior description – in this case, that of the Pasteurians as practical constructivists who know what they are doing even when they reify their facts. Even though Latour admits the need to part ways with his scientists and challenge their accounts, he does not seem prepared to challenge the positivist Pasteurians for their overwhelming belief in the 'unacceptable, intolerable, even immoral' myth of reason and science (1988a: 149) that he himself wishes to undermine. See his similar discussion of Pasteurian duplicity in Latour 1999a: 113ff.

13. See Lukàcs's conception of proletarian consciousness as 'the self-consciousness of the commodity' or 'the self-revelation of capitalist society' (1983: 168). See also Lawson 1985: 21.

14. See his characteristic rejoinder to accusations of narcissism, navel-gazing, and subjectivism, which in his view demonstrated 'an uneasiness with all efforts at self-knowing and self-reflection', and were merely 'false consciousness's effort to protect itself from change' (Gouldner 1973: 123–24).

15. This is effectively the same criticism as that brought against Mannheim by Harré, who has argued that we cannot assume 'any direction of influence from social position (situatedness) to intellectual product, since in the structurist view "social position" is as much an intellectual product as anything else' (1975: 282n). On the feminist argument, see extensively Prins 1997.

16. This generalized 'outsider within' position, it is irregularly admitted, more nearly describes the position of feminist intellectuals working in the (academic) centre (cf. Hill Collins 1986: 14–15; 1991: 12, 22; Harding 1991: 131–32; 1993: 65–66).

17. At the heart of all standpoint theory sits a persistent (but also fruitfully exploitable) ambiguity about the 'standpoint' as a category of lived experience and of socially assigned identity, or as an achieved, decisional category of *identification* and position-*taking*.

18. Cf. his remark that 'persons, at their most personal, are essentially the personification of exigencies actually or potentially inscribed in the structure of the field or, more precisely, in the position occupied within this field' (cit. Wacquant, in Bourdieu and Wacquant 1992: 44). Or see his characterization of *Homo Academicus* as an 'anti-biography' and his insistence that 'the most intimate truth of what we are, the most unthinkable unthought [...] is also inscribed in the objectivity and in the history of the social positions that we have held in the past and that we presently occupy' (Bourdieu and Wacquant 1992: 213).

19. See also Pels 2000: 218–20 and more extensively Chapter 5 above. Bourdieu's holistic 'field objectivism' has been criticized in similar terms by Jenkins (1992), Griller (1996), Kögler (1997a; 1977b), Bohman (1997) and Burkitt (1997). In this context, the criticisms by radical constructivists and semioticians of conventional sociological explanations should be taken much more seriously. See Woolgar's (1981) critical approach to sociological 'interest work', or Latour's view of 'society', 'social factors', and 'contexts' as the performative constructs of professional social scientists who strive

'to establish new types of calculations in their institutions', and try to render their definitions indispensable to as many people as possible (1988c: 161). See also Latour 1988d: 27–28: 'No matter how sociologists and historians love to put texts, ideas, and events in their context, this context is always made up of shifted characters inside another text. They can add one text to another, but not escape from it. We have access to co-texts not to context.'

20. In reversing the wheel of existentialist actionism, Bourdieu indeed repeats the circular loop of Sartrean *mauvaise foi*, which presumes that people misrecognize their 'essential' freedom and deceive themselves by acting as if they are things. People know they are free, they are aware of the essential distinction between people and things, but they still act as if the opposite were true. Bourdieu appears to reverse this logic by taking for granted an opposite philosophy of freedom which supports the ability to neutralize social determinations by objectifying them. Both critiques of reification, though diametrically opposite, 'suffer' from a form of circular reasoning that is not recognized as such.

21. Sometimes Bourdieu may appear close to ethnomethodology in adopting a descriptive notion of first-order, practical, or 'endogenous' reflexivity, according to which everyday performances of social reality do not necessarily escape the bonds of the natural attitude (cf. Pollner 1991; May 1998), and which is therefore quite compatible with everyday reification or everyday essentialism. But on the whole he tends to reserve the term 'reflexivity' for the self-referential and hence critical character of professional sociological analysis (cf. Bohman 1997: 177). My argument of course is that Bourdieuan second-order reflexivity to some extent repeats the reificatory 'natural attitude' of everyday world-making (insofar as and to the extent that it is indeed 'natural' and reificatory).

22. Warning to the reader: of course this 'essentialism' is once again cheerfully promoted as a circular statement of the second degree, i.e. it is projected from within the 'weak programme' of reflexive circularity of which I try to convince myself in this chapter – and which I hope may capture a few others of like mind and constitution.

23. In spite of many disclaimers that the sociologist should resist the temptation of taking up an absolute or 'regal' view (e.g. Bourdieu 1990a: 183; Bourdieu and Wacquant 1992: 68), Bourdieu therefore residually reverts to what he himself describes as the intellectualist or 'theoreticist' fallacy, the temptation of transcendental control by means of a sovereign policing overview. Perhaps the true greatness of a theory such as Bourdieu's (a compliment that can be also be applied to many other great – rather than grand – theories) is that it contains many productive contradictions, so that one can always argue with Bourdieu against Bourdieu (as one can always argue with Marx against Marx, with Weber against Weber, or with Mannheim against Mannheim).

24. An elaborate demonstration of how this radical reflexivity is set to work in the historical sociology of ideas, focusing upon the intriguing careers of the concepts of property and power, is provided in Pels 1998: 17, 254–59. The diamond-like schematization of the history of the concepts of fact and value presented in Chapter 4 above is of course another instance of combining a dialectical narrative with a reflexive form of circular reasoning.

25. Contra Bourdieu's anti-biographical asceticism, I tend to favour a softer and more ecumenical approach that distances itself from his imperious opposition between structuralist and personalistic styles of reflexivity. Perhaps one could clarify the issue by drawing three concentric 'orders' or 'rings' of reflexivity: an outer ring where traditional sociological variables operate, such as the 'holy trinity' of class, gender, and race; a middle ring of intellectual fields and interests such as those acknowledged by Bourdieu or Bauman; and an inner ring of autobiographical reflexivity that reimports a much-neglected social psychology or psychobiography of knowledge.

Notes

Notes to Chapter 8

1. Following Havelock's interpretation, Ong argues that Plato's analytic philosophy was possible only because of the effects that writing was beginning to have on mental processes. In Plato's time, more than three centuries after the introduction of the Greek alphabet, writing was finally diffused throughout the Greek population, and was sufficiently interiorized to affect thought processes more generally. Havelock has proposed that Plato's entire epistemology of trying to sort out *epistēme* from *doxa* 'was unwittingly a programmed rejection of the old moral, mobile, warm, personally interactive lifeworld of oral culture'. The Platonic notion of the Ideas or Forms is visually based; they are voiceless, immobile, devoid of all warmth, not interactive but isolated, not part of the human lifeworld at all but utterly above and beyond it (cf. Ong 1982: 80, 94; Goody and Watt 1968: 53). Echoing these writers, Harris notes that 'The careful organisation of a Socratic dialogue, let alone the organisation of connections between one dialogue and another, is the kind of organisation which not even a Plato can undertake without at least jotting down a few notes. The complexity of a long, articulated chain of rational argument cannot simply be carried in the head. The mind's eye needs to be able to scan discussion before it can plan discussion on that kind of scale, just as an architect needs drawings as soon as his projected buildings go beyond a certain level of simplicity, or a composer needs musical notation' (Harris 1986: 25).

2. See the parallel distinction between *otium* and *negotium*. Hobbes' criticism of rhetoric and the rhetoricians repeats these Platonic themes (Skinner 1996: 291).

3. Lit.: 'the flowing water urges him on'. Several categories of speech in Athenian lawcourts were timed by a water-clock (*Theaetetus* 173a transl. n. 2).

4. Cf. the famous anecdote about Thales who, while watching the stars, fell into a pit and was laughed at by a Thracian serving-girl (*Theaetetus* 174a, b). This anecdote inaugurates a long tradition featuring thinkers such as Burke, Tocqueville, Barrès or Schumpeter, who all disparage the intellectual for being impractical and standing apart and aloof from (political) reality. However, this distanciation and alienation precisely focuses the productive contribution of the 'professional strangers' that are intellectuals (Pels 2000: 222–23).

5. In all fairness, Callicles symmetrically admits that politicians will probably make fools of themselves when getting embroiled in philosophical discussions.

6. See also Deleuze's expression about 'creative stuttering', as cited in Rose 1999: 20n.

7. See also Latour's (1999a: 216–35) recasting of the debate between Socrates and Callicles as a contest between two rival forms of expertocracy and intellectual elitism. Latour successfully removes the epistemological obstacles to democratic scientific and political representation, but obfuscates the issue of their mutual autonomy. See Pels 2000: 198–201 for a contrasting argument that critique, or making a difference to the world, is initially not so much a matter of extending one's network and mobilizing allies but precisely of *diminishing* one's network by withdrawing from the *agora* and estranging oneself from the crowd.

8. Perfection, as the Cathars (who were also called the 'Perfects') remind us, was to be attained by catharsis or purification.

9. Cf. the Abbot Agatho's example of carrying a stone in his mouth for three years until he learned to be silent. One may also cite the pervasive silence of Trappist monasteries (or Cistercians of the Strict Observance), where speaking is not permitted in the cloister walk, the chapel, the library, and the monastic cell, and where 'The Great Silence' is observed from 8pm to 8am.

10. This means that I use the case of time rationalization in the monasteries differently from these interpretations, which instead emphasize the monasterial anticipation

of 'Weberian' labour rationalization as the *acceleration* of production in the time-saving spirit of capitalism (see also Thrift 1996: 193–96).

11. Did Descartes just follow a train of thought, or was he perhaps *writing* all the time? In the latter case, would his slogan 'I think, therefore I am', not need to be revised as follows: 'I think, therefore I write, which means that at least also the pen and paper on which I write exist, as well as the hand that holds the pen, which is integrally attached to my body, which hence also exists, as does the material world out there, which has the unfortunate effect of killing off my grounding dualisms of mind and body and mind and matter ...'

12. The older, pre-Cartesian, meaning of *askēsis* as a precondition of knowledge included a different conception of method as 'lived experience', as a *rite de passage*, a trial or experiment in the proper conduct of life (Szakolczai 1998: 12, 20). My additional point is of course that *askēsis* is a proper way of *slowing down* the pace of life.

13. Both views are developed in critical debate with the 'Weber thesis' and argue that the Protestant ethic was fully anticipated in the Catholic monasteries. Kaelber (1998: 61–63) interestingly argues that both the classical Weberian and these revisionist accounts tend to overestimate the impact of asceticism on the drive towards rationalization.

14. It is notable that, in Shapin's constructivist reading of Boyle, 'matters of fact' are generated by a specific multiplication (ideally: universalization) of the witnessing experience through a whole array of material, literary, and social technologies. Performing experiments in the social space of the laboratory or the public rooms of the Royal Society was explicitly contrasted with what occurred in the alchemist's closet precisely because of the semi-public nature of such locales. Another way of expanding the audience to distant but nevertheless direct witnesses was to facilitate replication by means of experimental protocols, as e.g. published in letters to other experimental philosophers. In addition, Boyle perfected a literary technology of 'virtual witnessing', producing clear images of the experimental scene in a writing style that was plain, modest, naturalistic and functional, diverting attention from the person and his reputation. In addition, it was iconographically and pictorially rich, suggesting that the images directly imitated reality, obviating the necessity for direct witnessing or replication (Shapin 1984). This technique of multiplying scientific authority by multiplying witnesses of course runs counter to the Socratic tactic of cornering 'just a few' witnesses in the pursuit of truth. But the experimentalists in the Royal Society still envisioned a *restricted* space for truth-finding, accessible only to gentlemen-scientists (Shapin 1994: 3ff.).

15. I have modified the translation in close comparison with the German original.

16. See Kant on the political role of the philosopher: 'It is not to be expected that kings will be philosophers or that philosophers will become kings; nor is it to be desired, however, since the possession of power inevitably corrupts the free judgment of reason. Kings or sovereign peoples [...] should not, however, force the class of philosophers to disappear or to remain silent, but should allow them to speak publicly. This is essential to both in order that light may be thrown on their affairs' (Kant 1991: 115).

17. This also implies that we need to reconcile liberal difference with the *emancipatory* purpose of the socialist critique of the liberal separations and neutralities as producing 'sham' liberties – a critique that has also been articulated on the radical right (e.g. Schmitt 1993).

18. See the longer-standing argument about the political implications of economies of increasing scale, as discussed in earlier literatures on the managerial revolution and neo-corporatism (Pels 1998: 164ff.). Enterprises change into polities and/or are incorporated into broader mixed-political structures of consultation.

19. Bourdieu's (1999; 2000b) 'discursive' definition of the political field (see

Notes

Chapter 5) and Foucault's conception of 'governmentality' (cf. Dean 1999) are usefully seconded by Fairclough's (2000) and Thompson's (2001) analyses of the new centrality of political communication and rhetoric in the current collusion between politics, government, and the mass media.

20. This is also why I think that resurrections of the 'idea of the idea of the university' (e.g. Rothblatt 1997) or defences of academic freedom that return to disinterestedness as their foundational principle (e.g. Filmer 1997; King and Webster 1997a; 1997b) are ultimately unconvincing. Robins and Webster rightly insist, against postmodernist subversions of the university's distinctiveness, that 'there are *hierarchies of difference* within and between universities, hierarchies which are underlain by criteria of judgement which, at root, define the university'. If this takes care of relativism, the fact that universities still retain a monopoly right to bestow academic credentials once again appears to beg the question rather than to resolve the problem of specificity and autonomy. For Robins and Webster, the legitimacy of academic qualifications 'hinges on public confidence that the teaching that takes place there, and the research that accompanies it, are guided by ideals, and maintained by standards, higher than those of the commercial and instrumental. These ideals – disinterestedness, critical inquiry, open debate, rigorous examination of evidence, and the like – are a crucial element of university life which, if under some strain, remain defining features' (Robins and Webster 1999: 216–17). My argument throughout has been that disinterestedness must be recast as a lack of interest in quick solutions and quick rewards (indeed, of a commercial and/or instrumental nature), and that critical inquiry, open debate, and rigorous research all require the pragmatic advantage, if not of disinterestedness in the conventional sense, than at least of a decelerated time-frame.

21. Etzkowitz and Webster discuss this normative shift in attitudes towards intellectual property, in which the norm of capitalization displaces that of disinterestedness. Gaining credibility in science is in their view increasingly tied to the ability to generate exploitable knowledge, which makes scientists more akin to economic entrepreneurs (1995: 503, 487).

22. Cf. the classical riposte of Durkheim to Lagardelle taking the 'standpoint of the working class': 'We know. You are the mass, you are a force. So what? Is that all you can say in justification'? (1970: 289).

23. Apart from the speed of exposure that is imposed by fast media, another link between celebrity status and haste is that celebrities must instantaneously deal with a great many people whom they don't know because all these people 'know' them and want a piece of them.

24. As a high-tech, quick-service product which students expect to operate like a bank, a cybermall, or a fast-food restaurant, 'McUniversity' dramatically transforms the times and spaces associated with higher education (Ritzer 1998: 151ff.).

Bibliography

Adam, B. (1990) *Time and Social Theory*. Cambridge: Polity Press.
— (1995) *Timewatch: The Social Analysis of Time*. Cambridge: Polity Press.
Ágh, A. (1989) 'The "Triangle Model" of Society and Beyond', in V. Gáthy (ed.) *State and Civil Society: Relationships in Flux*. Budapest: Institute of Sociology of the Hungarian Academy of Sciences.
Alexander, J. (1981) 'Looking for Theory: Facts and Values as the Intellectual Legacy of the 1970s', *Theory and Society* 10: 279–92.
Anrich, E. (1964) *Die Idee der Deutschen Universität: die fünf Grundschriften aus der Zeit ihrer Neubegründung durch klassischen Idealismus und romantische Realismus*. Darmstadt: Wissenschaftliche Buchgesellschaft.
Archer, M, R. Bhaskar, A. Collier, T. Lawson and A. Norrie (eds)(1998) *Critical Realism: Essential Readings*. London and New York: Routledge.
Armitage, J. (1999) 'Paul Virilio: An Introduction' in *idem* (ed.) (1999).
Armitage, J. (ed.) (1999) *Paul Virilio: From Modernism to Hypermodernism and Beyond*. London: Sage.
Ashmore, M. (1989) *The Reflexive Thesis: Wrighting Sociology of Scientific Knowledge*. Chicago: The University of Chicago Press.
— (1993) 'The Theatre of the Blind', *Social Studies of Science* 23(1): 67–106.
— (1996) 'Ending Up on the Wrong Side', *Social Studies of Science* 26(2): 305–22.
Austin, J.L. (1962) *How to Do Things With Words*. Oxford: Oxford University Press.

Banning, W. (1958) *Typen van zedeleer*. Haarlem: Bohn.
Barnes, B. (1974) *Scientific Knowledge and Sociological Theory*. London and Boston: Routledge & Kegan Paul.
— (1976) 'Natural Rationality: A Neglected Concept in the Social Sciences', *Philosophy of the Social Sciences* 6: 115–26.
— (1977) *Interests and the Growth of Knowledge*. London, Boston and Henley: Routledge & Kegan Paul.

— (1982) *T.S. Kuhn and Social Science*. London and Basingstoke: Macmillan.
— (1988) *The Nature of Power*. Urbana, IL: University of Illinois Press.
— (1994) 'Cultural Change – The Thought Styles of Mannheim and Kuhn', *Common Knowledge* 3(2): 65–78.
— (2001) 'The Micro/Macro Problem and the Problem of Structure and Agency', in G. Ritzer and B. Smart (eds) *The Handbook of Social Theory*. London: Sage.
Barry, A., T. Osborne and N. Rose (eds) (1996) *Foucault and Political Reason*. London: UCL Press.
Barthes, R. (1977) *Image Music Text*. London: Fontana.
Bauman, Z. (1987) *Legislators and Interpreters*. Cambridge: Polity Press.
— (1991) *Modernity and Ambivalence*. Cambridge: Polity Press.
— (1992a) *Intimations of Postmodernity*. London: Routledge.
— (1992b) 'Life-World and Expertise: Social Production of Dependency', in Stehr and Ericson (eds) (1992).
— (1993) *Postmodern Ethics*. Cambridge: Polity Press.
— (1997) 'Universities: Old, New, and Different', in Smith and Webster (eds) (1997).
Beck, U. (1992) *Risk Society*. London: Sage.
— (1996) 'World Risk Society as Cosmopolitan Society?', *Theory, Culture and Society* 13(4): 1–32.
— (1997) *The Reinvention of Politics*. Cambridge: Polity Press.
Beck, U., A. Giddens and S. Lash (1994) *Reflexive Modernization*. Cambridge: Polity Press.
Benjamin, W. (1973 [1939]) *Illuminations*. Ed. H. Arendt. London: Fontana.
Bessin, M., and G. Gasparini (2000) 'An Introduction', in *idem* (eds) 'Symposium: Speed and Social Life', *Social Science Information* 39(2) and 39(3): 195–200.
Bhabha, H.K. (1994) *The Location of Culture*. London and New York: Routledge.
Bhaskar, R. (1986) *Scientific Realism and Human Emancipation*. London: Verso.
Bijker, W. (1993) 'Do Not Despair: There is Life After Constructivism', *Science, Technology and Human Values* 18: 113–38.
Black, M. (1969) 'The Gap Between "Is" and "Should"', in Hudson (ed.) (1969).
Bloor, D. (1973) 'Wittgenstein and Mannheim on the Sociology of Mathematics', *Studies in the History and Philosophy of Science* 4(2): 173–91.
— (1981) 'The Strengths of the Strong Programme', *Philosophy of the Social Sciences* 11: 199–213.
— (1983) *Wittgenstein: A Social Theory of Knowledge*. New York: Columbia University Press.
— (1991 [1976]) *Knowledge and Social Imagery*. Chicago and London: The University of Chicago Press.
— (1992) 'Left and Right Wittgensteinians', in Pickering (ed.) (1992).
Bogen, D. (1996) 'The Allure of a "Truly General Theory of Knowledge and Science": A Comment on Pels', *Sociological Theory* 14(2): 187–94.

Bogen, D., and M. Lynch (1990) 'Social Critique and the Logic of Description', *Journal of Pragmatics* 14: 505–21.
Bohman, J. (1997) 'Reflexivity, Agency, and Constraint: The Paradoxes of Bourdieu's Sociology of Knowledge', *Social Epistemology* 11(2): 171–86.
Boltanski, L., and L. Thévenot (1991) *De la justification: Les économies de la grandeur*. Paris: Gallimard.
Bourdieu, P. (1981 [1975]) 'The Specificity of the Scientific Field and the Social Conditions of the Progress of Reason', in C. Lemert (ed.) *French Sociology: Rupture and Renewal Since 1968*. New York: Columbia University Press.
— (1984) *Distinction*. London and New York: Routledge and Kegan Paul.
— (1986) 'The Forms of Capital', in J.G. Richardson (ed.) *Handbook of Theory and Research for the Sociology of Education*. New York: Greenwood Press.
— (1988a) *Homo Academicus*. Cambridge: Polity Press.
— (1988b) 'Vive la crise! For Heterodoxy in Social Science', *Theory and Society* 17(5): 773–87.
— (1989) 'The Corporatism of the Universal: The Role of Intellectuals in the Modern World', *Telos* 81: 99–110.
— (1990a) *In Other Words*. Cambridge: Polity Press.
— (1990b) 'Animadversiones in Mertonem', in J. Clark, C. Modgil and S. Modgil (eds) *Robert K. Merton: Consensus and Controversy*. New York: The Falmer Press.
— (1991) *Language and Symbolic Power*. Cambridge: Polity Press.
— (1993a) *Sociology in Question*. London: Sage.
— (1993b) *The Field of Cultural Production*. Cambridge: Polity Press.
— (1993c) 'Narzisstische Reflexivität und wissenschaftliche Reflexivität', in E. Berg and M. Fuchs (eds) *Kultur, Soziale Praxis, Text*. Frankfurt: Suhrkamp.
— (1995) 'La Cause de la science', *Actes de la recherche en sciences sociales* 106/107: 3–10.
— (1996a) *The Rules of Art*. Cambridge: Polity Press.
— (1996b) 'Understanding', *Theory, Culture and Society* 13(2): 17–37.
— (1996c) 'On The Family as a Realized Category', *Theory, Culture and Society* 13(3): 19–26.
— (1997) *Les usages sociaux de la science*. Paris: INRA.
— (1998a) *Practical Reason: On the Theory of Action*. Cambridge: Polity Press.
— (1998b) *On Television and Journalism*. London: Pluto Press.
— (1999) 'Scattered Remarks', *European Journal of Social Theory* 2(3): 334–40.
— (2000a) *Pascalian Meditations*. Cambridge: Polity Press.
— (2000b) *Propos sur le champ politique*. Lyon: Presses Universitaires de Lyon.
— (2001) *Science de la science et réflexivité*. Paris: Éditions Raisons d'Agir.
Bourdieu, P., and L.J.D. Wacquant (1992) *An Introduction to Reflexive Sociology*. Chicago: University of Chicago Press.
Brubaker, R. (1984) *The Limits of Rationality*. London: Allen & Unwin.
Bucchi, M. (1998) *Science and the Media: Alternative Routes in Scientific Communication*. London and New York: Routledge.
Burkitt, I. (1997) 'The Situated Social Scientist: Reflexivity and Perspective in the Sociology of Knowledge', *Social Epistemology* 11(2): 193–202.

Bibliography

Butler, J. (1992) 'Contingent Foundations: Feminism and the Question of Postmodernism', in *idem* and J. Scott (eds) *Feminists Theorize the Political*. New York and London: Routledge.
Button, G. (1991) 'Introduction', in *idem* (ed.) (1991).
Button, G. (ed.) (1991) *Ethnomethodology and the Human Sciences*. Cambridge: Cambridge University Press.

Callebaut, W. (1993) *Taking the Naturalistic Turn*. Chicago: The University of Chicago Press.
Callon, M. (1986) 'Some Elements of a Sociology of Translation', in Law (ed.) (1986).
— (1995) 'Four Models for the Dynamics of Science', in Jasanoff et al. (eds) (1995).
Callon, M., and B. Latour (1981) 'Unscrewing the Big Leviathan', in K.Knorr-Cetina and A.V. Cicourel (eds) *Advances in Social Theory and Methodology*. London and Henley: Routledge and Kegan Paul.
— (1992) 'Don't Throw the Baby Out With the Bath School!', in Pickering (ed.) (1992).
Callon, M., J. Law and A. Rip (1986) *Mapping the Dynamics of Science*. London and Basingstoke: Macmillan.
Castells, M. (1996) *The Rise of the Network Society*. Oxford: Blackwell.
Chesneaux, J. (2000) 'Speed and Democracy: An Uneasy Dialogue', in Bessin and Gasparini (eds) (2000): 407–20.
Clegg, S., and G. Palmer (1996) 'Introduction: Producing Management Knowledge', in *idem* (eds) *The Politics of Management Knowledge*. London: Sage.
Clifford, J. (1988) *The Predicament of Culture*. Cambridge, MA: Harvard University Press.
Clifford, J., and G. Marcus (eds) (1986) *Writing Culture: The Poetics and Politics of Ethnography*. Berkeley: University of California Press.
Collins, H.M. (1981) 'Stages in the Empirical Programme of Relativism', *Social Studies of Science* 11: 3–10
— (1983) 'An Empirical Relativist Programme in the Sociology of Scientific Knowledge', in Knorr-Cetina and Mulkay (eds) (1983).
— (1986) 'Stages in the Empirical Programme of Relativism', *Social Studies of Science* 11: 3–10.
— (1990) *Artificial Experts: Social Knowledge and Intelligent Machines*. Cambridge, MA: The MIT Press.
— (1991a) *Changing Order*. Chicago and London: University of Chicago Press (2nd edn).
— (1991b) 'Captives and Victims: Comment on Scott, Richards, and Martin', *Science, Technology and Human Values* 16(2): 249–51.
— (1996) 'In Praise of Futile Gestures', *Social Studies of Science* 26(2): 229–44.
Collins, H.M., and S. Yearley (1992) 'Epistemological Chicken', in Pickering (ed.) (1992).
Collins, R. (1986) *Weberian Sociological Theory*. Cambridge: Cambridge University Press.

Comte, A. (1975 [1830–42]) *Cours de philosophie positive. Leçons 1 à 45.* Ed. M. Serres, F. Dagognet and A. Sinaceur. Paris: Hermann.
Corner, J., and D. Pels (eds) (forthcoming) *Media and the Restyling of Politics: Consumerism, Celebrity and Cynicism.* London: Sage
Coulter, J. (1989) *Mind in Action.* Cambridge: Polity Press.
Cozzens, S. (1990) 'Autonomy and Power in Science' in *idem* and Gieryn (eds) (1990).
Cozzens, S., and T. Gieryn (eds) (1990) *Theories of Science in Society.* Bloomington and Indianapolis: Indiana University Press.
Crook, S., J. Pakulski and M. Waters (1992) *Postmodernization: Change in Advanced Societies.* London: Sage.
Currie, J. (1998) 'Introduction', in *idem* and J. Newson (eds) *Universities and Globalization: Critical Perspectives.* London: Sage.
Czyzewski, M. (1994) 'Reflexivity of Actors versus Reflexivity of Accounts', *Theory, Culture and Society* 11(4): 161–68.

Dahlgren, P. (1992) 'Introduction', in *idem* and Sparks (eds) (1992).
Dahlgren, P., and C. Sparks (eds) (1992) *Journalism as Popular Culture.* London: Sage.
Dahrendorf, R. (1968) *Essays in the Theory of Society.* London: Routledge & Kegan Paul.
Darmon, G. (1986) 'The Asymmetry of Symmetry', *Social Science Information* 25(3): 743–52.
Dean, M. (1999) *Governmentality: Power and Rule in Modern Society.* London: Sage.
Dehue, T. (1995) *Changing the Rules: Psychology in the Netherlands 1900–1985.* Cambridge: Cambridge University Press.
Delanty, G. (2001) *Challenging Knowledge: The University in the Knowledge Society.* Buckingham: Open University Press.
Derber, C., W.A. Schwartz and Y. Magrass (1990) *Power in the Highest Degree: Professionals and the Rise of a New Mandarin Order.* New York: Oxford University Press.
Descartes, R. (1968 [1637]) *Discourse on Method and the Meditations.* London: Penguin Books.
De Man, H. (1928) *The Psychology of Socialism.* New York: Scribner's.
De Vries, G. (1992) 'Consequences of Wittgenstein's Farewell to Epistemology', in *L'étude sociale des sciences: Communications de la journée d'étude du 14 mai 1992,* Paris: CRHST.
Diamond, S. (1974) 'Anthropology in Question', in Hymes (ed.) (1974).
Doeser, M. (1986) 'Can the Dichotomy of Fact and Value be Maintained?', in *idem* and J.N. Kraay (eds) *Facts and Values.* Dordrecht, Boston and Lancaster: Martinus Nijhoff.
Douglas, M. (1986) *How Institutions Think.* London: Routledge & Kegan Paul.
Du Gay, P. (1996) *Consumption and Identity at Work.* London: Sage.
Du Gay, P., and M. Pryke (eds) (2002) *Cultural Economy: Cultural Analysis and Commercial Life.* London, Sage.
Durkheim, E. (1958) *Socialism.* Ed. A.W. Gouldner. London: Collier-Macmillan.

Bibliography

— (1965 [1892]) *Montesquieu and Rousseau*. Ann Arbor: University of Michigan Press.
— (1970) *La science sociale et l'action*. Ed. J.-C. Filloux. Paris: PUF.
— (1972) *Selected Writings*. Ed. A. Giddens. Cambridge: Cambridge University Press.
— (1982 [1895]) *The Rules of Sociological Method*. Ed. Steven Lukes. London and Basingstoke: Macmillan.
Dwyer, K. (1979) 'The Dialogic of Ethnology', *Dialectical Anthropology* 4: 205–24.

Eckersley, R. (1992) *Environmentalism and Political Theory*. Albany, NY: SUNY Press.
Edel, A. (1980) *Exploring Fact and Value*. New Brunswick and London: Transaction Books.
Edge, D. (1995) 'Reinventing the Wheel', in Jasanoff et al. (eds) (1995).
Elias, N. (1978a) *The Civilising Process*. Vol I. Oxford: Blackwell.
— (1978b) *What is Sociology?* London: Hutchinson.
— (1987) *Involvement and Detachment*. Oxford: Blackwell.
— (1990) *Norbert Elias über sich selbst*. Frankfurt: Suhrkamp.
— (1991) *The Symbol Theory*. Ed. R. Kilminster. London: Sage.
Elzinga, A. (1985) 'Research, Bureaucracy and the Drift of Epistemic Criteria', in B. Wittrock and A. Elzinga (eds) *The University Research System*. Stockholm: Almqvist & Wiksell.
Eriksen, T.H. (2001) *Tyranny of the Moment: Fast and Slow Time in the Information Age*. London: Pluto Press.
Etzkowitz, H. (1983) 'Entrepreneurial Scientists and Entrepreneurial Universities in American Academic Science', *Minerva* 21: 198–233.
— (1994) 'Academic–Industry Relations: A Sociological Paradigm for Economic Development', in L. Leydesdorff and P. van den Besselaar (eds) *Evolutionary Economics and Chaos Theory: New Directions in Technology Studies*. London: Pinter.
Etzkowitz, H., and L. Leydesdorff (eds) (1997) *Universities in the Global Economy: A Triple Helix of University–Industry–Government Relations*. London and Washington: Pinter.
Etzkowitz, H., and A. Webster (1995) 'Science as Intellectual Property', in Jasanoff et al. (eds) (1995).
Etzkowitz, H., A. Webster and P. Healey (eds) (1998) *Capitalizing Knowledge: New Intersections of Industry and Academia*. Albany, NY: SUNY Press.
Eyerman, R., and A. Jamison (1991) *Social Movements: A Cognitive Approach*. Cambridge: Polity Press.

Fabian, J. (1983) *Time and the Other: How Anthropology Makes its Object*. New York: Columbia University Press.
— (1990) 'Presence and Representation: The Other in Anthropological Writing', *Critical Inquiry* 16: 753–72.
— (1991) 'Dilemmas of Critical Anthropology', in Nencel and Pels (eds) (1991).

Fairclough, N. (1991) 'What Might we Mean by "Enterprise Discourse"?', in Keat and Abercrombie (eds) (1991).
— (1995) *Critical Discourse Analysis*. London and New York: Longman.
— (2000) *New Labour, New Language?* London and New York: Routledge.
Featherstone, M. (1991) *Consumer Culture and Postmodernism*. London: Sage.
Feyerabend, P. (1970) 'Consolations for the Specialist', in Lakatos and Musgrave (eds) (1970).
— (1975) *Against Method*. London: Verso.
— (1977) 'Experts in a Free Society', *Amsterdams Sociologisch Tijdschrift* 3(4): 389–405.
— (1978) *Science in A Free Society*. London: New Left Books.
Filmer, P. (1997) 'Disinterestedness and the Modern University', in Smith and Webster (eds) (1997).
Finnegan, R. (1988) *Literacy and Orality: Studies in the Technology of Communication*. Oxford: Blackwell.
Fiske, J. (1992) 'Popularity and the Politics of Information', in Dahlgren and Sparks (eds) (1992).
Fleck, L. (1980 [1935]) *Entstehung und Entwicklung einer wissenschaftlichen Tatsache*. Frankfurt: Suhrkamp.
— (1983) *Erfahrung und Tatsache: Gesammelte Aufsätze*. Frankfurt: Suhrkamp.
Foucault, M. (1980) *Power/Knowledge: Selected Interviews and Other Writings 1972–1977*. Ed. Colin Gordon. New York: Pantheon Books.
— (1985) *Ervaring en waarheid (interview met Duccio Trombadori)*. Nijmegen: TEU.
— (1994a [1983]) 'Critical Theory/Intellectual History', in Kelly (ed.) (1994).
— (1994b) 'Kritiek en Verlichting', *Krisis: Tijdschrift voor Filosofie*. 56: 64–79.
— (2002 [1969]) *The Archaeology of Knowledge*. London and New York: Routledge.
Fox Keller, E. (1985) *Reflections on Gender and Science*. New Haven: Yale University Press.
Freidson, Elliot (1986) *Professional Powers: A Study of the Institutionalization of Formal Knowledge*. Chicago and London: The University of Chicago Press.
Freyer, H. (1931) *Revolution von Rechts*. Düsseldorf and Cologne: Diederichs.
— (1932) 'Die Universität als hohe Schule des Staates', *Die Erziehung* 7: 520–37, 669–89.
— (1964 [1930]) *Soziologie als Wirklichkeitswissenschaft*. Stuttgart: Teubner.
Frisby, D. (1991) *The Alienated Mind: The Sociology of Knowledge in Germany 1918–1933*. London and New York: Routledge.
Fuchs, S. (1992) *The Professional Quest for Knowledge*. Albany, NY: SUNY Press.
Fuller, S. (1988) *Social Epistemology*. Bloomington: Indiana University Press.
— (1989) *Philosophy of Science and its Discontents*. Boulder, CO: Westview Press.
— (1992) 'Social Epistemology and the Research Agenda of Science Studies', in Pickering (ed.) (1992).
— (1993) *Philosophy, Rhetoric, and the End of Knowledge*. Madison: The University of Wisconsin Press.

— (2000a) *The Governance of Science*. Buckingham: Open University Press.
— (2000b) 'Why Science Studies Has Never Been Critical of Science', *Philosophy of the Social Sciences* 30(1): 5–32.

Gadamer, H. (1979) *Truth and Method*. London: Sheed and Ward.
Garfinkel, H. (1984) *Studies in Ethnomethodology*. Cambridge: Polity Press.
Gaudemar, J.-P. de (1979) *La mobilisation générale*. Paris: Editions du Champ Urbain.
Gergen, K.J. (1988) 'Feminist Critique of Science and the Challenge of Social Epistemology', in M. McCanney Gergen (ed.) *Feminist Thought and the Structure of Knowledge*. New York and London: New York University Press.
Gibbons, M., C. Limoges, H. Nowotny, S. Schartzman, P. Scott and M. Trow (1994) *The New Production of Knowledge*. London: Sage.
Giddens, A. (1982) *Profiles and Critiques in Social Theory*. London and Basingstoke: Macmillan.
— (1984) *The Constitution of Society*. Cambridge: Polity Press.
— (1987) *Social Theory and Modern Sociology*. Cambridge: Polity Press.
— (1990) *The Consequences of Modernity*. Cambridge: Polity Press.
— (1991) *Modernity and Self-Identity*. Cambridge: Polity Press.
Gieryn, T. (1983) 'Boundary-Work and the Demarcation of Science from Non-Science: Strains and Interests in Professional Ideologies of Scientists', *American Sociological Review* 48: 781–95.
— (1995) 'Boundaries of Science', in Jasanoff et al. (eds) (1995).
— (1999) *Cultural Boundaries of Science: Credibility on the Line*. Chicago and London: The University of Chicago Press.
Glasersfeld, E. von (1991) 'Knowing without Metaphysics: Aspects of the Radical Constructivist Position', in Steier (ed.) (1991).
Goehring, J.E. (1999) *Ascetics, Society, and the Desert: Studies in Early Egyptian Monasticism*. Harrisburg, PA: Trinity Press.
Goldman, H. (1994) 'From Social Theory to Sociology of Knowledge and Back: Karl Mannheim and the Sociology of Intellectual Knowledge Production', *Sociological Theory* 12(3): 266–78.
Goody, J. (1977) *The Domestication of the Savage Mind*. Cambridge: Cambridge University Press.
— (1987) *The Interface Between the Written and the Oral*. Cambridge: Cambridge University Press.
Goody, J., and I. Watt (1968) 'The Consequences of Literacy', in J. Goody (ed.) *Literacy in Traditional Societies*. Cambridge: Cambridge University Press.
Gordon, C. (1980) 'Afterword', in Foucault (1980).
Goudsblom, J. (1987) *De sociologie van Norbert Elias*. Amsterdam: Meulenhoff.
Gouldner, A.W. (1970) *The Coming Crisis of Western Sociology*. London: Heinemann.
— (1973) *For Sociology*. London: Allen Lane.
— (1976) *The Dialectic of Ideology and Technology*. New York: Oxford University Press.
— (1985) *Against Fragmentation: The Origins of Marxism and the Sociology of Intellectuals*. New York: Oxford University Press.

Grafstein, R. (1981) 'The Institutional Resolution of the Fact-Value Dilemma', *Journal in the Philosophy and Sociology of Science* 11: 1–14.
Graham, L. (1981) *Between Science and Values*. New York: Columbia University Press.
Griller, R. (1996) 'The Return of the Subject: The Methodology of Pierre Bourdieu', *Critical Sociology* 22(1): 3–28.
Grint, K., and S. Woolgar (1997) *The Machine at Work*. Cambridge: Polity.
Grit, Kor (2000) *Economisering als probleem: Een studie naar de bedrijfsmatige stad en de ondernemende universiteit*. Assen: Van Gorcum.
Gurvitch, G. (1964) *The Spectrum of Social Time*. Dordrecht: Reidel.
Gyarmati, G. (1975) 'The Doctrine of the Professions: Basis of a Power Structure', *International Social Science Journal* 27(4): 629–54.

Habermas, J. (1971) 'Technology and Science as "Ideology"', *Toward a Rational Society*. London: Heinemann.
— (1979) *Communication and the Evolution of Society*. London: Heinemann.
— (1986) *The Theory of Communicative Action*. Vol. I. Cambridge: Polity Press.
— (1987) *The Theory of Communicative Action*. Vol. II. Cambridge: Polity Press.
— (1989) *The New Conservatism*. Cambridge: Polity Press.
— (1990) *Moral Consciousness and Communicative Action*. Cambridge: Polity Press.
— (1992) *Post-Metaphysical Thinking: Philosophical Essays*. Cambridge: Polity Press.
— (1996) *Between Facts and Norms: Contribution to a Discourse Theory of Law and Democracy*. Cambridge: Polity.
Hacking, I. (1984) 'Wittgenstein Rules', *Social Studies of Science* 14: 469–76.
Hahn, R. (1971) *The Anatomy of a Scientific Institution: The Paris Academy of Sciences, 1666–1803*. Berkeley: University of California Press.
Hammersley, M. (2000) *Taking Sides in Social Research*. London and New York: Routledge.
Haraway, D. (1991) *Simians, Cyborgs, and Women*. London: Free Association Books.
Harding, S. (1991) *Whose Science? Whose Knowledge?* Milton Keynes: Open University Press.
— (1992) 'After the Neutrality Ideal: Science, Politics, and "Strong Objectivity"', *Social Research* 59(3): 567–87.
— (1993) 'Rethinking Standpoint Epistemology: What is "Strong Objectivity"?', in L. Alcoff and E. Potter (eds) *Feminist Epistemologies*. New York: Routledge.
Harré, R. (1975) 'Images of the World and Societal Icons', in K. Knorr-Cetina, H. Strasser and M. Zilian (eds) *Determinants and Controls of Scientific Development*. Dordrecht: Reidel.
Harris, R. (1986) *The Origin of Writing*. London: Duckworth.
Hartsock, N. (1983) The Feminist Standpoint', in S. Harding and M.B. Hintikka (eds) *Discovering Reality*. Dordrecht: Reidel.
Haslam, C., and A. Bryman (eds) (1994) *Social Scientists Meet the Media*. London and New York: Routledge.

Havel, V. (1988) 'Anti-Political Politics', in Keane (ed.) (1988).
Heidegger, M. (1972 [1927]) *Sein und Zeit*. Tübingen: Max Niemeyer Verlag.
Heilbron, J. (1995) *The Rise of Social Theory*. Cambridge: Polity Press.
Hekman, S. (1986) *Hermeneutics and the Sociology of Knowledge*. Cambridge: Polity Press.
Hennis, W. (1991) 'The Pitiless "Sobriety of Judgement": Max Weber between Carl Menger and Gustav von Schmoller – the Academic Politics of Value Freedom', *History of the Human Sciences* 4(1): 27–59.
— (1994) 'The Meaning of Wertfreiheit: On the Background and Motives of Max Weber's "Postulate"', *Sociological Theory* 12(2): 113–25.
Herrnstein Smith, B. (1988) *Contingencies of Value: Alternative Perspectives for Critical Theory*. Cambridge, MA: Harvard University Press.
— (1997) *Belief and Resistance: Dynamics of Contemporary Intellectual Controversy*. Cambridge, MA: Harvard University Press.
Hesse, M. (1980) 'Theory and Value in the Social Sciences', *Revolutions and Reconstructions in the Philosophy of Science*. Brighton: The Harvester Press.
Hetherington, K. (1997) *The Badlands of Modernity*. London and New York: Routledge.
Hetherington, K., and R. Munro (eds) (1997) *Ideas of Difference: Social Spaces and the Labour of Division*. Oxford: Blackwell.
Hetherington, K., and N. Lee (2000) 'Social Order and the Blank Figure', *Environment and Planning D: Society and Space* 18: 169–84.
Hill, S., and T. Turpin (1995) 'Cultures in Collision. The Emergence of a New Localism in Academic Research', in M. Strathern (ed.) *Shifting Contexts: Transformations in Anthropological Knowledge*. London and New York: Routledge.
Hill Collins, P. (1986) 'Learning From the Outsider Within: The Sociological Significance of Black Feminist Thought', *Social Problems* 33(6): 14–32.
— (1991) *Black Feminist Thought*. New York and London: Routledge.
Hirst, P. (1994) *Associative Democracy*. Cambridge: Polity Press.
Hörning, K.H., A. Gerhard and M. Michailow (1995) *Time Pioneers: Flexible Working Time and New Lifestyles*. Cambridge: Polity Press.
Hobbes, T. (1968 [1651]) *Leviathan*. Harmondsworth: Penguin Books.
Hollis, M. (1944) *The Philosophy of Social Science: An Introduction*. Cambridge: Cambridge University Press.
Hudson, W.D. (ed.) (1969) *The Is–Ought Question*. London and Basingstoke: Macmillan.
Hume, D. (1978 [1739–40]) *A Treatise of Human Nature*. Oxford: Clarendon Press.
Hutchby, I. (2000) *Conversation and Technology*. Cambridge: Polity Press.
Hymes, D. (ed.) (1974) *Reinventing Anthropology*. New York: Vintage Books.

Jameson, F. (1991) *Postmodernism, or the Cultural Logic of Late Capitalism*. London and New York: Verso.
Jasanoff, S. (1995) *Science at the Bar: Law, Science, and Technology in America*. Cambridge, MA: Harvard University Press.

Jasanoff, S., G.E. Markle, J.C. Petersen and T. Pinch (eds) (1995) *Handbook of Science and Technology Studies*. London: Sage.

Jauréquiberry, F. (2000) 'Mobile Telecommunications and the Management of Time', in Bessin and Gasparini (eds) (2000): 255–68.

Jayyusi, L. (1991) 'Values and Moral Judgment: Communicative Praxis as Moral Order', in Button (ed.) (1991).

Jenkins, R. (1992) *Pierre Bourdieu*. London and New York: Routledge.

Johnson, T. (1972) *Professions and Power*. London and Basingstoke: Macmillan.

Kant, I. (1991) *Political Writings*. Ed. H. Reiss. Cambridge: Cambridge University Press.

— (1992 [1798]) *The Conflict of the Faculties*. Lincoln, NE and London: University of Nebraska Press.

Kaelber, L. (1998) *Schools of Asceticism: Ideology and Organization in Medieval Religious Communities*. University Park, PA: Pennsylvania State University.

Keane, J. (1988) 'Introduction', in *idem* (ed.) (1988).

Keane, J. (ed.) (1988) *Civil Society and the State: New European Perspectives*. London and New York: Verso.

Keat, R. (1991) 'Introduction: Starship Britain or Universal Enterprise?' in *idem* and Abercrombie (eds)(1991).

Keat, R., and N. Abercrombie (eds) (1991) *Enterprise Culture*. London and New York: Routledge.

Keenan, T. (1987) 'The "Paradox" of Knowledge and Power: Reading Foucault on a Bias', *Political Theory* 15(1): 5–37.

Kellner, D. (1999) 'Virilio, War and Technology. Some Critical Reflections', in Armitage (ed.) (1999).

Kelly, M. (ed.) (1994) *Critique and Power: Recasting the Foucault/Habermas Debate*. Cambridge, MA: MIT Press.

Kelsen, H. (1967 [1934]) *Pure Theory of Law*. Berkeley and London: University of California Press.

Kern, S. (1983) *The Culture of Time and Space 1880–1918*. London: Weidenfeld & Nicolson.

Kettler, D., and V. Meja (1994) '"That Typically German Kind of Sociology which Verges Towards Philosophy": The Dispute about *Ideology and Utopia* in the United States', *Sociological Theory* 12(3): 279–303.

Kettler, D., V. Meja and N. Stehr (1990) 'Rationalizing the Irrational: Karl Mannheim and the Besetting Sin of German Intellectuals', *American Journal of Sociology* 95(6): 1441–73.

Kilminster, R. (1993) 'Norbert Elias and Karl Mannheim: Closeness and Distance', *Theory, Culture and Society* 10(3): 81–114.

— (1998) *The Sociological Revolution*. London and New York: Routledge.

Kim, K.-M. (1994) 'Natural versus Normative Rationality: Reassessing the Strong Programme in the Sociology of Knowledge', *Social Studies of Science* 24: 391–403.

— (1999) 'The Management of Temporality: Ethnomethodology as Historical Reconstruction of Practical Action', *The Sociological Quarterly* 40(3): 505–23.

Bibliography

Knorr-Cetina, K. (1977) 'Producing and Reproducing Knowledge: Descriptive or Constructive?' *Social Science Information* 16(6): 669–96.
— (1981) *The Manufacture of Knowledge*. New York: The Pergamon Press.
— (1982) 'Scientific Communities or Transepistemic Arenas of Research? A Critique of Quasi-Economic Models of Science', *Social Studies of Science* 12: 101–30.
— (1983) 'The Ethnographic Study of Scientific Work: Towards a Constructivist Interpretation of Science', in *idem* and Mulkay (eds) (1983).
— (1993) 'Strong Constructivism', *Social Studies of Science* 23(3): 555–63.
— (1999) *Epistemic Cultures: How the Sciences Make Knowledge*. Cambridge, MA: Harvard University Press.
Knorr-Cetina, K., and M. Mulkay (eds) (1983) *Science Observed*. London: Sage.
Kögler, H.H. (1997a) 'Alienation as Epistemological Source: Reflexivity and Social Background after Mannheim and Bourdieu', *Social Epistemology* 11(2): 141–64.
— (1997b) 'Reconceptualizing Reflexive Sociology: A Reply', *Social Epistemology* 11(2): 223–50.
Kohlberg, L. (1971) 'From Is to Ought: How to Commit the Naturalistic Fallacy and Get Away with It in the Study of Moral Development', in T. Mischel (ed.) *Cognitive Development and Epistemology*. New York and London: Academic Press.
Kolakowski, L. (1972) *Positivist Philosophy*. Harmondsworth: Penguin.
— (1977) 'The Myth of Human Self-Identity', in *idem* and S. Hampshire (eds) *The Socialist Idea: A Reappraisal*. London: Quartet Books.
— (1978) *Main Currents of Marxism*. 3 vols. Oxford: Clarendon Press.
Konrád, G. (1984) *Anti-Politics*. San Diego: Harcourt Brace Jovanovich.
— (1990) *Langzame opmerkingen in een snelle tijd*. Amsterdam: Van Gennep.
Kuhn, T. (1970a) *The Structure of Scientific Revolutions*. Chicago: The University of Chicago Press (2nd edn).
— (1970b) 'Logic of Discovery or Psychology of Research?', in Lakatos and Musgrave (eds) (1970).
— (1979) 'Foreword', in L. Fleck, *The Genesis and Development of a Scientific Fact*. Chicago: The University of Chicago Press.
Kumar, K. (1997) 'The Need for Place', in Smith and Webster (eds) (1997).

Lakatos, I. (1970) 'Falsification and the Methodology of Scientific Research Programmes', in *idem* and Musgrave (eds)(1970).
Lakatos, I., and A. Musgrave (eds) (1970) *Criticism and the Growth of Knowledge*. Cambridge: Cambridge University Press.
Lash, S. (1990) *Sociology of Postmodernism*. London: Routledge.
Lash, S., and J. Urry (1994) *Economies of Sign and Space*. London: Sage.
Latour, B. (1980) 'Is It Possible to Reconstruct the Research Process? Sociology of a Brain Peptide', in K. Knorr, R. Krohn and R. Whitley (eds) *The Social Process of Scientific Investigation*. Dordrecht and Boston: Reidel.
— (1983) 'Give Me a Laboratory and I Will Raise the World', in Knorr and Mulkay (eds) (1983).

— (1986) 'The Powers of Association', in Law (ed.) (1986).
— (1987) *Science in Action*. Milton Keynes: The Open University Press.
— (1988a) *The Pasteurization of France*. Cambridge, MA: Harvard University Press.
— (1988b) 'Mixing Humans with Non-Humans: Sociology of a Door-Closer', *Social Problems* 35: 298–310.
— (1988c) 'The Politics of Explanation', in Woolgar (ed.) (1988).
— (1988d) 'A Relativistic Account of Einstein's Relativity', *Social Studies of Science* 18: 3–44.
— (1989) 'Clothing the Naked Truth', in H. Lawson and L. Appignanesi (eds) *Dismantling Truth: Reality in the Post-Modern World*. London: Weidenfeld & Nicolson.
— (1990) 'Postmodern? No, Simply Amodern! Steps Towards an Anthropology of Science', *Studies in the History and Philosophy of Science* 21(1): 145–71.
— (1991) 'The Impact of Science Studies on Political Philosophy', *Science, Technology and Human Values* 16(1): 3–19.
— (1992) 'One More Turn After the Social Turn', in E. McMullan (ed.) *The Social Dimensions of Science*. Notre Dame, IN: University of Notre Dame Press.
— (1993a) *We Have Never Been Modern*. Cambridge, MA: Harvard University Press.
— (1993b) *La clef de Berlin et autres leçons d'un amateur de sciences*. Paris: La Découverte.
— (1999a) *Pandora's Hope: Essays on the Reality of Science Studies*. Cambridge, MA: Harvard University Press.
— (1999b) *Politiques de la nature: Comment fair entrer les sciences en démocratie*. Paris: La Découverte.
— (1999c) 'On recalling ANT', in Law and Hassard (eds) (1999).
— (2001) 'Réponse aux objections ...', *Revue du Mauss* 17(1): 137–51.
— (2002) 'Morality and Technology: The End of the Means', in Pels, Hetherington and Vandenberghe (eds) (2002).
Latour, B., and P. Fabbri (1977) 'La rhétorique du discours scientifique', *Actes de la recherche en sciences sociales* 13: 81–95.
Latour, B., and S. Woolgar (1986 [1979]) *Laboratory Life: The Construction of Scientific Facts*. Princeton: Princeton University Press.
Laudan, L. (1981) 'The Pseudo-Science of Science?', *Philosophy of the Social Sciences* 11: 173–98.
— (1984) *Science and Values*. Berkeley: University of California Press.
— (1987) 'Progress or Rationality? The Prospects for Normative Naturalism', *American Philosophical Quarterly* 24(1): 19–31.
— (1990) 'Normative Naturalism', *Philosophy of Science* 57: 44–59.
Lauretis, T. de (1990) 'Eccentric Subjects: Feminist Theory and Historical Consciousness', *Feminist Studies* 16(1): 115–50.
Law, J. (1986) 'Editor's Introduction: Power/Knowledge and the Dissolution of the Sociology of Knowledge', in *idem* (ed.)(1986).
— (1991) 'Introduction: Monsters, Machines, and Sociotechnical Relations,'

in *idem* (ed.) *A Sociology of Monsters: Essays on Power, Technology, and Domination*. London and New York: Routledge.
— (1999) 'After ANT: Complexity, Naming and Topology', in *idem* and Hassard (eds) (1999).
— (2002) 'Objects and Spaces', in Pels, Hetherington and Vandenberghe (eds) (2002).
Law, J. (ed.) (1986) *Power, Action, and Belief: A New Sociology of Knowledge?* London: Routledge.
Law, J., and J. Hassard (eds) (1999) *Actor Network Theory and After*. Oxford: Blackwell.
Lawson, H. (1985) *Reflexivity: The Post-Modern Predicament*. London: Hutchinson.
Layder, D. (1986) 'Social Reality as Figuration: A Critique of Elias's Conception of Sociological Analysis', *Sociology* 20(3): 367–86.
Lee, N., and S. Brown (1994) 'Otherness and the Actor Network: The Undiscovered Continent', *American Behavioral Scientist* 37: 772–90.
Lemaire, T. (1976) *Over de waarde van kulturen*. Baarn: Ambo.
Leydesdorff, L. (1995) 'The Triple Helix as a Social Achievement: Evolutionary Origins and Consequences of Triadic Communications', *EASST Review* 14(1): 70–73.
Liedman, S.-E. (1993) 'In Search of Isis: General Education in Germany and France', in Rothblatt and Wittrock (eds) (1993).
Luhmann, N. (1982) *The Differentation of Society*. New York: Columbia University Press.
— (1990) *Die Wissenschaft der Gesellschaft*. Frankfurt: Suhrkamp.
Lukàcs, G. (1983) *History and Class Consciousness*. London: Merlin Press.
Lukes, S. (1985) *Marxism and Morality*. Oxford: Clarendon Press.
Lynch, M. (1985) *Art and Artifact in Laboratory Science*. London: Routledge & Kegan Paul.
— (1992a) 'Extending Wittgenstein: The Pivotal Move from Epistemology to the Sociology of Science', in Pickering (ed.) (1992).
— (1992b) 'From the "Will to Theory" to the Discursive Collage: A Reply to Bloor's "Left and Right Wittgensteinians"', in Pickering (ed.) (1992).
— (1992c) 'Going Full Circle in the Sociology of Knowledge: Comment on Lynch and Fuhrman', *Science, Technology and Human Values* 17: 228–33.
— (1993) *Scientific Practice and Ordinary Action*. Cambridge: Cambridge University Press.
— (2000) 'Against Reflexivity as an Academic Virtue and Source of Privileged Knowledge', *Theory, Culture and Society* 17(3): 26–54.
Lynch, W.T. (1994) 'Ideology and the Sociology of Scientific Knowledge', *Social Studies of Science* 24: 197–227.
Lynch, W.T., and E. Fuhrman (1991) 'Recovering and Expanding the Normative: Marx and the New Sociology of Scientific Knowledge', *Science, Technology and Human Values* 16: 233–48.
— (1992) 'Ethnomethodology as Technocratic Ideology: Policing Epistemic Boundaries', *Science, Technology and Human Values* 17: 234–36.

MacIntyre, A. (1969), 'Hume on "Is" and "Ought"', in Hudson (ed.) (1969).
— (1984) *After Virtue*. Notre Dame, IN: University of Notre Dame Press.
Mackie, J.L. (1977) *Ethics: Inventing Right and Wrong*. Harmondsworth: Penguin Books.
Manin, B. (1997) *The Principles of Representative Government*. Cambridge: Cambridge University Press.
Mannheim, K. (1943) *Diagnosis of Our Time*. London: Routledge & Kegan Paul.
— (1952) *Essays in the Sociology of Knowledge*. London: Routledge & Kegan Paul.
— (1953) *Essays on Sociology and Social Psychology*. London: Routledge & Kegan Paul.
— (1968 [1936]) *Ideology and Utopia*. London: Routledge & Kegan Paul.
— (1982 [1922]) *Structures of Thinking*. London, Boston and Henley: Routledge & Kegan Paul.
Marinetti, F.T. (1972) *Selected Writings*. Ed. R.W. Flynt. London: Secker & Warburg.
Marshall, P.D. (1997) *Celebrity and Power: Fame in Contemporary Culture*. Minneapolis: University of Minnesota Press.
Martin, B. (1993) 'The Critique of Science Becomes Academic', *Science, Technology and Human Values* 18(2): 247–59.
— (1996) 'Sticking a Needle into Science: The Case of Polio Vaccines and the Origin of AIDS', *Social Studies of Science* 26(2): 245–76.
Martin, B., E. Richards and P. Scott (1991) 'Who's a Captive? Who's a Victim? Response to Collins's Method Talk', *Science, Technology and Human Values* 16(2): 252–55.
Marx, K. (1972 [1844]) 'Letter to Ruge' and 'Introduction to the Critique of Hegel's Philosophy of Right', in *The Marx–Engels Reader*. Ed. R.C. Tucker. New York: Norton.
Massey, D. (1994) *Space, Place and Gender*. Cambridge: Polity Press.
Maturana, H.R., and F.J. Varela (1984) *The Tree of Knowledge: The Biological Roots of Human Understanding*. Boston: Shambala.
May, J., and N. Thrift (2001) 'Introduction', in *idem* (eds) *Timespace: Geographies of Temporality*. London and New York: Routledge.
May, T. (1998) 'Reflexivity in the Age of Reconstructive Social Science', *International Journal of Social Research Methodology* 1(1): 7–24.
McClelland, C. (1980) *State, Society and University in Germany 1700–1914*. Cambridge: Cambridge University Press.
McHoul, A. (1988) 'Language and the Sociology of Mind', *Journal of Pragmatics* 12: 339–86.
— (1990) 'Critique and Description: An Analysis of Bogen and Lynch', *Journal of Pragmatics* 14: 523–32.
— (1994) 'Towards a Critical Ethnomethodology', *Theory, Culture and Society* 11(4): 105–26.
McLuhan, M. (2001 [1964]) *Understanding Media*. London and New York: Routledge.
Meja, V., and N. Stehr (eds) (1982) *Der Streit um die Wissenssoziologie*. 2 vols. Frankfurt: Suhrkamp.

Mennell, S. (1989) *Norbert Elias: Civilization and the Human Self-Image.* Oxford: Blackwell.
Merton, R. (1941) 'Karl Mannheim and the Sociology of Knowledge', *The Journal of Liberal Religion* 2: 125–47.
— (1973) *The Sociology of Science.* Chicago and London: The University of Chicago Press.
Meyrowitz, J. (1985) *No Sense of Place: The Impact of Electronic Media on Social Behavior.* New York: Oxford University Press.
Millar, J., and M. Schwartz (eds) (1998) *Speed – Visions of an Accelerated Age.* London: The Photographers' Gallery.
Mol, A., and J. Law (1994) 'Regions, Networks and Fluids: Anaemia and Social Topology', *Social Studies of Science* 26: 641–71.
Mommsen, W. (1989) *The Political and Social Theory of Max Weber.* Cambridge: Polity Press.
Monbiot, G. (2000) *Captive State: The Corporate Takeover of Britain.* London: Macmillan.
Montefiore, A. (1990) 'The Political Responsibility of Intellectuals', in I. MacLean, A. Montefiore and P. Winch (eds) *The Political Responsibility of Intellectuals.* Cambridge: Cambridge University Press.
Moore, G.E. (1993 [1903]) *Principia Ethica.* Cambridge: Cambridge University Press.
Mouton, J., and J. Muller (1995) 'A Typology of Shifts in Intellectual Formations in South Africa', in U.J. van Beek (ed.) *South Africa and Poland in Transition: A Comparative Perspective.* Pretoria: HSRC Publishers.
Mulkay, M. (1979) *Science and the Sociology of Knowledge.* London: George Allen & Unwin.
Mumford, L. (1967) *The Myth of the Machine: Technics and Human Development.* New York: Harcourt, Brace & World.

Neckel, S., and J. Wolf (1994) 'The Fascination of Amorality: Luhmann's Theory of Morality and its Resonances among German Intellectuals', *Theory, Culture and Society* 11(2): 69–99.
Nencel, L., and P. Pels (eds) (1991) *Constructing Knowledge: Authority and Critique in Social Science.* London: Sage.
Nietzsche, F. (1980 [1906]) *Der Wille zur Macht.* Stuttgart: Kröner.
— (1990 [1886]) *Beyond Good and Evil.* Harmondsworth: Penguin.
— (1996 [1887]) *On the Genealogy of Morals.* Oxford: Oxford University Press.
Nowotny, H. (1994) *Time: The Modern and Postmodern Experience.* Cambridge: Polity Press.

Oakes, G. (1988) 'Rickert's Value Theory and the Foundations of Weber's Methodology', *Sociological Theory* 6: 38–51.
Ong, W.J. (1982) *Literacy and Orality: The Technologization of the World.* London and New York: Methuen.
Ornstein, M. (1938) *The Role of Scientific Societies in the Seventeenth Century.* Chicago: University of Chicago Press.

Parsons, T. (1967) *Sociological Theory and Modern Society*. New York: The Free Press.
— (1968) 'Professions', *International Encyclopaedia of the Social Sciences*. Ed. D.L. Sills. New York: Macmillan and The Free Press.
Pasteur, L. (1922) *Oeuvres*. Vol. VII. Paris: Masson.
Pels, D. (1990) 'De natuurlijke saamhorigheid van feiten en waarden', *Kennis & Methode* 14(1): 14–43.
— (1991a) 'Values, Facts, and the Social Theory of Knowledge', *Kennis & Methode* 15(3): 274–84.
— (1991b) 'Elias and the Politics of Theory', *Theory, Culture and Society* 8(2): 177–83.
— (1993) *Het democratisch verschil: Jacques de Kadt en de nieuwe elite*. Amsterdam: Van Gennep.
— (1995a) 'Have We Never Been Modern? Towards a Demontage of Latour's Modern Constitution', *History of the Human Sciences* 8(3): 129–41.
— (1995b) 'The Politics of Critical Description: Recovering the Normative Complexity of Foucault's *pouvoir/savoir*', *American Behavioral Scientist* 38(7): 1018–41.
— (1996) 'Historical Positivism', *History of the Human Sciences* 9(1): 113–21.
— (1997) 'Mixing the Metaphors: Economics or Politics of Knowledge?', *Theory and Society* 26: 685–717.
— (1998) *Property and Power in Social Theory: A Study in Intellectual Rivalry*. London and New York: Routledge.
— (1999) 'The Critical Quadrangle: Anti-Essentialism in "Science" and "Common Sense"', conference paper, 'The Transformation of Knowledge', University of Surrey.
— (2000) *The Intellectual as Stranger: Studies in Spokespersonship*. London and New York: Routledge.
— (2002) 'Everyday Essentialism. Social Inertia and the Münchhausen Effect', in Pels, Hetherington and Vandenberghe (eds) (2002): 69–89.
Pels, D., K. Hetherington and F. Vandenberghe (eds) (2002) *Sociality/Materiality: The Status of the Object in Social Science*. Special issue of *Theory, Culture and Society* 19(5–6).
Phillips, D. (1977) *Wittgenstein and Scientific Knowledge: A Sociological Perspective*. London: Macmillan.
Pickering, A. (ed.) (1992) *Science as Practice and Culture*. Chicago: University of Chicago Press.
Pinch, T. (1993) 'Generations of SSK', *Social Studies of Science* 23(2): 370–72.
Pinch, T., and W.E. Bijker (1984) 'The Social Construction of Facts and Artifacts', *Social Studies of Science* 14: 399–441.
Pitkin, H.F. (1972) *Wittgenstein and Justice*. Berkeley: University of California Press.
Plato (1987a) *The Last Days of Socrates*. Harmondsworth: Penguin Books.
— (1987b) *Theaetetus*. Harmondsworth: Penguin Books.
— (1994) *Gorgias*. Oxford: Oxford University Press.
Pollner, M. (1987) *Mundane Reason: Reality in Everyday and Sociological Discourse*. Cambridge: Cambridge University Press.

— (1991) 'Left of Ethnomethodology: The Rise and Decline of Radical Reflexivity', *American Sociological Review* 56: 370–80.
Popkin, R. (1979) *The History of Scepticism from Erasmus to Spinoza*. Berkeley: University of California Press.
Popper, K. (1962) *The Open Society and its Enemies*. Vol. II. London: Routledge & Kegan Paul.
— (1970) 'Normal Science and its Dangers', in Lakatos and Musgrave (eds) (1970).
Portis, E.B. (1983) 'Max Weber and the Unity of Normative and Empirical Theory', *Political Studies* 31: 25–42.
Potter, J. (1996) *Representing Reality: Discourse, Rhetoric and Social Construction*. London: Sage.
Potter, J., and M. Wetherell (1988) 'Accomplishing Attitudes: Fact and Evaluation in Racist Discourse', *Text* 8(1–2): 51–68.
Power, M. (1997) *The Audit Society: Rituals of Verification*. Oxford: Oxford University Press.
Prins, B. (1997) 'The Standpoint in Question: Situated Knowledges and the Dutch Minorities Discourse'. PhD dissertation, University of Utrecht.
Proctor, R.N. (1991) *Value-Free Science? Purity and Power in Modern Knowledge*. Cambridge, MA: Harvard University Press.
Putnam, H. (1981) *Reason, Truth, and History*. Cambridge: Cambridge University Press.
— (1990) *Realism With a Human Face*. Cambridge, MA: Harvard University Press.
Putnam, R.A. (1985) 'Creating Facts and Values', *Philosophy* 60: 187–201.

Radder, H. (1996) *In and About the World*. Albany, NY: SUNY Press.
Radnitzky, G. (1973) *Contemporary Schools of Metascience*. Chicago: Regnery.
Rasch, W. (2000a) *Niklas Luhmann's Modernity: The Paradoxes of Differentiation*. Stanford: Stanford University Press.
— (2000b) 'Conflict as a Vocation: Carl Schmitt and the Possibility of Politics', *Theory, Culture and Society* 17(6): 1–32.
Rasch, W., and C. Wolfe (2000) 'Introduction: Systems Theory and the Politics of Postmodernity', in *idem* (eds) *Observing Complexity: Systems Theory and Postmodernity*. Minneapolis and London: University of Minnesota Press.
Ray, L. (1999) 'Social Differentiation, Transgression and the Politics of Irony', in *idem* and Sayer (eds) (1999).
Ray, L. and A. Sayer (eds) (1999) *Culture and Economy After the Cultural Turn*. London: Sage.
Readings, B. (1996) *The University in Ruins*. Cambridge, MA: Harvard University Press.
Restivo, S. (1995) 'The Theory Landscape in Science Studies', in Jasanoff et al. (eds) (1995).
Richards, E. (1991) *Vitamin C and Cancer: Medicine or Politics?* London: Macmillan.
— (1996) '(Un)Boxing the Monster', *Social Studies of Science* 26(2): 323–56.

Richards, E., and M. Ashmore (1996) 'More Sauce Please! Politics and SSK: Neutrality, Commitment and Beyond', *Social Studies of Science* 26(2): 219–28.

Richards, E., and J. Schuster (1989) 'The Feminine Method as Myth and Accounting Resource: A Challenge to Gender Studies and Social Studies of Science', *Social Studies of Science* 19: 697–720.

Richards, E., and M. Ashmore (eds) (1996) *The Politics of SSK: Neutrality versus Commitment*. Special issue, *Social Studies of Science* 26(2).

Ritzer, G. (1996) *The McDonaldization of Society*. London: Sage.

— (1998) *The McDonaldization Thesis: Explorations and Extensions*. London: Sage.

— (1999) *Enchanting a Disenchanted World*. London: Sage.

Robbins, B. (1991) 'Othering the Academy: Professionalism and Multiculturalism', *Social Research* 58(2): 355–72.

— (1993) *Secular Vocations: Intellectuals, Professionalism, Culture*. London and New York: Verso.

Robins, K., and F. Webster (1999) *Times of the Technoculture*. London and New York: Routledge.

Root, M. (1993) *Philosophy of Social Science*. Oxford: Blackwell.

Rose, N. (1999) *Powers of Freedom: Reframing Political Thought*. Cambridge: Cambridge University Press.

Roth, M. (1994) 'What Does the Sociology of Scientific Knowledge Explain?: or, When Epistemological Chickens Come Home to Roost', *History of the Human Sciences* 7(1): 95–108.

Rothblatt, S. (1997) *The Modern University and its Discontents*. Cambridge: Cambridge University Press.

Rothblatt, S., and B. Wittrock (eds) (1993) *The European and American University since 1800: Historical and Sociological Essays*. Cambridge: Cambridge University Press.

Rouse, J. (1987) *Knowledge and Power*. Ithaca, NY: Cornell University Press.

Ruschemeyer, D. (1983) 'Professional Autonomy and the Social Control of Expertise', in R. Dingwall and P. Lewis (eds) *The Sociology of Professions*. New York: St Martin's Press.

Santiso, J. (2000) 'Political Sluggishness and Economic Speed: A Latin American Perspective', in Bessin and Gasparini (eds) (2000): 233–53.

Sarfatti-Larson, M. (1977) *The Rise of Professionalism: A Sociological Analysis*. Berkeley: University of California Press.

Schäfer, L., and T. Schnelle (1980) 'Einleitung', in Fleck (1980).

— (1983) 'Die Aktualität Ludwik Flecks in Wissenschaftssoziologie und Erkenntnistheorie', in Fleck (1983).

Schaffer, S. (1991) 'The Eighteenth Brumaire of Bruno Latour', *Studies in the History and Philosophy of Science* 22(1): 174–92.

Schmitt, C. (1935) *Staat, Bewegung, Volk*. Hamburg: Hanseatische Verlagsanstalt.

— (1988) *Positionen und Begriffe im Kampf mit Weimar-Genf-Versailles*. Berlin: Duncker & Humblot.

— (1993) 'The Age of Neutralizations and Depoliticizations', *Telos* 96: 130–42.
— (1996) *The Concept of the Political*. Ed. G. Schwab and T.B. Strong. Chicago and London: University of Chicago Press.
Schmitt, F.C. (1994) 'Socializing Epistemology: An Introduction through Two Sample Issues', in *idem* (ed.) *Socializing Epistemology: The Social Dimensions of Knowledge*. London: Rowman & Littlefield.
Scholte, B. (1974) 'Toward a Reflexive and Critical Anthropology', in Hymes (ed.) (1974).
Schnädelbach, H. (1983) *Philosophie in Deutschland 1831-1933*. Frankfurt: Suhrkamp.
Schwartz, B. (1974) 'Waiting, Exchange, and Power: The Distribution of Time in Social Systems', *American Journal of Sociology* 79: 841—870.
Scott, A. (1987) 'Politics and Method in Mannheim's *Ideology and Utopia*', *Sociology* 21(1): 41–54.
— (1997) 'Between Autonomy and Responsibility: Max Weber on Scholars, Academics and Intellectuals', in J. Jennings and A. Kemp-Welch (eds) *Intellectuals in Politics: From the Dreyfus Affair to Salman Rushdie*. London and New York: Routledge.
Scott, A., and J. Street (2000) 'From Media Politics to E-Protest: The Use of Popular Culture and New Media in Parties and Social Movements', *Information, Communication and Society* 3(2): 215–40.
Scott, P., E. Richards and B. Martin (1990) 'Captives of Controversy: The Myth of the Neutral Social Researcher in Contemporary Scientific Controversies', *Science, Technology and Human Values* 15(4): 474–94.
Searle, J. (1969) *Speech Acts*. Cambridge: Cambridge University Press.
Serres, M. (1982) *Hermes: Literature, Science, Philosophy*. Baltimore: Johns Hopkins University Press.
Shapin, S. (1984) 'Pump and Circumstance: Robert Boyle's Literary Technology', *Social Studies of Science* 14: 481–520.
— (1994) *A Social History of Truth*. Chicago and London: University of Chicago Press.
Shapin, S., and S. Schaffer (1985) *Leviathan and the Air-Pump: Hobbes and Boyle and the Experimental Life*. Princeton: Princeton University Press.
Sharrock, W.W., and R.J. Anderson (1984) 'The Wittgenstein Connection', *Human Studies* 7: 375–86.
Shields, R. (1991) *Places on the Margin: Alternative Geographies of Modernity*. London and New York: Routledge.
Shumar, W. (1997) *College for Sale: A Critique of the Commodification of Higher Education*. London and Washington: Falmer Press.
Simons, H.W., and M. Billig (eds) (1994) *After Postmodernism: Reconstructing Ideology Critique*. London: Sage.
Skinner, Q. (1978) *The Foundations of Political Thought*. Vol I. Cambridge: Cambridge University Press.
— (1988) 'Some Problems in the Analysis of Political Thought and Action', in Tully (ed.) (1988).
— (1996) *Reason and Rhetoric in the Philosophy of Hobbes*. Cambridge: Cambridge University Press.

Slater, D., and F. Tonkiss (2001) *Market Society: Markets and Modern Social Theory*. Cambridge: Polity Press.
Slaughter, S., and G. Rhoades (1996) 'The Emergence of a Competitiveness R&D Policy Coalition and the Commercialization of Academic Science and Technology', *Science, Technology and Human Values* 21(3): 303–39.
Slaughter, S., and L. Leslie (1997) *Academic Capitalism: Politics, Policies and the Entrepreneurial University*. Baltimore and London: The Johns Hopkins University Press.
Sloterdijk, P. (1983) *Kritik der zynischen Vernunft*. 2 vols. Frankfurt: Suhrkamp.
— (1998) 'Modernity as Mobilisation', in Millar and Schwarz (eds) (1998).
Smith, A. (1976 [1776]) *An Inquiry into the Nature and Causes of the Wealth of Nations*. Ed. R.H. Campbell and A.S. Skinner. Oxford: Clarendon Press.
Smith, A., and F. Webster (1997a) 'Changing Ideas of the University' in *idem* (eds) (1997).
— (1997b) 'Conclusion: An Affirming Flame' in *idem* (eds) (1997).
Smith, A., and F. Webster (eds) (1997) *The Postmodern University? Contested Visions of Higher Education in Society*. Buckingham: SRHE/Open University Press.
Soja, E. (1996) *Thirdspace: Journeys to LA and Other Real-and-Imagined Places*. Oxford: Blackwell.
Squires, J. (ed.) (1993) *Principled Positions: Postmodernism and the Rediscovery of Value*. London: Lawrence and Wishart.
Stafford, B.M. (1994) *Artful Science*. Cambridge, MA: MIT Press.
Steding, C. (1932) *Politik und Wissenschaft bei Max Weber*. Breslau: Korn Verlag.
Stedman-Jones, S. (1988) 'Fact/Value', in C. Jencks (ed.) *Core Sociological Dichotomies*. London: Sage.
Stehr, N. (1981) 'The Magic Triangle: In Defense of a General Sociology of Knowledge', *Philosophy of the Social Sciences* 11: 225–29.
— (1994) *Knowledge Societies*. London: Sage.
Stehr, N., and V. Meja (1982) 'Zur gegenwärtigen Lage wissenssoziologischer Konzeptionen', in Meja and Stehr (eds) (1982).
Stehr, N., and R.V. Ericson (eds) *The Culture and Power of Knowledge*. Berlin and New York: De Gruyter.
Steier, F. (1991a) 'Introduction: Research as Self-Reflexivity, Self-Reflexivity as Social Process', in *idem* (ed.) (1991).
— (1991b) 'Reflexivity and Methodology: An Ecological Constructionism', in *idem* (ed.) (1991).
Steier, F. (ed.) (1991) *Research and Reflexivity*. London: Sage.
Stocking, G. (1982 [1968]) *Race, Culture and Evolution: Essays in the History of Anthropology*. Chicago: The University of Chicago Press.
Street, B.V. (1984) *Literacy in Theory and Practice*. Cambridge: Cambridge University Press.
Sturken, M., and L. Cartwright (2001) *Practices of Looking: An Introduction to Visual Culture*. Oxford: Oxford University Press.
Szakolczai, A. (1998) *Max Weber and Michel Foucault: Parallel Life-Works*. London and New York: Routledge.

Bibliography

Taylor, C. (1978) 'Neutrality in Political Science', in P. Laslett and W.G. Runciman (eds) *Philosophy, Politics and Society*. 3rd series. Oxford: Blackwell.
Tenbruck, F.H. (1981) 'Emile Durkheim oder die Geburt der Gesellschaft aus dem Geist der Soziologie', *Zeitschrift für Soziologie* 10(4): 333–50.
Teubner, G. (1993) *Law as an Autopoetic System*. Oxford: Blackwell.
Thompson, J. (1995) *The Media and Modernity*. Cambridge: Polity Press.
— (2001) *Political Scandal: Power and Visibility in the Media Age*. Cambridge: Polity Press.
Thrift, N. (1996) *Spatial Formations*. London: Sage.
— (1999) 'Capitalism's Cultural Turn', in Ray and Sayer (eds) (1999).
— (2002) 'Performance Cultures in the New Economy', in Du Gay and Pryke (eds) (2002).
Tully, J. (1988) 'The Pen is a Mighty Sword: Quentin Skinner's Analysis of Politics' in *idem* (ed.) (1988).
Tully, J. (ed.) (1988) *Meaning and Context: Quentin Skinner and his Critics*. Cambridge: Cambridge University Press.
Turner, S.P., and R.A. Factor (1984) *Max Weber and the Dispute over Reason and Value*. London: Routledge & Kegan Paul.
Turner, V. (1982) *From Ritual to Theatre: The Human Seriousness of Play*. New York: Performing Arts Journal Publications.

Urry, J. (1985) 'Social Relations, Space and Time', in D. Gregory and J. Urry (eds) *Social Relations and Spatial Structures*. London: Macmillan.
— (1990) *The Tourist Gaze*. London: Sage.
— (1995) *Consuming Places*. London and New York: Routledge.
— (2000) *Sociology Without Societies*. London and New York: Routledge.

Van El, C. (2002) *Figuraties en verklaringen: Stijlgebonden schoolvorming in de Nederlandse sociologie sinds '1968'*. Amsterdam: Aksant.
Vierhaus, R. (1977) 'Rankes Begriff der historischen Objektivität', in R. Koselleck (ed.) *Objektivität und Parteilichkeit*. Munich: Klett-Cotta..
Virilio, P. (1986) *Speed and Politics: An Essay on Dromology*. New York: Semiotext(e).
Von Beyme, K. (1991) *Theorie der Politik im 20. Jahrhundert*. Frankfurt: Suhrkamp.
Von Ferber, C. (1965) 'Der Werturteilsstreit 1909/1959', in E. Topitsch (ed.) *Logik der Sozialwissenschaften*. Cologne and Berlin: Luchterhand.

Walzer, M. (1984) 'Liberalism and the Art of Separation', *Political Theory* 12(3): 315–30.
Weber, M. (1949) *On the Methodology of the Social Sciences*. Ed. E. Shils and H.A. Finch. New York: The Free Press.
— (1970) 'Science as a Vocation', in *From Max Weber: Essays in Sociology*. Ed. H.H. Gerth and C.W. Mills. London and New York: Routledge.
— (1994) *Political Writings*. Ed. P. Lassman and R. Speirs. Cambridge: Cambridge University Press.

Webster, F. (1995) *Theories of the Information Society*. London and New York: Routledge.
Wernick, A. (1991) *Promotional Culture*. London: Sage.
Whimster, S., and S. Lash (1987) 'Introduction', in S. Lash and S. Whimster (eds) *Max Weber: Rationality and Modernity*. London: Allen & Unwin.
White, G.D., and F.C. Hauck (eds) (1998) *Campus, Inc.: Corporate Power in the Ivory Tower*. Amherst, NY: Prometheus Books.
Williams, R. (1989 [1958]) *Resources of Hope: Culture, Democracy, Socialism*. London: Verso.
Willis, P. (1978) *Profane Culture*. London: Routledge & Kegan Paul.
Winner, L. (1993) 'Upon Opening the Black Box and Finding it Empty', *Science, Technology and Human Values* 18(3): 362–78.
Wittgenstein, L. (1998 [1953]) *Philosophical Investigations*. Oxford: Blackwell.
Wittrock, B. (1993) 'The Modern University: The Three Transformations', in Rothblatt and Wittrock (eds) (1993).
Wolff, K. (1959) 'The Sociology of Knowledge and Sociological Theory', in L. Gross (ed.) *Symposium on Sociological Theory*. Evanston, IL: Row Peterson & Co.
Wood, Denis (1993) *The Power of Maps*. London: Routledge.
Woolgar, S. (1981) 'Interests and Explanation in the Social Study of Science', *Social Studies of Science* 11: 365–94.
— (1983) 'Irony in the Social Study of Science', in Knorr-Cetina and Mulkay (eds) (1983).
— (1988a) *Science: The Very Idea*. London: Tavistock.
— (1988b) 'Reflexivity is the Ethnographer of the Text', in *idem* (ed.) (1988).
— (1991) 'The Turn to Technology in Social Studies of Science', *Science, Technology and Human Values* 16: 20–50.
— (1992) 'Some Remarks about Positionism: A Reply to Collins and Yearley', in Pickering (ed.) (1992).
— (1997) 'Science and Technology Studies and the Renewal of Social Theory', in S.P. Turner (ed.) *Social Theory and Sociology: The Classics and Beyond*. Oxford: Blackwell.
Woolgar, S. (ed.) (1988) *Knowledge and Reflexivity: New Frontiers in the Sociology of Knowledge*. London: Sage.
Woolgar, S., and D. Pawluch (1985) 'Ontological Gerrymandering: The Anatomy of Social Problems Explanations', *Social Problems* 32(3): 214–27.

Young, M. (1988) *The Metronomic Society: Natural Rhythms and Human Timetables*. London: Thames and Hudson.
Young, M., and T. Schuller (eds) (1988) *The Rhythms of Society*. London and New York: Routledge.

Zerubavel, E. (1981) *Hidden Rhythms: Schedules and Calendars in Social Life*. Chicago: University of Chicago Press.
Žižek, S. (ed.) (1994) *Mapping Ideology*. London and New York: Verso.
Zolo, D. (1992) *Democracy and Complexity: A Realist Approach*. Cambridge: Polity Press.

Index

academic
  capitalism 2, 212, 215, 225
  freedom 1, 2, 23, 46, 112, 149, 179, 192, 193, 195, 207–16, 224, 245
acceleration 5–9, 14, 49–50, 151, 196, 214, 216, 219, 227, 244
agnosticism 14, 15, 21, 69, 71, 101, 103, 104, 106, 130, 131, 134, 143, 147, 148, 149, 156, 176, 195, 224, 230, 233
anti-essentialism, *see* essentialism
anti-politics 14, 22, 109, 119, 120, 122, 123, 126–29, 179, 200, 236
asymmetry 64, 76, 87, 90, 93, 131, 133, 135, 144, 146–50, 155, 197, 230, 232, 237, 238, 239
autonomy 2, 9, 10, 12, 13, 14, 20, 22, 23, 24, 26, 29, 30, 31, 36, 38, 42, 45, 47, 49–50, 81, 96, 97, 108ff., 136, 142, 147, 149, 177, 179ff., 224, 226, 227, 234, 235, 243

Bacon, F. 23, 28, 77, 78, 79, 93, 95, 184, 186, 187, 188, 194, 196, 197, 211
Barnes, B. 21, 52, 62, 63, 68, 91, 117, 130, 134, 197, 227, 229, 232, 236, 237
Beck, U. 27, 56, 198, 205, 206, 227, 240
Bildung 188
Bloor, D. 21, 52, 53, 55, 62, 64, 68, 71, 91, 94, 130, 133, 134, 169, 227, 228, 229, 230, 232, 236
boundary work 3, 26, 42, 150, 201, 208
Bourdieu, P. 1, 12, 13, 14, 18, 22, 23, 27, 46, 47, 48, 52, 56, 57–59, 65, 66, 67, 84, 93, 108ff., 149, 158, 167, 169–74, 199, 200, 215, 224, 225, 226, 228, 229, 230, 233, 234, 235, 241, 242, 244
Boyle, R. 23, 77, 78, 93, 102, 132, 137–43, 147–48, 237, 238, 244

Callon, M. 55, 71, 92, 132, 143, 137, 140, 142, 150, 197, 233, 238, 239
celebrity (fame) 2, 3, 9, 35, 38, 48, 50, 181, 182, 183, 185, 206, 207, 212, 213, 214, 218, 245
chronopolitics (politics of time) 4, 8, 14, 22, 45, 49, 150, 180, 196, 198, 200, 215
circularity 14, 16, 23, 75, 85, 116, 118, 155, 161, 164–66, 168, 169, 170, 172, 173, 174, 175, 176, 177, 178, 198, 201, 219,

269

220, 231, 233, 234, 238, 239, 240, 242
Collins, H. 21, 52, 55, 68, 71, 130, 132, 133, 134, 141, 143, 144, 146, 148, 227, 229, 230, 239
competition, intellectual 24, 27, 44, 60, 62, 78, 90–94, 112, 113–19, 125, 128, 147, 149, 175–78, 191, 209, 210, 211, 212, 232, 234, 235, 239
Comte, A. 28, 77, 79, 93, 95, 96, 157, 232, 239
constructivism 15, 19, 20, 21, 23, 27, 28, 52, 55, 57, 58, 65, 72, 74, 82, 83, 84, 91, 92, 101, 103, 104, 108, 109, 132, 137, 138, 139, 158, 161, 163, 164, 173, 174, 176, 223, 224, 228, 230, 237, 239, 240, 241, 244
critical phenomenology 8, 20, 30, 49, 207, 225
critique 9, 14, 20, 55, 71, 72, 73, 83, 89–90, 103, 107, 131, 143, 146, 161, 163, 165, 174, 175, 179, 197, 200, 201, 236, 238, 240, 243
  of ideology 15, 56, 104, 155, 156, 160, 166, 167, 224, 225, 228
culturalization 26, 37, 38, 80, 202, 203, 204–06

deceleration (delay) 32–33, 36, 45, 46, 49, 97, 126–31, 136, 151, 152, 153, 154, 156, 160, 164, 179ff., 225, 234, 235, 244, 245
(de-)differentiation 5, 8, 21, 24, 37, 38, 40–43, 48, 81, 89, 105, 108, 110, 112, 126, 152, 201, 202, 204, 206, 207, 208, 209, 226
delay, *see* deceleration
demarcation 1, 9, 11, 13, 17, 36, 37–38, 45, 62, 73, 79, 81, 84–85, 127, 131, 141, 147, 163, 176, 177, 179, 191, 199, 200, 212, 219, 224

Descartes, R. 3, 11, 12, 13, 14, 23, 78, 79, 160, 184–85, 187, 194, 195, 217, 244
detachment, *see* involvement
dialectics 15, 40–41, 82, 85, 88, 115, 120, 131, 202, 224, 233, 234, 242
diamond matrix 22, 82ff., 231, 234, 242
distance/distancing (*see also* involvement/detachment) 3, 71, 73, 82, 94, 97, 114, 136, 142, 146, 149, 152, 160, 162, 184, 196, 199, 208, 227, 243
doing/saying 31, 102, 104, 138, 167, 178, 225, 233
doubt 3, 12, 13, 14, 22, 23, 24, 129, 160, 179, 185, 194–201, 216, 217, 227
duality 22, 27, 28, 108–13, 115, 116, 118, 122, 123, 126, 127, 128, 193, 234
Durkheim, E. 4, 9–12, 14, 35, 40, 56, 61, 79, 93, 94, 95, 96, 111, 116, 196, 197, 198, 199, 202, 218, 222, 223, 229, 245

economization 37, 38, 48, 202, 203, 204, 210
Elias, N. 14, 22, 27, 52, 57, 65, 93, 94–97, 99, 105, 199, 228, 229, 230, 233
enterprise culture 2, 43, 50, 204, 210
essentialism 1, 5, 9, 20, 29, 38, 39, 76, 81, 83, 84, 103, 115, 120, 152, 174, 177, 179, 182, 187, 188, 193, 195, 201, 203, 210, 213, 220, 223, 225, 226, 228, 232, 234, 236, 242
estrangement 3, 128, 152, 161, 179, 180, 183–85, 195, 196, 199, 201, 215, 216, 227, 243
ethnomethodology 52, 70–73, 82, 104, 115–18, 139, 145, 153, 158, 165, 197, 199, 200, 208, 213, 227, 242

Index

fact/value 10, 11, 14, 21, 17, 51, 55, 68, 70, 71, 74ff., 115–18, 174, 198, 199, 231, 233, 234
fame, *see* celebrity
fascism 7, 18, 223, 224
Feyerabend, P. 25, 26, 93, 109, 110, 121, 131, 231, 234
feminism 16, 18, 27, 169–70, 204, 218, 241
Fleck, L. 52, 57, 60–61, 227, 229
fluidity (flow) 3, 4, 29, 37, 43, 141, 201
Foucault, M. 13, 15, 20, 27, 28, 52, 59, 84, 109, 110, 121, 149, 157, 177, 199, 204, 224, 225, 228, 230, 233, 235, 245
Freyer, H. 17, 18, 39, 129, 223, 224, 231

Garfinkel, H. 92, 219–20, 240
Giddens, A. 27, 52, 56, 98, 222, 227, 233, 240
globalization 5, 7, 50, 219
Gouldner, A. 52, 57, 59, 84, 87, 120, 121, 158, 167, 231, 235, 239, 241
Gurvitch, G. 4–5, 8

Habermas, J. 8, 14, 22, 37, 39, 40, 46, 52, 83, 84, 88, 93, 97–100, 106, 119, 121, 188, 191, 192, 193, 199, 200, 202, 213, 218, 226, 233, 234, 235, 236
Harding, S. 169–71, 174, 241
hybridity 5, 50, 59, 102, 103, 127, 133, 149, 197, 201, 202, 208, 212, 213
hyper-differentiation, *see* de-differentiation
Hobbes, T. 23, 102, 132, 137–43, 147–48, 157, 202, 237, 238, 243

ideal speech situation 14, 46, 100, 199, 213, 226
ideology critique, *see* critique
identity 29, 39, 43, 44, 141, 153, 159, 160, 162, 163, 171, 237, 239, 241
indifference 54, 70–73, 105, 136
image, *see* visual culture
individualization 5, 7, 41
inscription, *see* writing
involvement/detachment 94–97, 112, 114, 119, 127, 128, 130, 136, 146, 149, 181, 193, 195, 199, 201, 238
intellectuals 14, 16, 22, 31, 33, 45, 48, 56, 108–09, 112, 119, 127, 129, 135, 167, 168, 169, 171, 172, 173, 174, 175, 193, 200, 206, 207, 228, 234, 235, 236, 241, 243
is/ought, *see* fact/value

Kant, I. 9, 24, 40–41, 59, 74, 79, 80, 84, 87, 93, 157, 187, 189–92, 193, 194, 201, 215, 216, 225
(neo-)Kantianism 17, 63, 69, 80, 88, 89, 97, 116, 226, 244
Knorr-Cetina, K. 52, 57, 58, 59, 71, 84, 224, 238
knowledge politics 1, 13, 15, 20, 21, 22, 28, 29, 37, 63, 85–86, 87, 100, 102, 104, 108, 109, 110, 112, 113, 114, 118, 127, 128, 131, 132, 134, 135, 142, 146, 147, 163, 198, 200, 204, 217, 232, 234, 236
knowledge-political continuum 20, 28–32, 110, 124, 125, 128, 129, 150, 182, 200, 202, 215, 226
Kuhn, T. 21, 57, 59, 60, 63, 64, 91, 93, 101, 134, 227, 240

last instances 38, 41, 43, 48, 98, 202, 203, 206, 224, 226
Latour, B. 1, 12, 13, 18, 19, 22, 23, 28, 41, 52, 55, 57, 58–59, 65, 66, 67, 69, 71, 84, 91, 93, 101–05, 106, 108, 117, 119, 129, 132, 134, 137, 138, 139–43,

271

147, 148, 149, 150–56, 160–63, 176, 197, 198, 204, 209, 220, 223, 224, 229, 230, 233, 234, 236, 237, 238, 239, 240, 241, 243
lifeworld/system 98–99, 214
literacy, *see* writing
Luhmann, N. 40, 159, 203, 240
Lynch, M. 52, 53, 55, 65, 68–69, 70, 71, 72, 138, 164, 227, 228, 230, 237, 239, 240

Mannheim, K. 21, 23, 51ff., 167, 169, 183, 193, 227, 228, 229, 230, 241, 242
marginality, *see* estrangement
Marx, K. (Marxism) 6, 15, 16, 18, 19, 27, 37, 39, 43, 56, 59, 65, 67, 86, 88, 89–90, 98, 127, 146, 167, 169, 188, 201, 202, 204, 212, 213, 214, 218, 224, 232, 233, 234, 236, 242
materiality (materialization) 23, 32, 47, 154, 155, 226, 227
media(tization) 5, 38, 47, 48–49, 125, 199, 205–06, 207, 219, 227
methodism (or methodological voluntarism) 131, 133–35, 185, 195, 198, 199, 200, 201, 234, 236, 238
Merton, R. 29, 58, 116, 228, 229
micro/macro 55, 230
mobility 29
mobilization (total-) 5, 6, 7, 129, 151, 162, 200, 219

natural proximity
 of facts and values 14, 17, 18, 19, 20, 74ff., 117, 177, 230, 231
 of communicative and strategic action 100, 235
 of good and evil 111–12, 116, 129
 of involvement and detachment 96, 112, 114, 234

of knowledge and power (or reason and force) 12–13, 14–20, 27–28, 113–15, 129
naturalism (also anti-) 53, 54, 64, 66, 68, 69, 71, 78, 86, 89, 90–94, 101, 115, 134, 148, 155, 159, 160, 163, 166, 167, 169, 232, 233
naturalistic fallacy 76, 79, 83, 86–90, 91, 94, 98, 100, 103, 232, 233
neutrality 1, 15, 17, 18, 19, 20, 21, 24, 29, 64, 67, 68, 71, 72, 74, 81, 92, 94, 96, 97, 106, 114, 119, 125, 129, 130, 131, 132, 135, 136, 143–49, 151, 158, 168, 170, 171, 177, 187, 194, 195, 196, 197–98, 199, 200, 201, 208, 210, 211, 212, 213, 224, 232, 236, 238, 244
Nietzsche, F. 3, 9, 16, 27, 50, 115, 157, 199, 222, 234
normative complexity 21, 67–70, 152, 153
normative continuum 94–95, 233
normativism (anti-) 54, 68, 71, 86, 90–94, 115, 232
normativistic fallacy 86–90
'nothing special' 1, 25, 30, 38, 108, 109, 151, 188, 210, 213, 222

orality, *see* talk

partisanship 15, 16, 19, 36, 67, 72, 95, 132, 143–49, 212, 237, 238
performativity 14, 15, 16, 17, 19, 20, 28, 31, 54, 72, 73, 81, 83, 84, 85, 97, 99, 100, 103, 104, 106, 107, 115, 117, 120, 124, 127, 135, 139, 140, 141, 146, 159, 164, 165, 168, 171, 182, 195, 218, 223, 232, 235, 240, 241
performative contradiction 97–100, 166, 218, 232, 233
Plato 3, 9, 23, 25, 35, 50, 76, 78,

79, 125, 126, 127, 157, 158, 180–83, 187, 188, 193, 194, 213, 217
political correctness 2, 31, 50, 141, 212, 215, 231, 236, 243, 244
politicization 15, 30, 37, 38, 47, 48, 67, 74, 127, 147, 190, 192, 202, 203, 204, 205
politics
of knowledge, *see* knowledge politics
of time, *see* chronopolitics
Popper, K. 22, 29, 75, 76, 84, 87, 91, 92–93, 176, 201, 217
postliberalism 21, 24, 37, 42, 43, 202, 204, 209, 212
pragmatism 2, 9, 10, 12, 14, 20, 23, 28, 31, 44, 45, 49, 51, 83, 199, 200, 213, 215, 226, 232, 245
professionalism 66, 108–10, 112, 120–21, 149, 194, 234

rationality 8, 11, 12, 19, 21, 25–27, 33, 40, 41–42, 51, 54, 59, 61, 62, 63, 66, 68, 72, 74, 76, 90, 98, 130, 136, 196, 207, 220, 233, 235
reading (*see also* writing) 2, 31–35, 45, 49, 152–54, 180, 184, 226
realism 62, 72, 104, 138, 139, 145, 164, 174, 175, 237, 239, 240
anti- 56, 149, 228
circular 165, 175, 240
reflexive 240
reductionism 86–90, 95, 105–07, 115, 226, 232, 233
reflexivity 14, 20, 21, 23, 44, 55, 56, 57, 58, 61, 68, 71, 72, 75, 83, 90, 95, 96, 103, 105, 106, 115, 116, 117, 131, 146, 154, 155, 157ff., 179, 200, 201, 218, 234, 236, 238, 239
reification 23, 72, 73, 78, 79, 83, 92, 95, 99, 101, 102, 103, 104, 105, 106, 115, 117, 119, 138, 139, 141, 150, 153, 155, 156, 162, 163, 177, 180, 194, 195,
197, 200, 208, 219, 220, 223, 225, 226, 233, 239, 241, 242
relativism (*see also* value-free) 54, 62, 84, 85, 97, 99, 101, 130, 134, 143, 144, 149, 160, 167, 170, 175, 200, 218, 219, 229, 231, 240, 245
representation 3, 14, 30, 33, 40, 42, 47, 48, 49, 54, 56, 59, 67, 83, 101, 142, 173, 177, 205, 212, 237, 238, 239, 243
reputation (recognition) 27, 35, 47–48, 113, 118, 121, 122, 123, 124, 149, 209, 216, 212, 239, 240–41
rivalry, *see* competition

saying/doing, *see* doing/saying
science and technology studies (STS) 14, 22, 27, 51ff., 90–94, 130ff., 197–98, 224, 227–28, 230
self-interested science 25–28, 30–31, 46, 112, 122, 209
Schmitt, C. 17, 18, 39, 43, 97, 129, 156, 167, 168, 201, 202, 204, 224, 244
Sloterdijk, P. 6, 7, 176
slowdown, *see* deceleration
Smith, A. 6, 67, 119, 203
socialism 10, 11, 18, 39, 88, 198, 232, 244
social epistemology 56–57, 61, 65, 66, 68, 69, 72, 228
social triangle 20, 21, 24, 36–40, 42, 45, 48, 49, 110, 124, 147, 179, 198, 202–07, 209, 215
Socrates, *see* Plato
speech act theory 54, 84, 232, 233
spokespersonship 16, 23, 33, 110, 111, 121, 123, 128, 140, 142, 149, 150, 153, 155, 159, 162, 167, 168, 169, 171, 178, 206, 219, 223, 224, 225, 238, 240–41
standpoint theory 16, 27, 169–71, 175, 241, 245

strangerhood, *see* estrangement
Strong Programme 51, 52, 53, 55, 58, 62–64, 68, 92, 94, 119, 130, 169, 218, 229, 232, 236
symmetry 14, 15, 22, 23, 26, 52, 54, 56, 62, 64, 65, 69, 87, 92, 94, 130ff., 179, 195, 197–98, 200, 204, 229, 230, 236, 237, 238, 240
synchronization 5, 8, 198, 215, 216

talk 2, 20, 31–35, 46, 49, 50, 112, 153, 155, 180, 207, 214, 225, 226
third positions/spaces/ways 3, 29, 37, 43, 131, 133, 135, 136, 143–46, 148, 149, 201, 239
thought style 56, 60–61, 229
Thrift, N. 47, 206, 222
time-geography 4, 179, 222
time/space 3, 100, 122, 124, 131, 150, 152, 180, 201

uncertainty 105, 151, 153, 156, 177, 197, 198, 215, 219
universalism 1, 8, 9, 12, 16, 19, 24, 25, 26, 28, 29, 30, 56, 61, 109, 110, 115, 118–21, 127, 162, 165, 177, 187, 193, 194, 196, 204, 229, 233, 235, 237, 239, 240, 244
university 24, 179, 184, 188–94, 208–16, 245
 critical 210–11
 entrepreneurial 38, 210–11

value-freedom, *see* neutrality
value-free relativism 19, 21, 22, 54–57, 82, 92, 130–31, 201, 233
Virilio, P. 7, 8, 49, 129
visual culture 46, 48–49, 207, 242

weak
 autonomy 21, 30, 31, 43, 124, 179
 boundaries 8, 20, 29, 38, 50, 110, 131, 150, 196, 202, 208, 210, 215
 demarcation/differentiation 30, 43, 44, 150, 199, 238
 enemies 178, 198
 epistemology 12, 52, 198
 programme 53, 132, 242
 social theory 14, 22, 23, 36, 105, 146–50, 162–65, 169, 170, 178, 198, 200, 201, 217–21, 230, 231, 239
Weber, M. 6, 8, 15, 17, 18, 19, 21, 22, 40, 64, 65, 67, 71, 75, 76, 81, 83, 84, 86, 87, 88, 93, 97, 98, 114, 146, 177, 194, 201, 218, 223, 224, 231, 232, 233, 236, 242, 244
Wittgenstein, L. 17, 21, 51ff., 82, 92, 130, 201, 227, 228, 229, 230, 233
Woolgar, S. 28, 52, 62, 65, 68, 71, 84, 92, 101, 119, 132, 133, 134, 152, 153, 155, 164, 165, 169, 176, 224, 227, 228, 229, 233, 238, 239, 240, 241
writing (*see also* reading) 2, 20, 23, 31–35, 45, 49, 50, 112, 129, 150–56, 180, 207, 214, 225, 226, 244